The Smart Revolution Towards the Sustainable Digital Society

ADVANCES IN INFORMATION, COMMUNICATION AND ENTERTAINMENT MARKETS

Series Editor: Gary Madden, *Curtin University of Technology, Australia*

This series is intended to provide a vehicle for thorough and up-to-date analysis of the recent developments and revolutionary changes that are occurring in digital markets. The industrial organisation of broadband ISPs, Internet backbone providers and content/application providers is in turmoil. Furthermore, the traditional lines between these services are blurring, as each attempts to enter adjacent markets. We are seeing the emergence of powerful, competing platforms, linked in complex ways that challenge traditional economic analyses. Advances in Information, Communication and Entertainment Markets showcases cutting-edge research and emerging practice and policy, covering a wide range of topics relating to digital technologies and their impact on the media, law, the arts, business and other elements of society.

The Smart Revolution Towards the Sustainable Digital Society

Beyond the Era of Convergence

Edited by

Hitoshi Mitomo

Waseda University, Tokyo, Japan

Hidenori Fuke

Komazawa University, Tokyo, Japan

Erik Bohlin

Chalmers University of Technology, Gothenburg, Sweden

ADVANCES IN INFORMATION, COMMUNICATION AND ENTERTAINMENT MARKETS

EE Edward Elgar PUBLISHING

Cheltenham, UK • Northampton, MA, USA

Published by
Edward Elgar Publishing Limited
The Lypiatts
15 Lansdown Road
Cheltenham
Glos GL50 2JA
UK

Edward Elgar Publishing, Inc.
William Pratt House
9 Dewey Court
Northampton
Massachusetts 01060
USA

A catalogue record for this book
is available from the British Library

Library of Congress Control Number: 2015933352

This book is available electronically in the **Elgar**online
Economics subject collection
DOI 10.4337/9781784710040

MIX
Paper from
responsible sources
FSC® C013056

ISBN 978 1 78471 003 3 (cased)
ISBN 978 1 78471 004 0 (eBook)

Typeset by Servis Filmsetting Ltd, Stockport, Cheshire
Printed and bound in Great Britain by T.J. International Ltd, Padstow

Contents

Contributors

Hasnat Ahmad is a Research Officer at the Planning Commission of Pakistan in Lahore, Pakistan. He earned a Master of Philosophy (MPhil) degree in Econometrics and Quantitative Economics from the International Islamic University, Pakistan in 2008. In 2010 he enrolled in the International PhD Program at the Curtin Business School, Australia. His PhD, entitled 'Demographic dividend and economic performance: the case of Pakistan', is currently under examination. Hasnat, as a Research Assistant at the Communication Economics and Electronic Market Centre (CEEM), published several scholarly articles in journals including *Applied Economics Letters* and *Telecommunications Policy*.

Erik Bohlin is Professor in Technology Assessment at the Department of Technology Management and Economics at Chalmers University of Technology, Gothenburg, Sweden. He has published books and papers in a number of areas relating to the information society, including policy, strategy and management. He is Chair of the International Telecommunications Society and Chief Editor of *Telecommunications Policy*. He obtained his graduate degree in Business Administration and Economics at the Stockholm School of Economics (1987) and his PhD at Chalmers University of Technology (1995).

Teruyuki Bunno is Professor of Business Management, Faculty of Business Administration, Kinki University, Osaka, Japan. Major areas of specialty include innovation theory, life cycle of firms and new business creation. Current research focuses on human resources roles in the process of firms' growth. Publications include 'An empirical analysis of organizational innovation generated by ICT in Japanese SMEs', in S. Allegrezza and S. Dubrocard (eds), *Internet Econometrics* (Palgrave Macmillan, 2012); 'Business issues of venture business to overcome', in R. Takada, H. Ueno, T. Murakoso and K. Maeda (eds), *Contemporary Study of Small- and Medium-Sized Enterprises* (in Japanese), (Doyuukan, 2009); and *Venture Business Theory* (in Japanese), ed. with K. Ohta and K. Ikeda (Jikkyo Shuppan, 2007). He is a board member of the Japan Academy of Small Business Studies and the Kansai Association for Venture and Entrepreneur Studies.

Martin Cave is a Visiting Professor at Imperial College Business School, London, UK and an inquiry chair at the UK Competition and Markets Authority. He has been BP Centennial Professor at the London School of Economics, and a Professor at Warwick Business School. He holds BA, BPhil and DPhil degrees from Oxford University. He is an economist specialising in competition law and the regulation of network industries. He is co-author of *Essentials of Modern Spectrum Management* (Cambridge University Press, 2007) and *Understanding Regulation* (Oxford University Press, 2011); and co-editor of *The Handbook of Telecommunications Economics* (Elsevier, 2002, 2005). He has also written a number of journal articles; and has advised a number of governments and regulatory agencies on spectrum, telecommunications and other matters.

Matthias Ehrler studied Economics at the University of Muenster, Germany before working for the regional fixed network operator Net Cologne GmbH (2000–03) as Manager Regulatory Affairs. Between January 2004 and June 2007, he worked for O_2 Germany (now Telefónica Germany) as Senior Regulatory Counsel Economy and Strategy and as a certified project manager. At SBR, he focused on the strategic and economic aspects of the deployment of next generation networks (NGN/NGA), in particular fibre optic networks, and was responsible for several projects with an emphasis on market and competition analysis, strategic assessment and business development. He is currently involved in regulatory projects with national and international institutions and authorities.

Natascha Freund worked in the telecommunications, unfair competition and regulatory law department of Telekom Austria AG, Vienna, Austria after completing her law studies (1991–97) and a postgraduate course in European law (LLM). During this period, she handled telecommunications and unfair competition law, as well as regulatory agendas, and oversaw every aspect of the market presence of Telekom Austria AG. She also attended to all matters concerning the Act Against Unfair Competition. In 2006, she was awarded her Doctorate in Legal Science. Thereafter she worked as a trainee lawyer for law firm Fellner, Wratzfeld und Partner Rechtsanwälte GmbH and in 2008, she joined the competition regulation team of Rundfunk- und Telekom Regulierungs-GmbH, both in Austria. Since 2012, she has been working as a legal expert for SBR-net Consulting AG, Austria.

Hidenori Fuke is currently Professor at the Faculty of Global Media Studies, after serving the Dean of the Faculty and the Chairman of the Graduate School of Global Media, at Komazawa University, Tokyo,

Japan. He was a member of the board of directors of the International Telecommunications Society from 1998 to 2012. After working for Nippon Telegraph and Telephone Corporation on regulatory matters as a senior executive and a member of the board of directors at InfoCom Research, he joined Kansai University in Osaka as a Professor. He received his BA in Economics from the University of Tokyo, and his MLitt from the University of Glasgow. He also received a PhD in International Public Policy from Osaka University. His major areas of specialisation include the economic analysis of the telecommunications industry and regulatory policy. His current research focuses on changes in the structure of the info-communications industry brought about by the wide deployment of smartphone and broadband Internet, and their policy implications.

Teresa Garín-Muñoz is Associate Professor of Economics at the National Distance Education University (UNED), Madrid, Spain. She has been a Visiting Scholar at the University of California, San Diego (UCSD). She holds a Master of Arts in Economics from UCSD and a PhD in Economics from UNED. Her research interests are in microeconomics (demand modelling, consumer satisfaction, consumer protection). Most of her research has been devoted to the areas of tourism and telecommunications. She has published in international refereed journals, including *Telecommunications Policy*, *Information Economics and Policy*, *Tourism Management*, *Tourism Economics*, *International Journal of Tourism Research* and *Applied Economics*.

Covadonga Gijón holds a European PhD in Economics from the Complutense University of Madrid, Spain. She has been a Visiting Scholar at the Newcastle Business School at Northumbria University, UK. She also holds a Master of Arts in Economies of Telecommunications from the National Distance Education University (UNED). She previously held positions as Assistant Professor at the Antonio Nebrija University, Pontificia University – ICADE and Complutense University. While in that position she was involved in research at CeDInt from the Polytechnic University of Madrid. Her research covers topics in telecommunications economics, consumer satisfaction, broadband speed, the digital divide, personal information online, and model construction and selection.

Keiko Hatta is a Chief Consultant at Global Study Group, InfoCom Research, Tokyo, Japan, where she specialises in telecommunications regulations with a particular focus on the mobile sector. Prior to her current position, she was a Visiting Fellow at Oxford University where she worked on international comparative studies of regulatory systems. She studied at Waseda University and the Graduate School at the University of Tokyo. She has written a number of articles and book chapters on the topic of mobile regulation.

Anders Henten is Professor with Special Responsibilities at the Center for Communication, Media and Information Technologies (CMI) at the Department of Electronic Systems, Aalborg University, Denmark. He is a graduate in Communications and International Development Studies from Roskilde University in Denmark (1989) and holds a PhD from the Technical University of Denmark (1995). He has worked professionally in the area of communications economy and policy for more than 25 years; and has participated in numerous research projects financed by, for example, the European Community, the Nordic Council of Ministers, Danish Research Councils and Ministries, and in consultancies, financed by World Bank, UNCTAD, ITU, Danish Ministries, etc. He has published nationally and internationally – more than 250 academic publications in international journals, books, conference proceedings, etc.

Hiroki Idota is Professor of Management Information Systems, Faculty of Economics, Kinki University, Osaka, Japan. He is also a board member of the Japan Society for Information and Management and the Nippon Urban Management and Local Government Research Association. He received a BA in Economics from Kwansei Gakuin University, an MA in Informatics from Kansai University and a PhD in Economics from Osaka University. He also holds CISA (Certificated Information System Auditor). Major areas of specialty include management information systems and the management of technology. Current research focuses on innovation using ICT. Publications include 'An empirical analysis of organizational innovation generated by ICT in Japanese SMEs', in S. Allegrezza and S. Dubrocard (eds) *Internet Econometrics* (Palgrave Macmillan, 2012); 'Corporate productivity and the stage of ICT development' (*Information Management Technology*, **13**, 2012); and *Theory and Practice of Information Security Management* (in Japanese), (Hakuto-Shobo, 2004), which received the Telecom Social Science Encouragement Award of the Telecommunications Advancement Foundation in March 2005.

Toshiya Jitsuzumi is a Professor in the Faculty of Economics, Kyushu University, Fukuoka, Japan. He obtained an LLB from the University of Tokyo, an MBA from the Stern School of Business, New York University, and a DSc from the Graduate School of Global Information and Telecommunication Studies, Waseda University. Before assuming his current post, he worked for 18 years for the Ministry of Post and Telecommunications (now Ministry of Internal Affairs and

Communications), Japan. His main research topics include telecommunications policy and Internet economics, and his current focus is on network neutrality. During 2007–08, he was a Visiting Scholar at the Columbia Institute for Tele-Information (CITI), Columbia University, as an Abe Fellow sponsored by the Social Science Research Council/ Japan Foundation Center for Global Partnership. His recent publications include *Economics of the Telecom Industry 2.0* (Kyushu University Press, 2013); and *Economic Perspectives on Network Neutrality: Analyses of Broadband Quality of Service* (Kyushu University Press, 2013).

Norihiro Kasuga has been Professor of Industrial Organization in the Faculty of Economics, Konan University, Kobe, Japan since 2013. He received his BA in Economics from the University of Tokyo, his MA from Yale University and his PhD from Yokohama National University. After serving at the Ministry of Internal Affairs and Communications, he taught economics at Nagasaki University, Kobe University and Kinki University. During 2005–06, he was a Research Affiliate at the Yale Center for International Area Studies (YCIAS). His current major interests cover the method of regulatory impact assessment, the information distribution process by print and broadcasting media, and the impact of mergers and acquisitions on the media industry.

Mikio Kimura is Director of the Research Institute at the Japan Commercial Broadcasters Association (JBA), Japan. He holds a Master of International Business (Aoyama Gakuin University) and a PhD (the University of Tokyo). His research has covered quantitative analyses of innovation in ICT and empirical research of the media industry. He has conducted social research in media use behaviour as well. His current research focuses on developing new business models for the mass-media industry combined with the Internet or social media. He is also interested in the role of mass media during times of major disaster.

Chatchai Kongaut is currently a Doctoral Student at the Department of Technology Management and Economics, Chalmers University of Technology, Gothenburg, Sweden, with Professor Erik Bohlin as his main supervisor. He was born in Bangkok, Thailand. He received a BA in Economics from Thammasat University, Thailand (2006), and an MSc in Economics and Econometrics from the University of Nottingham, UK (2010). He currently also works as an economist at the National Broadcasting and Telecommunications Commission (NBTC), Thailand. His research interests focus mainly on the economic dimensions of ultra-fast Internet, the EU Digital Agenda and the impact of broadband policy.

Yu-Li Liu, PhD (Indiana University), is a Professor in the Department of Radio and TV at National Chengchi University (NCCU), Taipei, Taiwan. She served as one of the first-term Commissioners of the National Communications Commission from 2006 to 2008. She is the president of both the Chinese Communication Association and the Taiwan Communications Society. She has received grants from the US Fulbright Foundation and Taiwan's Ministry of Science and Technology for research (MOST) projects in the areas of convergence, broadband networks, digital TV and Over-the-Top TV. She was also granted an Excellent Research Award by MOST (2013 and 2014) and awarded 'Distinguished Professor' by NCCU for her research, teaching and service (August 2014–July 2017). She has published numerous books such as *Policy and Marketing Strategy for Digital Media* (co-edited with Prof. Robert Picard); *Multi-Channel TV and Audience*; *Cable TV Management and Programming Strategy*; *Cable TV Programming and Policy in China*; *Radio and TV and Telecommunications*, etc.

Rafael López is Assistant Professor of Economics at the Complutense University, Madrid, Spain. Previously he was Assistant Professor at Carlos III University in Madrid and Pompeu Fabra University in Barcelona. He has been a Visiting Scholar at the Alicante University and the Castilla-La Mancha University. He holds a PhD in Economics and Finance and an MSc in Economics from Pompeu Fabra University. His main fields of interest are game theory and microeconomics. His main line of research has been in consumer protection theory, particularly within the telecommunications sector. His research has been published in international refereed journals, including *Telecommunications Policy*.

Martin Lundborg holds a Master of Science in Business Administration from Jönköping International Business School (JIBS), Sweden. He joined SBR Juconomy Consulting in 2004, where he has been responsible for several projects with an emphasis on cost modelling, cost assessments, cost accounting, quantitative analyses (including business statistics), regulation of the telecommunications industry, market analysis, carrier management and spectrum administration. These projects have been conducted for regulatory authorities and communication ministries in Europe, the Middle East and the Pacific as well as for operators and industry organisations in Europe and the Middle East.

Gary Madden is Professor of Economics at Curtin University, Perth, Australia. He is an Affiliated Researcher (and Faculty Advisor) with the Center for Tele-Information at Columbia University, New York, USA. Madden specializes in the economics, business, and regulation of communications and high-tech industries. His work analyses complex communications policy matters and provides quantitative strategic advice to cope with changes in competition, regulation, and technology. His interests blend academic rigor with an understanding developed from his service on the Board of Directors of the International Telecommunications Society. Madden is an expert in empirically modeling telecommunications, specializing in wireline, wireless, cable, and emerging technologies. He has worked spectrum auctions, regulatory matters, assessment of competitive market conditions, convergence and business strategies, and demand forecasting and pricing. Dr. Madden has authored numerous articles and books on wireless and wireline competition and strategies and frequently lectures in these areas at industry conferences, and at universities. His involvement includes co-supervising the PhD student, and the specification and conduct of the econometric analysis. He will contribute to the specification and estimation of the econometric work, development of a specification test and analyse the implications for policy.

Hitoshi Mitomo is Professor of Telecommunications Economics and Policy at the Graduate School of Asia-Pacific Studies (GSAPS), Waseda University, Tokyo, Japan. He was a Visiting Professor at the Center for Information and Communication Research, Stockholm School of Economics in 2007. He has been serving as the president of Japan's Society of Information and Communication Research since 2013. He is a council member of the Ministry of Internal Affairs and Communications and has been involved in policy-making processes related to ICT and the media. He holds a Master's degree in Environment Science (University of Tsukuba) and a Doctorate in Engineering (Toyohashi University of Technology). His research covers a wide range of socio-economic and regulatory issues of ICT deployment and utilisation. He is also interested in applying behavioural economics to the assessment of media access behaviour.

Kyoung-Youn Na is a Research Fellow in the Construction Policy Division at the Construction and Economy Research Institute of Korea (CERIK). He earned his PhD in Economics at Korea University in 2013. Further, in 2011 he won the Korean Economics Association (KEA) excellence award for a young economist. He has published in the areas of industrial organisation, infrastructure policy and applied microeconomics, his articles appearing in *Telecommunications Policy*, *Annals of Regional Science*, *Review of Applied Economics*, *Journal of the Korean Official Statistics*, etc. His current research interests include public policy surrounding ICT and antitrust.

Akihiro Nakamura is a Professor at the School of International Management in Yokohama City University, Japan. He holds a BA, an MA and a PhD in Business and Commerce from Keio University in Tokyo as well as an MA in Statistics from Yale University. His research interests focus on industrial organisation and applied econometrics, and he currently oversees several telecommunications research and transportation projects. His articles have appeared in several journals such as *Telecommunications Policy*, *Information Economics and Policy*, and *Socio-Economic Planning Sciences*.

Tokio Otsuka is Associate Professor of the Department of English and Information Management, Shumei University, Yachiyo, Japan. He received a Doctor of Science in Global Information and Telecommunication Studies from Waseda University. He previously worked as a Research Associate at Waseda University and as Assistant Professor at Shumei University. His research focuses on the socio-economic impact of ICT.

Teodosio Pérez-Amaral is Associate Professor and Head of the Department of Economics at Complutense University, Madrid, Spain. He previously held the position of staff economist at the Bank of Spain. He has been a Visiting Scholar at the University of California, San Diego. He has also been a consultant to DIW of Berlin, Deutsche Telekom and Telefónica. Professor Pérez-Amaral holds a PhD in Economics from the University of California, San Diego. He has published in academic journals, including *Journal of Econometrics*, *Oxford Bulletin of Economics and Statistics*, *Econometric Theory*, *Applied Economics*, *Journal of Forecasting*, *Telecommunications Policy* and *Information Economics and Policy*. His research covers topics in telecommunications economics, finance, and model construction and selection. He is Associate Editor of *Telecommunications Policy* and *Mathematics and Computers in Simulation*.

Ernst-Olav Ruhle, after completing his studies in Economics at the University of Hanover in 1993, worked at the renowned Scientific Institute for Economics (WIK) in Bad Honnef until 1995. Following a period with the Hannover-Hildesheim Chamber of Industry and Commerce and after finishing his Doctorate in Economics in 1996, he was employed with RWE Telliance in Essen until September 1998. Between 1998 and 2001, he was Managing Director for Regulatory Affairs for Telekom Austria AG in Vienna. Between 2001 and 2002, he was Director for Interconnection, Roaming and Regulatory Affairs with 3G Mobile Telecommunications GmbH. He was a member of Piepenbrock Schuster Attorneys (now SBR Attorneys-at-Law) between 2002 and 2004. Since the foundation of

Piepenbrock Schuster Consulting (now: SBR-net Consulting), Austria, he has been the Managing Director of this company.

Naoko Sakurai is a Research Fellow at the Institute of Asia-Pacific Studies, Waseda University, Tokyo, Japan. She received her Bachelor's degree in Literature from Waseda University. She then worked for Waseda University as an administrative staff member for 35 years until 2013. She received her Master's degree (2006) and her Doctor of Science (2009) in Global Information and Telecommunication Studies from Waseda University. Her research interest is in the measurement of the value of information security, as well as the protection of personal information. She is a member of Japan's Society of Security Management and Society of Public Utility Economics. She has published many articles including 'The value and evaluation of information security' (in Japanese), 2011.

Manabu Shishikura is Professor at the Faculty of Economics, Nagasaki University, Japan. He received a PhD from Keio University. Prior to starting his academic career, he served for 11 years at the Ministry of Internal Affairs and Communications, Japan. His main areas of research are public economics, industrial organisation, and corporate strategies in media and network industries.

Minoru Sugaya is Professor of Media Policy and Economics at Keio University, Tokyo, Japan. After graduating from Keio University, he received an MA in Public Administration from the International Christian University, an MA in telecommunications from Michigan State University, and a PhD from the International Christian University. He was Visiting Researcher at the Kennedy School at Harvard University in 1991 and at Michigan State University in 1992. In 1998, he was a member of the Department of Communication at the University of Hawaii as a Fulbright Senior Researcher. Along with his abundant works in Japanese, he has also published many English articles including 'Media convergence policy in Japan: coping with multimedia' (*Keio Communications Review*, 2000); and 'Cable Television and Government Policy in Japan' (*Telecommunications Policy*, 1995). He has served in several roles as a government committee member and chairperson including universal service, cable television, and media convergence. He is also the former president of the Japan Association of Information and Communication Research.

Reza Tadayoni is Associate Professor at the Center for Communication, Media and Information Technologies (CMI) at Aalborg University, Denmark. He holds an MScEE from DTU (Danish Technical University) specialising in broadband communication, and holds a PhD from DTU in the field of media convergence. His main research focus is on ICT development and media convergence. He has participated in a number of Danish, European and international research projects, has supervised a number of PhD projects as well as Master and Bachelor theses, and has published 100+ scientific papers and research reports.

Keisuke Takachi is currently a Principal Economist at the Japan Center for Economic Research (JCER). In this role, he leads a study project on how the ongoing rapid progress in information and communications technologies (ICT) is affecting the long-term economic growth of Japan. He originally joined the Ministry of Internal Affairs and Communications of Japan in 1991 (then the Ministry of Posts and Telecommunications of Japan) and he led efforts such as the evaluation of competition in the Japanese telecommunications business field, the international deployment of Japanese digital television standard, ISDB-T, and the coordination of ICT policies in local governments. His academic research is mainly concerned with the analysis of competition around economic platforms in broadband industries and empirical studies of the consumer benefits of telecommunications services. He received a PhD in economics from Kyushu University, an MPA from the University of San Francisco and an LLB from the University of Tokyo.

Masatsugu Tsuji is Professor of Economics, Graduate School of Applied Informatics, University of Hyogo, Kobe City and Professor Emeritus of Osaka University, Japan. He received a BA from Kyoto University in 1965; an MA from Osaka University in 1967; and a PhD in Economics from Stanford University in 1976. He served as a Visiting Professor at Carnegie Mellon University, US and National Cheng Kung University, Taiwan; as a board director at the International Telecommunications Society and CPRsouth; on the editorial board of the *Journal of the International Society of Telemedicine and eHealth*, and *Smart Homecare Technology and TeleHealth*; as coordinator of e-Health Economics, the International Society for Telemedicine and e-Health. Current research focuses on identifying factors promoting innovation in Japan and ASEAN economies, in particular the role of ICT in the innovation process. Publications include *Industrial Clusters: Upgrading and Innovation in East Asia* (Edward Elgar, 2011); *From Agglomeration to Innovation: Upgrading Industrial Clusters in Emerging Economies* (Palgrave Macmillan, 2009); and *Industrial Agglomeration and New Technology* (Edward Elgar, 2007).

Chang-Ho Yoon is Professor Emeritus of Economics at Korea University, Seoul, Korea. He served as Professor of Economics (1982–2014), Dean of the College of Political Science and Economics, Korea University (2008–2010), Commissioner (non-standing) of the Korea Fair Trading Commission (2000–2006), and Member of the Policy Council, Ministry of Information and Communication, Korea (2004–2005). He has coedited several volumes on competition policy, ICT policy in Korea and North Korea in transition.

Preface

Information and communication technologies (ICT) have enabled us to enjoy a variety of applications and services through a diverse range of network technologies. With innovative progress in wired and wireless broadband technologies and devices, we are now greatly benefitting from new forms of services such as social networking and cloud computing services. We can easily access an abundant array of information that helps us seek out ways to live in comfort. In addition, as a consequence of the advancement of ICT applications, our society is ready to plunge into a new phase in which big data and open data are formed and processed so as to create innovative changes in value. 'Smart revolution' is defined as a term that symbolises such a dramatic change.

The objective of this book is to present a comprehensive evaluation of the smart revolution, including its social and economic impacts. It proposes a modern framework to help assess how recent progress in ICT can contribute to societies as a whole. The authors offer a guide to how advanced network societies have been realised through market mechanisms with the assistance of governmental policies that complement the deficiencies of this mechanism and foster the penetration of ICT. Through ICT, we are connected with each other more at both a personal and a global level, and this will ultimately herald a new era of ICT that will shape the 'digital society'.

This book addresses the following key questions and seeks to help readers understand what aspects are important in the digital society:

- How does the smart revolution impact economies and people's way of life?
- How has the ICT ecosystem been developed, and should changes be in accordance with the smart revolution?
- How are consumers' interests threatened by the smart revolution, and how should they be protected?
- How does the smart revolution create new environments for businesses and how does it affect stakeholders?
- How should the smart revolution enhance innovation and in so doing what kind of policies and deregulation should accompany it?

This book highlights a variety of issues including policies, economics, business and social issues that form the smart revolution, based on global experiences accumulated through the activities of the International Telecommunications Society (ITS). It is expected that it will stimulate future discussions on the smart revolution and contribute to the reformation of info-communications policies and business strategies.

While the planning of this book is rooted in the Silver Lining Conference of the ITS held in Tokyo in 2010, it presents an entirely original treatment, including reflections on dramatic events such as policies for emergency ICT systems against the background of the 2011 Japan earthquake. Similarly, other content is devoted to many of the recent innovative changes in ICT.

The publication plan of this book was able to overcome difficulties caused by the Great East Japan Earthquake in March 2011. Given the fact that, even after a two-year absence, the contributors to this book willingly responded to the book proposal, the editors would like to express heartfelt gratitude to them. The editors also express their warm thanks to PhD candidates Maria Massaro, Chalmers University of Technology, Nahoko Mitsuyama and Soonbo Nicholas Woo, Waseda University. Their editorial support work was extremely helpful.

Hitoshi Mitomo, Hidenori Fuke and Erik Bohlin

Introduction

Erik Bohlin, Hidenori Fuke and Hitoshi Mitomo

1. INTRODUCTION

Technology development in information and communication is increasingly gathering pace. In particular, smartphone technology is rapidly advancing and we are seeing new breakthroughs every day. The Internet and mobility, combined together, are driving massive changes in our societies, boosting what has been called the 'smart revolution'.

The smart revolution is so spectacularly pervasive that it is blurring boundaries between the real and the virtual worlds. The birth of intelligent technology marked the shift from the concept of mobile phones to that of smartphones. It is no longer simply about handheld devices that allow users to communicate with each other; it is about providers of highly advanced applications and functions, which rely on an ever astonishing ability to fit computing into our pockets.

Moreover, the emerging smart revolution is advancing the process of technology convergence between media, telecommunications and information technology, which were once distinct universes. The prerequisite for technology convergence is digitalisation, which has introduced a standardised information digital structure, enabling a whole new set of communication capabilities and opportunities. Predictions of technological singularity, meaning the unfolding of extra-human intelligent artificial technologies, leads us to understand that the implications of technological convergence are even broader than first predicted. Moreover, although first conceived of as a merging of previously separated fields related to technology, convergence is a multidimensional concept that goes beyond merely the field of technology. Indeed, it transforms economics, social, cultural and regulatory aspects of our societies.

The smart revolution needs to be modelled and analysed in order to understand its unique features and the roles played by the different entities and their interactions. To this end, the concept of a smart ecosystem shows promise. Within such a system, there are several roles to be played by different groups of entities that interact with each other and with their environment. A main point is that there is not only competition, but also different kinds of cooperation, symbiosis and mutual interdependence between the various markets and players.

Moreover, the smart revolution cannot be separated from globalisation, in which economies, societies, cultures and policy-making processes have become integrated through the global network of trade and transactions, communication, migration and movement. Information and communication technologies (ICTs) have lowered the cost of and increased access to digitalised information and, thus, accelerated the diffusion of innovation and promoted growth and productivity. Indeed, the smart revolution is now pushing forward the globalisation process, further shortening both time and distance, and advancing mobility, dynamism and integration at full steam. A long-term and fundamental transformation of our societies is taking place, and it is therefore of great importance for us to reform and improve existing policy settings in light of these new realities.

In broader terms, the information society is the queen of globalisation. And this queen – known by many names (digital society, knowledge society, network society, post-industrial society, etc.) – has many suitors. A policy race towards leadership of the information society has appealed to Europe, the United States, Japan and other strong economies, over time. The 1980s saw Japan leading the scene, bringing forward the Integrated Services Digital Network (ISDN), with a vision of a nationwide ISDN network that would evolve into B-ISDN afterwards, to include broadband. In the early 1990s, the United States developed the vision of constructing National Information Structures and Information Superhighways, under the Clinton-Gore Administration. In those same years, the European commitment towards an information society got off the ground. Several policy initiatives were adopted, with the main goal of fostering access to broadband connectivity.

During the first years of this century, Japan consolidated its supremacy in the mobile Internet technology field, with the tremendous success of the i-mode service. However, the smart revolution is overturning the rules of the game, and both Japan and Europe have lost important leadership and initiative in this new phase. In fact, Europe, who it was believed was leading the mobile economy, is underperforming not only relative to other advanced world powers, such as the United States, but also to some countries from the Asia-Pacific region, which are emerging as very strong forces.

Against this backdrop, policies, strategies and research must be conducted on the relationships between ICT technologies, global policy and regulation in this new emerging reality, in order to reap the full benefits of the smart ecosystem. We are experiencing a very long-term and fundamental transformation of our societies and therefore it is of great importance that we reform and improve existing policy settings based on these new realities, and conduct research to support such developments. Moreover, it is important for us to look even beyond the smart revolution. This book is one contribution to such an end.

2. AN OVERVIEW OF THE BOOK

This introduction will provide a brief summary of the following chapters in the book, and take up some pivotal issues for future research and policy pursued from a number of different perspectives towards and beyond the smart revolution. The book is divided into six parts, each focusing on different themes and aspects, as follows:

- Part I The ICT Ecosystem in Transformation;
- Part II What Did the Smart Revolution Achieve for Business and Society?;
- Part III New Issues Created by the Smart Revolution;
- Part IV Regulation Required for the Smart Revolution;
- Part V New Features in the Media Industry;
- Part VI The Smart Revolution in Countries.

Part I The ICT Ecosystem in Transformation

In Chapter 1, Anders Henten and Reza Tadayoni claim that the IT industry is leading the process of convergence between media, telecommunications and information technology. In support of this claim, the authors propose a theoretical framework building upon the basic concepts of convergence based on substitution and convergence based on complementation. To this, they add the structuralist concept of articulation to analyse the relationships between converging ICT sectors and the resource-based view (RBV) of companies with an extended sectoral perspective. Alongside this theoretical contribution to convergence theory, empirical evidence is examined corroborating the claim that IT is dominating the converging ICT sector. First, there is a comparison of the basic technological properties of Internet communications and telecoms. Furthermore, two cases are explored: one on the smartphone market and one on OTT TV. The two cases have an exemplary character illustrating the increasing dominance of the IT area in relation to telecoms and broadcasting, respectively.

Chapter 2, by Keisuke Takachi and Toshiya Jitsuzumi, calls for competition policy actions to cope with the changing market rules in the mobile industry in Japan. The authors start from acknowledging the widespread adoption of mobile services, and the prominent role played by content provision in terms of revenue. They focus on the behaviour of smartphone makers in relation to content aggregation, while assuming a monopoly in the smartphone manufacturing market, for simplicity of analysis. Thus, by means of an economic model to analyse profit maximisation of content aggregator(s) and monopolistic smartphone markers, the authors demonstrate how separation between smartphone makers and content aggregators and, consequently, promoting competition among content aggregators generates greater benefits for society as a whole. The authors believe these results are not restricted to Japan, but can also be applied to other markets where smartphone use is rapidly proliferating.

Chapter 3, by Minoru Sugaya, describes the transformation of the ICT ecosystem focusing on the shift of economic power traditionally stemming from communication networks, to digital technologies. This kind of power, known as 'digital power' finds its best expression in the smart device age. The implications of this shift in terms of industry structure, business models and policy responses are outlined. The development of the IT industry is examined from the perspective of economic power and categorised into three periods: the age of network, the age of platform and the age of smart devices. Each telecommunication regulator has been organised within the age of voice telephone and analogue broadcasting and their function in each period is re-examined. Through this process, the development of government and business relationships is clarified. The conclusive aim of this chapter is to illuminate the future development trends in the ICT industry discussing whether traditional regulators are able to cope with environmental changes.

Part II What Did the Smart Revolution Achieve for Business and Society?

Chapter 4, by Hiroki Idota, Teruyuki Bunno and Masatsugu Tsuji, makes out the case for greater consideration of social media as valuable enablers of product innovation, in particular in service industries. The authors attempt to examine the relationship between the role of social media inside the firm and product innovation and focus on the following research questions: 1) whether social capital influences the use of social media; 2) whether social media promotes product innovation; and 3) how the effects of social media on product innovation differ between the manufacturing and service industries. The authors argue that social media such as blogs (short form of 'web logs') and SNSs (social network services/social network sites) ease the flow of information between final consumers and firms' related functions/units, thus promoting the sharing of information that eventually contributes to product and process innovation. The analysis clarifies that well-informed social capital is a key factor for taking advantage of direct contact with customers by means of social media. For instance, if employees understand the objectives of the firm and collaborate with each other, SNSs can function well.

In Chapter 5, Kyoung-Youn Na and Chang-Ho Yoon report the results from an econometric analysis of complementarity between ICT and transportation infrastructures. By using dynamic panel data of OECD member countries, the authors empirically demonstrate the positive impact of widespread ICT use in its contribution to motorway infrastructure and production efficiency in developed countries. More specifically, the convergence of ICT and transportation technology has led to the development of the so-called intelligent transportation system (ITS), whose performance increases over time only if the ICT infrastructure grows concomitantly beyond a certain threshold level. The authors state that although the empirical analysis is carried out in OECD member countries, the pattern of complementarity can be applied to developing countries as well, where ICT convergence also takes place. They also point out the importance of measuring with accuracy the productivity effects of infrastructure investment.

The analysis of Akihiro Nakamura in Chapter 6 focuses on the role of ICT in disaster management, measuring the Japanese demand for countermeasures against natural disasters in the telecommunications field. The data are retrieved from an online survey that was conducted in collaboration with the Information Communications Research Institute of Japan in 2012, after the Great East Japan Earthquake hit the northern part of Japan, in March 2011. The results show that there is a very high willingness-to-pay (WTP) for provisions such as non-interruption of communications services and restoration of services within a 24-hour timeframe, with higher priority on mobile (voice) services when formulating appropriate backup plans for communication services during times of disasters. Moreover, the analysis has highlighted the national consensus on the necessity for countermeasures to be taken against the failure of telecommunications services. The novelty of this study relates to the absence of empirical studies for estimating the WTP for countermeasures in the telecommunication field.

In Chapter 7, relying on the same online survey used in the previous chapter, Hitoshi Mitomo, Tokio Otsuka and Mikio Kimura explore the 'Pythagorean effect' of media information in terms of people's contribution to and participation in the post-quake recovery process after the Great East Japan Earthquake. The data were collected from across Japan excluding the three heavily damaged prefectures and included more than

2000 samples aged from people in their teens to the over sixties. A two-fold approach was adopted. A deterministic approach was applied to construct a model to assess the contribution of media information and direct experience to post-earthquake behaviour. The study reveals that media information triggers information gathering while damage from the earthquake provokes countermeasures against possible aftermath events. An exploratory factor analysis is adopted to examine how people are impressed by media information and reveals different behaviour after the earthquake. In addition, the study shows that the role played by the Internet in providing information is still behind the substantive presence of television broadcasting.

Part III New Issues Created by the Smart Revolution

Chapter 8, by Rafael López, Teodosio Pérez-Amaral, Teresa Garín-Muñoz and Covadonga Gijón explores the problem of low-quality customer service in the mobile telecommunications industry. The authors provide theoretical support to the issue by using simple game theoretical models. Low-quality customer service levels are obtained as part of an equilibrium strategy profile for the firms. This is mainly owing to the existence of incomplete information: when signing a mobile phone contract, consumers cannot observe (ex ante) the operator's customer service quality and therefore firms may lack incentives to provide (ex post) high quality customer service levels. As a benchmark, the simple case of a one-shot monopoly market is investigated, showing that incomplete information leads to an inefficient outcome. Furthermore, the analysis of theoretical models with repeated interaction shows that the problem persists despite the introduction of competition. In terms of policy implications, this result suggests that inefficiencies should be solved through regulation via consumer protection.

Chapter 9, by Naoko Sakurai, addresses the problem of personal information leakage favoured by the adoption of ICT and the need to strengthen data protection systems. The study examines a sample comprising over 5200 incidents that occurred between January 2000 and December 2011 in a number of industry sectors. Specific attention was focused on the differences that existed among the types of industry that experienced leaks of personal information. The analysis reveals that particular sectors such as 'financial and insurance' and 'medical institutions' struggle the most with management of personal information. In addition, the results suggest that intentionally caused incidents tended to involve the leaking of large amounts of personal information. The final aim of this chapter is to discuss the problems and offer some concrete data that could be used for devising efficient security measures and to contribute to the further refinement, improvement or revision of current laws.

Part IV Regulation Required for the Smart Revolution

In Chapter 10, Martin Cave and Keiko Hatta engage in a deep exploration of two regulatory challenges that have arisen as a result of the proliferation of smart devices. First, they focus on the regulatory implications of fixed/mobile convergence, which is challenging the traditional assumption made by regulators that fixed and mobile retail and wholesale services do not fall within the same markets. There is a growing number of mobile only consumers that consider fixed and mobile services as substitutes. However, there is also a growing complementarity between fixed and mobile that is likely to delay the abandonment of fixed regulation. The second topic is cloud computing. Here, the greatest regulatory challenge is to time interventions directed at promoting interoperability and permitting portability of data, while avoiding the suppression of innovation and competition. Because suppliers of cloud services are also providers of devices and their operating systems, vertical issues may also arise. Overall, this chapter highlights continuing tensions between the promotion of competition and innovation and continuing to protect end users in an ever-changing regulatory regime.

Chapter 11, by Hidenori Fuke, reports on the increasing number of consumers complaining of unexpectedly high international mobile roaming services (IMRS) charges in Japan, as a consequence of the diffusion of smart devices. The analysis of the IMRS system in Japan is taken as an example to demonstrate that excessive deregulation works against users' interests. In fact, the author argues that one factor behind the increasing problems with IMRS is the excessive deregulation of cellular services. The IMRS market is not currently competitive and, if market forces are not working, some degree of regulation is necessary in order to protect users. The main conclusion is that the absence of regulation in the cellular market has gone too far, and thus works to the detriment of users' interest. The author also denounces the drawbacks of an extremely complicated IMRS structure that is not easily understandable for users. The main aim of the author is to initiate discussion on IMRS in Japan towards a reconsideration of deregulatory measures.

Part V New Features in the Media Industry

Manabu Shishikura and Norihiro Kasuga in Chapter 12 provide an empirical analysis of user demand for variety in the broadcasting platform. The author derives the demand function for channels and total TV watching time considering time and budget constraint, and shows that watching time is proportional to a user's available time while wages and price do not affect a user's total TV watching time. On the other hand, the demand for channels depends on the real wage rate, and a rise in wages increases a user's demand for the number of channels. Given the basic assumption that the viewing time of each channel is divided equally by the number of channels, the time spent watching each channel decreases as the total number of available channels increases. The authors empirically demonstrate the positive effect on the user of the increase in the number of channels. In addition, the result is that there is a difference in the marginal WTP for an increase in the number of channels between television advertising, public broadcasting and pay-TV. Moreover, the marginal WTP for additional channels of television advertising and public broadcasting is greater than that for pay-TV.

In Chapter 13, Yu-Li Liu suggests 'co-opetition' as the most viable strategy for the traditional and new media platforms to coexist in the smart ecosystem. Focusing on the media sector in Taiwan, the author discusses the competition among the players in the related TV market and analyses the impact of the newly emerging media on the traditional media platforms. The author argues that examples of cooperation and competition between different players such as new media, traditional platforms and content providers can often be observed in the media value chain. So, while newly emerging media services such as IPTV and OTT video services have begun to be popularised among users, by providing consumers with more choices, flexibility and convenience, traditional media have also found ways to match the new media platforms, adopting innovative business models characterised by time-shifting, on-demand and app delivery with different pricing options. With the impact of convergence and the structural change taking place in the media value chain, the players in the new ecosystem not only compete, but also cooperate.

Part VI The Smart Revolution in Countries

Chapter 14, by Gary Madden and Hasnat Ahmad, presents a snapshot of the Australian digital market and discusses evolving opportunities and challenges offered to consumers, businesses and the government. Consumers are changing, extensively using broadband for searches, gaming, media entertainment, social networking and shopping, and relying on smartphones and tablets for these purposes. This is stimulating retailers to develop digital strategies to meet customer expectations. For businesses, e-commerce provides opportunities to reduce costs and enhance sales. Indeed $189 billion of online orders were received during 2010–11. Importantly, 43 per cent of businesses report an online presence via a website, eBay store or social media page. This digital revolution is also encouraging online service delivery and engagement by the government. Examples of these developments include the Digital Local Government Programs that help communities in NBN deployment areas to develop planning applications, consultations and reports.

In Chapter 15, Ernst-Olav Ruhle, Matthias Ehrler, Natascha Freund and Martin Lundborg focus on the puzzling development of NGA networks in the European Union. Despite the fact that the European Union has undertaken massive initiatives, for example by introducing the digital agenda for Europe, thereby aiming at the implementation of higher bandwidths for end-users, investment activity is still rather low and therefore, information and communication infrastructures as the basis for a sustainable digital society are not as far developed as in other continents. The authors elaborate the reasons why the European Union is still relying on copper networks to a great extent for technological as well as for commercial reasons by describing general trends as well as by looking at specific case studies in four national markets, i.e. Germany, the UK, the Netherlands and Switzerland. A comparison between these four countries is provided and some specifics in comparison to other regions of the world are highlighted.

Chapter 16, by Chatchai Kongaut and Erik Bohlin, makes a contribution in the area of regulation by giving an overview of the problems of market definition analysis and significant market power frameworks in the EU. The focus is on the European telecommunications sector, where economic issues along with service convergence are bringing new challenges to both operators and regulators. For example, the two-sided market in mobile telephony affects the mobile termination market when defining relevant markets. Similarly, platform competition and substitution between fixed and mobile need to be considered when defining relevant markets for wholesale broadband services. The analysis of this study is based on document analysis of previous studies and reports, including the public consultation in 2013 on the revision of the Recommendation on relevant markets. The authors conclude by emphasising the importance for the market definition and SMP analysis to balance policy between economics and law, static and dynamic, and consider market definition case by case, depending on the priority and situation in each member state perspective.

3. BEYOND THE SMART REVOLUTION: TOWARDS A SUSTAINABLE DIGITAL SOCIETY

The smart ecosystem poses new challenges, but also new opportunities. There is a continuous shifting and interdependence between global and regional interests that calls for a comprehensive approach towards the building of an information society. From a policy perspective, there is an increasing need to coordinate several policy areas, making ICT a more integral part of our society, given that the use of ICTs when conducting activities makes the economy more transparent and efficient, favouring sustainable growth and human development. For instance, there is wide agreement on the importance of accessible and usable telecommunications infrastructures in disaster management.

The bridges between the information society, global market and national competitiveness, and the critical and still unresolved concerns of sustainable development are growing in importance. The issue of sustainability has been rising on the world scene as one of the most significant issues for the future of mankind. First thought to be exclusively related to environmental preservation, it actually encompasses several social, economic and cultural aspects. Moreover, there has been a shift from a static perspective towards the preservation of existing settings to a dynamic approach towards the creation of future opportunities.

The key formula is the implementation of a sustainable information society. The information society is considered a strategic organising principle, instrumental for the long-term well-being of any nation or region. Policies and strategies need to be defined for the future development path of the information society, as pursuing development without attention to sustainability will turn out to be very problematic for humanity in the long run. Sustainability is a very long-term challenge, as it means providing non-declining future welfare, balancing both present and future needs in a global context. How to implement a sustainable information society is all the more important in the current globalised environment sought after by several world powers such as the United States, China, Japan, India, Russia and Europe, to mention just a few, which seek to stand out, while developing strategies for mutual consensus and collaboration necessary for their long-term prosperity and security. As the links between the information society and sustainability are not yet well developed, researchers from different disciplines are expected to focus their studies on this phenomenon.

Sustainability concerns the path towards growth and prosperity, based on ubiquitous access to information and the production of a non-exhaustible resource: knowledge. However, it must be clear that the information society may generate both positive and negative trends; it is therefore important to evaluate the net effects of ICTs on sustainability. For instance, studies need to be focused on security issues, as cyber intrusions can have disastrous effects for humanity and, thus, need to be properly managed. Moreover, it needs to be understood to what extent the implementation of an information society impacts employment structures, labour skills and the development of various sectors such as public, health care, education and energy. Moreover, the consumption of ICTs greatly contributes to environmental pollution. From a cultural perspective, the information society might favour cultural homogenisation. It is still unclear whether this is critical for sustainable development or whether existing cultural identities must be protected. In the context of social sustainability, an important goal is to understand the impact of ICTs on lifestyles. Mobile Internet is considered a disruptive technology, given its potential to give rise to new ways of understanding, perceiving and interacting with and within the world. Concerns relate to the capability of living a life that can be sustained over time without compromising future generations.

An interesting ongoing debate centres on the concept of the well-being of a country. The idea pushed forward is that the traditional measure in terms of income or goods and services consumed by individuals partially represents the overall progress of a nation or region. The information society unavoidably impacts on the economy, but there is a psychological or subjective dimension of well-being that goes beyond the mere economic resources. It concerns measures of quality of life in its non-economic aspects.

Clearly, the pursuit of a sustainable information society has to face many challenges. The role of research is to improve our understanding of the relationship between the information society and sustainability in a geopolitical context, so that opportunities can be seized, while constraints can be restricted. This book is one step towards this ambitious goal.

PART I

The ICT ecosystem in transformation

1. The dominance of the IT industry in a converging ICT ecosystem

Anders Henten and Reza Tadayoni

1. INTRODUCTION

The aim of this chapter is to examine the increasing dominance of IT companies in the converging ICT industry and, on the basis of this development, to contribute to extending the theoretical understanding of market and industry convergence in the ICT area. Convergence is one of these concepts that will not go away. The relevance of the concept has been put into question time and again (Lind, 2004). However, convergence developments are increasingly affecting developments within the IT, telecoms and media sectors.

At a technical level as well as at a business and market level, IT, telecoms and media have increasingly been converging during the past decades based technologically on digitisation, in a broad sense, and on Internet, more specifically. In the last decade, it has become still clearer that IT is developing into being the dominant business area in this convergence. Relating to telecoms, the major problems of Nokia in the mobile market are an example of this development, where IT-based companies like Apple and Google presently are prominent players in the smartphone market. With respect to broadcasting, the progress of Over-The-Top (OTT) TV is a recent example of a similar development with IT rising as a dominant field in the convergence processes. Public service broadcasters are trying to counter the increasingly prominent position of IT-based companies in the broadcast area, for instance, with HBBTV (Hybrid Broadband Broadcast TV) initiatives; but they are on the defensive.

The term convergence is often used, but relatively little theory concerning convergence has been developed. At best, definitions of the concept have been suggested and the interrelationships between technical convergence and company, market and industry convergences, and policy and regulatory convergences have been discussed (Saxtoft, 2008). Furthermore the convergence of genres and the implications of the technology and market convergences for content production have been analysed from media research points of views (Jenkins, 2006). In the present chapter, emphasis is on industry and market aspects with technology developments as important drivers of convergence.

Greenstein and Khanna (1997) differentiated between convergence based on substitutes and convergence based on complements. Pennings and Puranam (2001) extended this framework of analysis with a supply and demand dimension, establishing a 2x2 framework, where Stieglitz (2003) later re-labelled supply (technology) and demand (product). These contributions still constitute the basic theoretical understanding of industrial convergence. The theory aim of our chapter is to further extend this framework towards being able to explain why and how new dominance relations develop in converging industries.

The approach used in the chapter is to confront the existing theoretical framework with recent developments in the ICT area. We are not witnessing a convergence on 'equal terms', but a convergence where one of the sectors clearly dominates and sets the conditions for the processes of convergence. This poses the question as to what the mechanisms for such an 'unequal' convergence are and how dominance is structured. An examination of these issues will allow for an extension of the existing theory on industry convergence.

The empirical material discussed is primarily the two exemplary cases of, respectively, the smartphone development, with Nokia as the big loser and the IT companies Apple and Google as big winners, and the video and TV development, with OTT companies on the way forward and traditional TV broadcasters on the defensive.

In order to analyse convergence from a point of view concerning the interplay between technology and market developments, the technology and market aspects will first be presented separately. In the section following the introduction, there is a presentation of technology developments, where emphasis is on the implications of the location of intelligence in the networks at the edges with the Internet end-to-end form of communication. The telecommunication operators have traditionally placed the intelligence inside the networks and have thus been in control of the streams of communication. This has changed with the Internet and the implications of this have become all the more visible with the increasing importance of the Internet.

In the subsequent section, theories regarding industry and market convergences and a theoretical approach to the dominance of one type of player or element vis-à-vis the others in a converging environment are presented. Theories regarding industry and market convergence take their point of departure in the abovementioned publications, primarily the one by Greenstein and Khanna (1997). The theoretical approach concerning the dominance relationship is based on the theory of the resource base of companies (Barney et al., 2011), on the one hand, and on the theory of articulation (Barthes, 1972), on the other.

The following two sections examine the cases of, respectively, Nokia vs Apple and Android-based producers in the smartphone markets, and OTT vs HBBTV in the video markets. Finally there is a summary and conclusion extending the theoretical understanding of convergence processes regarding IT, telecoms and media.

2. TECHNOLOGY DEVELOPMENTS

The emergence of the Internet is the most radical innovation in the communication field in the last decades. From a technological point of view, this innovation is driven primarily by the digitisation of communication technologies and platforms. The Internet is a 'network of networks' based on the TCP/IP protocol and is designed in a way that enables a radically different environment for innovation, development and organisation of services/applications and infrastructure platforms as compared to telecommunication networks. The history of the Internet, its design principles and the differences between the Internet and regular telecommunication networks are discussed in a number of papers including Saltzer et al. (1981), Clark (1988), Doyle et al. (2002), Willinger and Doyle (2002) and Leiner et al. (2009).

The Internet is engineered in such a manner that it has the potential to integrate all communication services in one and the same network. Some of the key people behind the design of the Internet were asked to write its brief historical account. They start their account by stating: 'The Internet is at once a worldwide broadcasting capability, a mechanism for information dissemination, and a medium for collaboration and interaction between individuals and their computers without regard for geographic location' (Leiner et al., 2009). This has proven to be a true statement, but only when certain conditions are fulfilled – as discussed further in the present section of the chapter. It is important to emphasise that when these conditions are fulfilled, the Internet is the only existing infrastructure that has the ability to integrate all communication services. This is an important issue in our discussion in this chapter concerning the dominance in the convergence process.

A number of design issues in the original development of the Internet provide the Internet with a comparative advantage over other infrastructures (broadcast and telecoms) and thereby give the IT sector a comparative advantage over the media and telecommunication sectors.

Clark (1988) divided the design principles (goals) into two categories: 1) the fundamental goals regarding the interconnection of different networks ('network of networks') using packet-based technologies; and 2) the 'second level goals', which are defined as a) robustness, b) ability to deliver multiple types of service, c) ability to accommodate a variety of networks, d) distributed architecture, e) cost efficiency, f) host connection with low efforts and g) accountability. Other important parameters that have been crucial for the development of the Internet have been scalability and flexibility, i.e. adaptation to the explosive growth and a design that can and actually does evolve over time in response to changing conditions (Willinger and Doyle, 2002). These design principles have resulted in an 'end-to-end' architecture, where the intelligence is moved from the core to the edges of the network.

Furthermore, the separation of the underlying network technology and the services has been vital in order to remove entry barriers for the service providers when entering the network. The only precondition for service provision is access to the network. This has created huge dynamics in service developments on the Internet. Denton (1999, p. 3) stated that '[t]he end-to end Internet model promises to do away with the idea that anyone would have a monopoly in the definition of services. If the Internet open-architecture model prevails, a telephone company of the future will still be able to define services, but so will every other user of the communication system capable of writing good code.'

One can say that packet-based switching, the robustness criteria implying a distributed architecture, and the service and network heterogeneity requirements (i.e. being able to deliver multiple types of service and being able to connect multiple networks without requiring changes in their design) were some of the design parameters that enabled the Internet to be a valid candidate to integrate all the communication services in one and the same network. However, when it comes to services that require specific quality of service (QoS) parameters, there are challenges that have worked against this integration.

The problems with QoS stem from the fact that the Internet is based on a 'best effort' principle, which means that the network, in its original design, does not guaranty a timely, reliable or secure communication service. Saltzer et al. (1981) analysed the security issue and argue that designing a network with high levels of security is a waste of resources as many services do not require high levels of security, and that the security issue can be dealt with at the edges of the network for those services that require security. Also the reliability issues are taken care of at the edges using the TCP transport protocol by retransmitting the 'lost packages'. Furthermore, the congestion control mechanism of TCP helps regulate from 'the edges' the use of resources in the core of the network in a sophisticated manner (e.g. Schiller, 2003), which dramatically reduces the need for traffic control and management in the core network.

The 'end-to-end argument' also applies to other QoS parameters. The end-to-end architecture enables the applications at the edges of the network to deal with the QoS issues. As the technology develops we can see more and more efficient ways of dealing with QoS parameters at the application level, which supports the provision of 'multiple types of service' on the Internet. 'The layered model of the Internet allows for many applications, each of which would have its own quality of services', suggested Denton (1999, p. 3). There are, however, issues related, in particular, to real time services (like voice and video telephony) that are highly sensitive to end-to-end delays and delay variations and services with high capacity demand (like video/TV) that have had a difficult time being delivered over the Internet. This is why we see that dedicated networks for telephony and TV are still coexisting with the Internet. This situation is, nevertheless, changing because issues related to, for example, delay variation are dealt with at the edge of the network and because the bandwidth capacity/throughput is increasing in the broadband networks.

The real integration of voice/video telephony and TV/video on the Internet is only possible when the development of broadband infrastructures and coding/compression technologies is at a level that makes it possible to deliver these services at the demanded quality over the Internet. As the capacity/throughput is a limited and costly resource, even when the access networks are capable of carrying these services, it has been important for TV/video to implement content delivery networks (CDNs) in the core network to bring the content as close to the users as possible. The CDNs have been instrumental for delivering TV/video over the Internet using different mechanisms including OTT provisions. In recent years, the broadband infrastructures are developing rapidly and CDNs are finding their way into many networks in the advanced markets. This creates good possibilities for the delivery of TV services like Netflix and the like over the Internet. However, in the short- and middle-term perspectives, we will still have dedicated networks for TV as it takes time to deploy broadband networks that can deliver the capacities needed for Ultra HDTV or 3D TV and the like.

As we can see in this brief account, the Internet is based on design principles that make it the prime candidate for integrating all communication services from the outset. The convergence process has in reality been dominated by a process of delivery of more and more communication services over the Internet. The process has been slow because of other circumstances, in particular the fact that broadband networks must be in place. But the tendency is clear: All services have shown the potential to be delivered over the Internet or the managed IP networks. In the discussion of the dominance in the convergence process, it becomes clearer why the IT/Internet sector dominates this process. In a geopolitical context, it also becomes clearer that the IT sector in the US has comparative advantages over anyone else in the world owing to its massive involvement in Internet design and development. This also applies to CDNs, which are highly US dominated.

3. THEORIES ON CONVERGENCE, ARTICULATION AND THE RESOURCE BASE OF COMPANIES

A large number of publications have been written on convergence from a technology point of view and from market, company, content or regulatory points of views – or combinations thereof (Saxtoft, 2008). However, the issue of convergence itself is relatively little theorised (Lind, 2004; Weaver, 2007). The interplay between technology, market, policy and cultural developments is often put forward as an overall analytical framework. Such an analytical framework provides a good general starting point for understanding the forces promoting convergence at a macro level. However, it does not, for instance, contribute much to the conception of the strategic action possibilities of companies at a micro level, nor does it in any sufficient manner provide a framework for understanding the dominance relationships at the meso and micro levels that we are interested in here.

3.1 Substitutes and Complements

Greenstein and Khanna's 1997 paper along with others in the same edited volume was a basic and successful attempt at putting the convergence concept into a theoretical context allowing for a deeper understanding of the business implications of convergence developments. The theoretical framework that Greenstein and Khanna (1997) presented differentiates between convergence based on substitution and convergence based on complementation. If, for instance, one service substitutes for another based on technology convergence – say VoIP substitutes for POT (plain old telephony) – this is a convergence based on substitutes. If, on the other hand, different services are provided on the same network, where they formerly would have been delivered on separate networks, we are dealing with a convergence based on complementation.

However, as pointed out by Greenstein (1999) in a paper summarising and discussing the findings of the 1997 Greenstein and Khanna paper, there is often substitution and complementation at the same time. The VoIP service mentioned will, for example, often be delivered as part of a multi-play service package in connection with IPTV and general Internet access. It is also possible that convergence processes will lead to a combination of substitution and complementation resulting in distinctly new services. Herzhoff (2009) found that there are five archetypes of convergence conceptualisation in the information systems (IS) literature: alignment, interoperability, optimisation, recombination and correspondence. Although these conceptualisations emanate from the IS research tradition and not from the business and management research tradition in which the Greenstein and Khanna paper is positioned, there is a considerable degree of overlap between the conceptualisations of convergence, but also inspiration to be found in the archetypes from the IS research tradition.

Pennings and Puranam (2001) extended the substitution/complementation framework with a supply and demand dimension, constructing a 2 × 2 framework of analysis. A substitution-based convergence, for instance, may be initiated from the demand side when 'the needs of different consumer sets become similar' and a complementation-based convergence may take its point of departure in 'different technologies brought together to create new kinds of technology' (Pennings and Puranam, 2001). Furthermore, Stieglitz (2003) re-labels the supply side as the technology variable and the demand side as the product variable in order to specify the primary factors affecting convergence developments.

Other contributions have been made, e.g. Mueller (1999), Lei (2000) and Hacklin et al. (2004). However, the contributions of Greenstein and Khanna (1997), Pennings and Puranam (2001) and Stieglitz (2003) still constitute the basic theoretical point of departure for understanding industrial convergence. Based on the original work by Rosenberg (1963) on the convergence developments in the machine tool industry from 1840 to 1910, another issue in the convergence literature is concerned with the differentiation between upstream technology convergence and downstream industry convergence (e.g. Gambardella and Torrisi, 1998). Moreover, the issue of convergence versus divergence plays a prominent role in the convergence literature, as industry developments point at the two opposite directions as the 'two sides of a coin' in dynamic industry developments caused by boundary changes in industries (e.g. Christensen, 2010).

In the present chapter, however, there is emphasis on the differentiation between substitution and complementation, as this most precisely points at the issue of dominance relationships between IT, telecoms and media, where these different sectors are distinct but yet converge as complements to a large extent but also, to some extent, as substitutes.

3.2 Articulation

The articulation concept has several different denotations. In the present paper there is focus on a conceptualisation that is not often seen presently and which takes its point of departure in the application of the concept of articulation by Barthes (1972). Articulation is an important concept in structuralist theory, where it is used to explain how different units are related to one another. Barthes committed a large part of his work to linguistics and semiotics and used the term articulation to signify an activity in which the rules of association between units in texts are discovered and established. Barthes (1972) identified two basic activities in structuralism, dissection and articulation where dissection means the finding of fragments engendering a certain meaning, and articulation means the activity of establishing the relationships between these fragments.

These ideas were taken up by Althusser and Balibar (1969) in their structuralist inspired interpretation of Marxist theory. Balibar introduced the notion that a 'social formation' is an articulation of different 'modes of production' and this notion was further developed by, for instance, Rey (1973) and a long range of other social scientists in the 1970s and 1980s (e.g. Taylor, 1979; Hindess and Hirst, 1977). The idea is that in specific societies, i.e. 'social formations', there can be different 'modes of production' existing at the same time, where 'modes of production' means the way in which goods and services are produced including the basic economic and social relations between classes in society. The modes of production are articulated in specific manners in relation to each other so that they will not only coexist but also be subject to complex interrelationships, where one mode of production will dominate over the others.

Henten and Tadayoni (2002) attempted a transposition of the concept of articulation for the purpose of analysing the interrelationships between traditional broadcasting and video on the Internet. At that point in time, video via Internet was only in its infancy. However, since then IPTV has developed and spread to a much larger extent and OTT TV has developed strongly. Therefore it seems timely to take up the discussion once again on a much broader basis encompassing the whole convergence process of IT, telecoms and media. In addition to the concept of articulation, the dominance relationships emphasised by the use of the concept in relation to the interplay between different modes of production are important to include when analysing the relationships between the fields of IT, telecoms and media. Not only do these fields converge; there are also relationships involving dominance between them.

This constitutes an add-on to the idea of convergence as complementation or substitution or combinations thereof. The fields of IT, telecoms and media complement and substitute for each other, subject to conditions of dominance. Other theoretical concepts such as business ecosystems (Moore, 1993) could also be helpful as analytical frameworks for examining the convergence developments. However, the ecosystem concept is better for studying relationships between leading companies – whether 'keystones' or 'dominators' as denoted by Iansiti and Levien (2004) – and the partners with whom they cooperate in smaller scale industrial settings. Specific ecosystems will be the results of certain articulations of IT, telecoms and media. However, for examining the broader picture of the convergence of IT, telecoms and media, the combined concepts of articulation and dominance are seen as a superior framework for analysis.

3.3 The Resource-based View

The next step to discuss is why and how the specific conditions of dominance develop. The question that needs to be answered is why it is that companies coming from the telecoms tradition or the media side in the convergence processes do not seem to be able to obtain a dominant position in the convergence of IT, telecoms and media. Many attempts have been made – some of them with success, for instance the i-mode service of NTT DOCOMO – however, most of the attempts with little success. The general picture is that convergence developments increasingly lead in the direction of IT as the dominant field in the convergence of IT, telecoms and media. The question is why companies from telecoms and media do not, to a sufficient degree, move into the IT field and subsume the IT field under their mode of operation.

An important and basic starting point for this discussion is the technology development as presented in the second section of the paper on Internet as an end-to-end network. This development constitutes the technological contribution to the dominance of IT companies in their articulation with companies from the telecoms or media sectors. However, this technological contribution is not sufficient for understanding the dominance of IT in the ICT convergence. Industries and companies are far more than their basic technological platforms. They also encompass the broader sets of resources and capabilities that industries and companies have, including knowledge, labour power, managerial competences and traditions, etc., and they also comprise the roles and positions that these industries and companies have acquired during their lifespan and development. This is where the resource-based view (RBV) of companies comes in as a useful framework – or rather the 'dark side' of RBV – as the resources that, in certain contexts, have been the basis for competitive advantages become a barrier in new market contexts. This is a phenomenon that, for instance, Christensen (1997) with the theory of disruptive innovations has pointed at.

The overall aim of the theoretical contributions on the resource base of companies is to determine why and how it is that some companies are able to obtain sustainable competitive advantages as compared to other companies in the same industry. As opposed to the theory of disruptive innovations, RBV focuses on the resources and capabilities (Amit and Schoemaker, 1993) of companies, allowing them to maintain competitive advantages. RBV focuses on the success stories and why and how companies obtain incumbent positions in industries, while disruptive innovations focus on the failures of incumbents.

The fundamental contribution of RBV to the understanding of the position of companies in markets is the focus on the resources that companies have and develop instead of the traditional neoclassical focus on the products – their qualities and prices. The emphasis of RBV is thus on the internal resources and processes in the companies, while neoclassical theory focuses on the external environment – the markets in which the products are sold. According to RBV, the competitive advantage of companies is based on heterogeneous bundles of tangible and intangible assets (Hewitt-Dundas, 2006) comprising knowledge assets, human assets, technical assets, economic assets, etc.

In a summary of the developments of RBV, Barney et al. (2011) provided an overview of the most important contributions to the theory of the resource base of companies and the discussions and criticism of the theory, beginning with Penrose (1959), who presented a starting point for the later contributions on RBV, giving emphasis to the constraints that inadequate resources can put on the development of companies. The term 'resource-based view' was coined by Wernerfelt (1984), and in 1991 Barney developed the characteristics of resources that would provide companies with competitive advantages: The resources must be valuable, rare, inimitable and non-substitutable. Amit and Schoemaker (1993) differentiated between resources and capabilities, where the resources are the companies' assets, and the capabilities are the competences of the companies to leverage those resources. Additionally, Teece, Pisano and Shuen (1997) introduced the idea of dynamic capabilities.

These contributions mostly emphasise the resources and capabilities of companies leading to competitive advantages, except for Penrose (1959), giving emphasis to constraints. The question could then be why it is that companies are not able to further develop their 'dynamic capabilities' and why it is that they lose their leading positions, as industries and technologies develop. This is where the theory on disruptive innovations provides some explanatory power (Christensen, 1997). The theory on disruptive innovations builds on the idea of the centrality of the resources and capabilities of companies, but focuses on why it is that companies fail in spite of their former dominant position. The explanation is, in a very brief version, that the resources and capabilities of incumbent companies are too path dependent on the hitherto dominant technologies and business models and, therefore, have great difficulties in changing direction when technology and market circumstances change. This is, to a large extent, what is happening with telecoms and media companies in the new convergent environments. The business models of companies in the telecoms area and in the broadcasting and, in broader terms, media areas have existed for a long time and have developed a certain degree of rigidity in their approach to markets.

4. NOKIA VERSUS APPLE AND GOOGLE

The Nokia case illustrates clearly the issues related to the dominance relationships in the convergence processes and in particular the dominance of the IT sector, represented in this case by Apple and Google.

From 1998 to 2011, Nokia was the biggest producer of mobile handsets in the world. Nokia reached almost 50 per cent of the worldwide mobile handset market and was successful in economically more developed markets as well as developing economies. There were a number of reasons for Nokia's success. In terms of the handsets and GUI, Nokia developed better GUIs than other handset producers and their GUIs were recognisable from generation to generation. Furthermore, the terminals were highly robust and of high quality.

If we go further back into history and look for reasons why Nokia became a success from the beginning, it was the timing of its decision to change from analogue to digital, which was of great importance: 'In early 1992, there were many who did not really believe in digital technology as analogue technology was very mature and many established companies would stick to analogue. Nokia chose to believe in the digital technology.' This is a statement taken from an interview with one of the former directors of the company. Other circumstances have been vital for this success, among them: the liberalisation of the European telecom market; the European GSM decision; the reorientation of the Finnish economy towards Western Europe after the downfall of the Soviet Union; the early liberalisation of the Finnish telecommunication market; the tradition that came from NMT (Nordic Mobile Telephone); a strong engineering culture in Nokia; and successful strategic buyouts.

The question is then why Nokia was not able to follow up on its previous success in the mobile handset market. This question has a demand-side as well as a supply-side dimension. On the demand side, there is a range of different issues relating, e.g., to the usability of mobile handsets, where iPhones, when being launched, simply had a superior usability compared to any other smartphone at the time. There is also the branding issue, where Apple has been the leader since the release of the first iPhones and where Nokia has been the loser. Furthermore, there is the issue of the relative ease of switching from one handset to another for the users. The mobile handset market is a very volatile market, which, for instance, Motorola witnessed before Nokia with the great success and downfall of their Razr phones.

On the supply side, two interrelated sets of factors are relevant for answering the question regarding Nokia's problems in the smartphone market. One set of factors has to do with the overall market developments – in this case, the growth of the smartphone market and the increasing dominance of IT companies in the converging ICT markets. This set of factors could be termed the 'external environment' but it is no more external than allowing for the huge influence that Nokia has had on market developments. This is why one has to see the two sets of factors as interrelated. The other set of factors could be considered as internal in the sense that they are centred on the individual company and its resource base – in this case, Nokia Mobile. The main issues relating to the resources of the company is the fact that Nokia, to a large extent, was an engineering driven company and, furthermore, that it is a telecommunication hardware company. Nokia obviously also had marketing functions and design functions, but the basic DNA of the company is its engineering functions. Moreover, it is basically a telecommunication hardware company. Software is clearly extremely important for mobile handset providers – and even more so with smartphones. Developing and implementing operating systems is a crucial aspect of Nokia's development strategy as can be seen in the emphasis on the development of Symbian and, later, the implementation of Microsoft mobile operating software. However, a major element in explaining the problems Nokia is encountering in the smartphone market is likely to be related to the fact that software is not Nokia's core business.

With respect to business model and ecosystem modelling, Nokia misjudged the dominant business models in the smartphone market. Nokia stayed with the idea that it would be the mobile network operators that would organise the upcoming mobile Internet markets. This, however, turned out not to be the case. The organisers and dominant players in the mobile Internet and smartphone markets have turned out to be IT companies – Apple and Google. This entails quite different business models and ecosystems, which Nokia was not able to catch up on and consequently the company was forced into a partnership with Microsoft, which ended up in the total acquisition of Nokia Mobile (Nokia's Devices and Services unit) by Microsoft. What went wrong for Nokia was that the company connected its business model/ecosystem closely with telecommunication operators. In short, Nokia delivered handsets to the telecommunication industry, not putting enough emphasis on the fact that the convergence process changes the dynamics in the mobile market. As soon as the mobile terminals became IP enabled, either through the mobile networks or through Wi-Fi or the like, the stakeholders from IT and Internet could enter and change the business radically. This was what happened with the introduction of the iPhone in 2007 – the first version ever with only Wi-Fi.

When Apple came with iPhone in 2007, Nokia had been in the smartphone market a couple of years and the Nokia N95 was a popular smartphone that sold quite well. But what made iPhone and the later Google Android based ecosystems different was that they used the Internet protocol as the network infrastructure and by that took advantage of the separation between services and network infrastructures by moving the service creation and development from the core to the edge and from the mobile operators to the entrepreneurs at the edge of the network. Consequently, they had the possibility to get shares in the value propositions involving millions of producers and billions of consumers.

In February 2014 just before the final acquisition of Nokia Mobile by Microsoft, the company launched a new family of devices named Nokia X, based on a version of Android OS. These devices are based on Nokia's map (HERE) and their services are based on Microsoft services (Skype, Outlook, etc.) so generally the look and feel is like a Microsoft phone. This could be seen as a signal that even when Nokia becomes an integrated part of Microsoft there remains the possibility of using Android OS in parts of Microsoft's mobile device portfolio. In this chapter we don't discuss what could have happened if Nokia had made this decision to use Android earlier in the process of competition between standards for smartphones, but the decision to develop devices based on Android OS shows even more clearly the superiority and dominance of IT companies like Google in the convergence process.

5. THE CASE OF TV OVER INTERNET – FOCUS ON OTT

As discussed in the section concerning the Internet, Internet can be used as the network infrastructure for delivering TV when certain circumstances are in place and certain requirements are met. There are different ways of delivering TV using the Internet Protocol: IPTV (Internet Protocol TV), HBBTV (Hybrid Broadband Broadcast TV) and OTT TV (Over-The-Top TV). As seen from the following, OTT TV illustrates the IT dominance in the convergence process.

IPTV denotes the provision of TV through broadband networks using the Internet Protocol (IP). IPTV can be provided both as managed services and as a best effort service. However, the term IPTV generally denotes the delivery of TV through managed broadband networks. In such a managed provision, the IPTV provider guarantees a specific degree of quality of service (QoS), while in the case of best effort it is offered through the general Internet. There are a number of different IPTV technologies available on the market, and ITU (the International Telecommunication Union) has specific focus on developing standards for IPTV. IPTV as a managed service uses only the Internet Protocol as the delivery mechanism and does not change the traditional broadcasting business model radically.

Recently, however, the TV sector has witnessed changes following two different tracks: 1) a wave of changes integrating broadcast with the Internet world under the heading of Hybrid Broadband Broadcast TV (HBBTV); and 2) a pure Internet-based approach under the heading of Over-The-Top (OTT) TV technologies and platforms.

The idea of HBBTV platforms is to connect smart TVs to broadcast as well as broadband networks, enabling the broadcasters to develop advanced services related to their content. The HBBTV idea is to make it possible for the broadcasters to retain control of the platforms and the associated applications. EBU (the European Broadcasting Union) has been one of the main promoters of this development.

With OTT, the Internet is seen as the generic platform for the provision of broadcast content and the associated advanced services regarding, for instance, the integration with social networking and community services taking advantage of the proven dynamics and huge innovation potentials of the Internet. This track is US centric with companies like Google and Apple the main players. OTT television is a relatively new term that has arisen partly for technological reasons and partly for business and regulatory reasons. The technological part is concerned with the fact that technological solutions have been developed to ensure a higher degree of

QoS of television/video delivered over the general Internet. This applies, for instance, to the HLS (HTTP Live Streaming) standard promoted by Apple, the WebM technology developed by Google or the Silverlight Smooth Streaming solution by Microsoft.

The recent technological developments have facilitated the development of business approaches, delivering streaming video and audio content over the general Internet in a sufficiently good quality for the users to utilise it as a source of their consumption of video and audio content. Internationally known examples of such business enterprises are Netflix and Hulu. But there are already a large number of such business companies, and in the coming years there will be a huge variety of companies with a narrower, more national focus. These companies are basically content aggregators assembling different kinds of content (movies, series, and other video and audio programs), which are marketed to the users utilising different types of business model: for example, subscription or advertising.

These types of company are often called third party providers, as they are neither content producers nor network providers. They are content aggregators, who provide content packages over the general Internet – though some of them will certainly expand into neighbouring areas such as content production. The fact that they are third parties using the general Internet as the means of transportation also touches upon the regulatory reason for the development of the OTT term. The situation is that ISPs that have been providing managed triple play services lose control of content provided over their networks and become mere 'bit pipes' for the OTT providers. Not only do they lose a business opportunity (triple play), but they are also supposed to deliver the network capacity, which is necessary for the Internet users to get access to the OTT services. In the long run, this will require investments in infrastructure facilities, which the network providers are not certain that they can get compensated by their subscribers with decreasing broadband access prices and flat rate subscriptions. The OTT term pinpoints this situation.

The OTT development must be seen in light of the general relationships between companies coming from the IT world versus companies from the broadcasting and telecommunication world. These different spheres are increasingly converging, but it seems that it is the IT world that is increasingly dominant. As seen in the Nokia case this has already, to a large extent, happened in the mobile communication area, where Androids (Google) and iPhones (Apple) have become dominant. In addition, it also seems to be happening in the television and video/audio area – Apple's and Google's interest here is to get into this new realm of the daily life of users and to combine the video services with their other Internet services. Microsoft has been working on being there for a long time with the Xbox as an example. But other large IT companies are also moving into this sphere.

6. SUMMARY AND CONCLUSION

An important aim of this chapter has been to attract attention to the emerging relations of dominance in the convergence of IT, telecoms and media. In the context of the shifting relations of dominance internationally in the mobile communications area, Henten, Falch and Vandrup (2010) discussed the increasing US dominance in the mobile Internet area as opposed to earlier phases in mobile developments, where Europe dominated the 2G phase with GSM and Japan forged ahead with NTT DOCOMO's i-mode service at the very beginning of the 21st century. Since the middle of the first decade of the 21st century, companies from the US have increasingly come to the fore, which Henten, Falch and Vandrup (2010) explain as being based on the strong position of the US regarding IT and Internet.

The chapter has examined two kinds of empirical evidence in order to discuss the emerging relations of dominance in the ICT area. First, there was a comparison of the basic technological properties of Internet communications and telecoms. Furthermore two cases have been discussed: one on the smartphone market and one relating to OTT TV. The two cases clearly illustrate the increasing dominance of the IT area in relation to telecoms and broadcasting, respectively.

The two cases illustrate a general development trend. Digitisation is the very basic technological prerequisite for ICT convergence. However, the Internet constitutes the more specific digital infrastructure that today is a fundamental defining basis for the actual convergence developments. How digital convergence would have developed without Internet is, to a large extent, a speculative endeavour, although prior but terminated technology developments could illustrate other development paths. One example of such a prior development path is ATM-based network technologies initiated by the telecoms field, which have lost importance with the proliferation of TCP/IP. Another example is the development of videotex in the late 1980s and early 1990s. This was a system developed by telecoms companies that preceded the graphical interface developed in connection with the Internet World Wide Web. However, with the development of the WWW, videotex quickly disappeared except in France, where it vanished more slowly, as the French incumbent telecoms operator had invested enormously in its videotex system, Teletel (Bouwman and Christoffersen, 1992).

Internet is thus the core defining technological complex, and although Internet did not come out of private IT companies, IT companies – and especially US-based IT companies – were much faster to incorporate Internet technologies into their products and services. Internet with its end-to-end property was basically in line with the networking interests of IT companies. IT companies in the US had worked since the 1960s on getting into networking, including having the right to use the telecommunication networks for computer networking. What the IT companies needed was for the networks to work as dump bit-pipes for their computer communication and this collided with the interests of the telecommunication operators who wanted to maintain control of the networks, which entailed keeping the intelligence in the networks.

The basic philosophy of the Internet was well in line with the interests of IT companies. Furthermore, IT companies were the new 'kids on the block'. The IT field was the new dynamic business area as opposed to telecommunication and broadcasting, which had existed for approximately 100 years in the case of telecoms and approximately 50 years in the case of broadcasting. This has meant that telecoms and broadcasting are embedded in longstanding modes of operation that have constrained their possibilities for adopting the new Internet mode of operation. Part of this is also that telecoms as well as broadcasting have traditionally been tightly regulated sectors, which adds an additional element to their path dependent behaviour.

The increasing emergence of IT in ICT convergence has thus been on its way for quite some time. However, development of the relationships between IT, telecoms and media has far from ended. It is a continuous process, and the purpose of this chapter has been to analyse how these processes of convergence evolve in terms of complementation, substitution and dominance.

The differentiation between complementation and substitution proposed by Greenstein and Khanna (1997) constitutes a good point of departure for analysing how convergence between IT, telecoms and media evolves in terms of new products and services as a result of novel combinations and the closing down and establishment of other product and service areas. This provides a valuable framework for analysing the strategic options of companies in such markets. However, it does not provide a sufficient framework for understanding the conditions for dominance between the different types of market players. The articulation concept provides the foundation for such a framework. While the concept of articulation has been used for how 'modes of production' are combined, it can also be used for the combination of the modes of operation of different market players, coming from IT, telecoms and media, respectively.

Additionally, in order to understand the mode of operation of the different types of company and the different kinds of sector, the resource-based view of companies with its emphasis on the resources and capabilities that promote or constrain the development paths of the companies in question becomes the third element in a theoretical framework that helps in analysing the evolving dominance relationships in ICT convergence.

Adding the articulation of modes of operation, the resource-based view and its kindred theory on disruptive innovations makes it possible to examine the relations of dominance in ICT convergence, where there are strong indications that IT is emerging as the dominant field. The manner in which IT, telecoms and media are articulated allows for companies from the IT field to subsume the interest of companies from the telecoms and media fields under their strategic interests. Therefore, the framework proposed is a combination of the complementation/substitution framework with the articulation concept used on modes of operation and the resource-based view on companies extended to a sector perspective and the path dependent developments of the IT, telecoms and media sectors.

REFERENCES

Althusser, L. and E. Balibar (1969), *Lire 'Le Capital'*, Paris: Maspero.

Amit, R. and P.J.H. Schoemaker (1993), 'Strategic assets and organizational rent', *Strategic Management Journal*, **14** (1), 33–46.

Barney, J.B. (1991), 'Firm resources and sustained competitive advantage', *Journal of Management*, **17** (1), 99–120.

Barney, J.B., D.J. Ketchen Jr. and M. Wright (2011), 'The future of resource-based theory: revitalization or decline?', *Journal of Management*, **37** (5), 1299–1315.

Barthes, R. (1972), 'The structuralist activity', in R. Barthes, *Critical Essays*, Evanston, IL: Northwestern University Press.

Bouwman, H. and M. Christoffersen (eds) (1992), *Relaunching Videotex*, Dordrecht: Kluwer Academic Publishers.

Christensen, C.M. (1997), *The Innovator's Dilemma: When New Technologies Cause Great Firms to Fail*, Boston: Harvard Business School Press.

Christensen, J.F. (2010), *Towards a Framework of Industry and Product Market Convergence*, DRUID, Imperial College London Business School, 16–18 June 2010, available at http://www2.druid.dk/conferences/viewpaper.php?id=501467&cf=43 (accessed 11 April 2013).

Clark, D.D. (1988), 'The design philosophy of the DARPA Internet protocols', SIGCOMM '88 Symposium Proceedings on Communications Architectures and Protocols, ACM, New York, pp. 106–114.

Denton, T.M. (1999), *Netheads versus Bellheads: Research into Emerging Policy Issues in the Development and Deployment of Internet Protocols*, final report for the Federal Department of Industry, Contract Number U4525-9-0038, available at http://www.tmdenton.com/images/papers/bellheads.pdf (accessed 26 February 2015).

Doyle, J.C., S.H. Low, F. Paganini, G. Vinnicombe, W. Willinger and P. Parrilo (2002), *Robustness and the Internet: Theoretical Foundations*, available at http://netlab.caltech.edu/publications/RobustInternetII-Doyle-2002.pdf (accessed 2 May 2013).

Gambardella, A. and S. Torrisi (1998), 'Does technological convergence imply convergence in markets? Evidence from the electronics industry', *Research Policy*, **27** (5), 445–63.

Greenstein, S. and T. Khanna (1997), 'What does industry convergence mean?', in D.B. Yoffie (ed.), *Competing in the Age of Digital Convergence*, Boston, MA: Harvard Business School Press, pp. 201–26.

Greenstein, S. (1999), *Technological Convergence*, mimeo, available at http://www.kellogg.northwestern.edu/faculty/greenstein/images/htm/Research/articles/convergence-handbook-final.pdf (accessed 2 May 2013).

Hacklin, F., V. Raurich and C. Marxt (2004), 'How incremental innovation becomes disruptive: the case of technology convergence', *Proceedings of the IEEE International Engineering Management Conference*, Singapore, 2004, available at http://ieeexplore.ieee.org/stamp/stamp.jsp?tp=&arnumber=1407070 (accessed 5 April 2013).

Henten, A. and R. Tadayoni (2002), *Constitution of Markets for Internet TV*, CTI Working Papers No.71, Lyngby: DTU.

Henten, A., M. Falch and K. Vandrup (2010), 'Mobile Internet developments in Europe, East Asia and the US', in A. Gentzoglanis and A. Henten (eds), *Regulation and the Evolution in the Global Telecommunications Industry*, Cheltenham: Edward Elgar Publishing, pp. 317–39.

Herzhoff, J. (2009), 'The ICT Convergence Discourse in the Information Systems Literature: A Second-Order Observation', 17th European Conference on Information Systems, 8–10 June 2009, Verona, Italy, available at http://digital-infrastructures.org/publications/Herzhoff2009.pdf (accessed 28 February 2014).

Hewitt-Dundas, N. (2006), 'Resource and capability constraints to innovation in small and large plants', *Small Business Economics*, **26**, 257–77.

Hindess, B. and P. Hirst (1977), *Mode of Production and Social Formation*, London: MacMillan.

Iansiti, M. and R. Levien (2004), 'Strategy as ecology', *Harvard Business Review*, **82** (3), 68–78.

Jenkins, H. (2006), *Convergence Culture*, New York: New York University Press.

Lei, D.T. (2000), 'Industry evolution and competence development: the imperatives of technological convergence', *International Journal of Technology Management*, **19** (7/8), 699–738.

Leiner, B.M., V.G. Cerf, D.D. Clark, R.E. Kahn, L. Kleinrock, D.C. Lynch, J. Postel, L.G. Roberts and S. Wolff (2009), *Brief History of the Internet*, available at http://www.internetsociety.org/internet/what-internet/history-internet/brief-history-internet (accessed 31 May 2013).

Lind, J. (2004), *Convergence: History of Term Usage and Lessons for Firm Strategies*, available at http://www.userpage.fu-berlin.de/~jmueller/its/. . ./1_LIND.doc (accessed 20 May 2013).

Moore, J.F. (1993), 'Predators and prey: a new ecology of competition', *Harvard Business Review*, **71** (2), 75–86.

Mueller, M.L. (1999), 'Digital convergence and its consequences', *The Public*, **6** (3), 11–28.

Pennings, J.M. and P. Puranam (2001), *Market Convergence and Firm Strategy:*

New Directions for Theory and Research, paper presented at the ECIS Conference on 'The Future of Innovation Studies', 20–23 September, Eindhoven, the Netherlands.
Penrose, E. (1959), *The Theory of the Growth of the Firm*, New York: Wiley.
Rey, P.-P. (1973), *Les Alliances des Classes*, Paris: Maspero.
Rosenberg, N. (1963), 'Technological change in the machine tool industry, 1840–1910', *The Journal of Economic History*, **23** (4), 414–43.
Saltzer, J.H., D.P. Reed and D.D. Clark (1981), 'End-to-end argument in system design', *Proceedings of the Second International Conference on Distributed Systems*, Paris, France, 8–10 April 1981, IEEE Computer Society, pp. 509–12.
Saxtoft, C. (2008), *Convergence: User Experience, Communications Enablers and Business Opportunities*, Hoboken, NJ: Wiley.
Schiller, J.H. (2003), *Mobile Communications*, 2nd edn, Boston, MA: Addison-Wesley.
Stieglitz, N. (2003), 'Digital dynamics and types of industry convergence: the evolution of the handheld computers market in the 1990s and beyond', in J.F. Christensen and P. Maskell (eds), *The Industrial Dynamics of the New Digital Economy*, Cheltenham, UK and Northampton, MA: Edward Elgar Publishing, Chapter 7.
Taylor, J.G. (1979), *From Modernization to Modes of Production: A Critique of the Sociologies of Development and Underdevelopment*, London: Macmillan.
Teece, D., G. Pisano and A. Shuen (1997), 'Dynamic capabilities and strategic management', *Strategic Management Journal*, **18** (7), 509–34.
Weaver, B. (2007), 'Industry convergence: driving forces, factors and consequences', *19th NFF Conference*, Bergen, 9–11 August 2007, available at http://www.nhh.no/Files/Filer/institutter/for/conferences/nff/papers/weaver (accessed 24 May 2013).
Wernerfelt, B. (1984), 'A resource-based view of the firm', *Strategic Management Journal*, **5** (2), 171–80.
Willinger, W. and J.C. Doyle (2002), *Robustness and the Internet: Design and Evolution*, available at http://netlab.caltech.edu/publications/JDoylepart1_vers42002.pdf (accessed 20 March 2013).

2. Model analysis of the two-sided market of the mobile broadband business: in Japan and worldwide

Keisuke Takachi and Toshiya Jitsuzumi

1. BACKGROUND AND OBJECTIVE

Mobile broadband services are widely used in Japan: the Telecommunications Carriers Association of Japan states that Japan's 129 million mobile subscribers are long-term evolution or third generation (3G) mobile service users as of the end of 2012.[1] The use of smartphones is rapidly proliferating: according to MIC (2013), it is expected that 84 per cent of the approximately 39 million mobile terminals shipped in Japan in 2012 are smartphones. An increasing proportion of the profits of mobile operators is derived from the smartphone business, e.g. from the sale of content[2] distribution, billing to content usage and the sale of high-spec smartphones; this shift is now influencing these companies' business strategies. Not only mobile operators but also all other players currently in the mobile broadband business – such as content aggregators, smartphone makers and consumers – must adapt to these new market conditions.

The Japanese government is no exception; it needs to consider whether it is necessary to take new policy action to regulate these new market conditions. The Ministry of Internal Affairs and Communications (MIC), the regulatory authority in Japan, has been paying attention to the market conditions of the mobile broadband business and one of its foci is whether mobile operators' exclusive provision of network functions relevant to content distribution (e.g. authorisation, billing and management of content menu) has any effect on the development of the mobile broadband ecosystem. For example, the MIC (2009) defined those functions as comprising the telecommunications platform (p. 5) and proposed a policy initiative to enhance diversity and interoperability among mobile operators. The MIC's main target was NTT DOCOMO's i-mode service; at that time DOCOMO's market share was about 50 per cent and the company was very powerful in the market, since the company was able to fully control the services and the specifications of feature phones in the i-mode business ecosystem. The MIC became seriously concerned that DOCOMO would intervene in the rapidly growing content market, by leveraging its market power in the mobile broadband market.

There has been much policy-oriented discussion on a monopolist's strategic behaviour in adjacent markets. According to Rey and Tirole (2007), such strategic behaviour is called foreclosure; it refers to a dominant firm's denial of proper access to an essential good it produces, with the intent of extending monopoly power from that segment of the market (e.g. the 'bottleneck segment') to an adjacent segment (e.g. the potentially competitive segment). There is an argument wherein Farrell and Weiser (2003) use the term internalising complementary efficiencies; according to that argument, even a monopolist has incentives to provide access to an essential good it produces, or its platform, when it is efficient to do so, and to deny access only when access is inefficient. Whinston (1990) and Economides (1998) nonetheless conclude that a monopoly does have incentives to discriminate downstream complements in cases where the platform monopoly makes a commitment to a certain technology or has rivals to its subsidiary. Van Schewick (2007) argues that even a competitive platform provider may foreclose its downstream market to increase switching costs.

Today, in the age of smartphones, regulatory authorities need to reconsider this subject. In the current mobile broadband business, Apple and Samsung are dominating the value chain and controlling the quality of customer experience, rather than mobile operators. Moreover, companies such as Google, Facebook and Skype, which are considered 'over the top' services, are providing substitutes for traditional telecommunications services. Changes in the mobile broadband business have pushed the MIC to expand its policy interest, from vertical integration by mobile operators to the whole of the mobile broadband business. For example, the MIC (2011) stated that 'it is possible that goods and services of non-telecommunications service layers – in which the attractiveness of mobile terminals or content is important to consumer decision-making – significantly influence competition among mobile operators' (p. 27); as such, the MIC said that it would closely observe market trends.

The objective of this chapter is to examine two different models of the current mobile broadband business as well as their influence on enterprise profits and the necessity for policy intervention. In this chapter, we provide the latest information on the Japanese mobile broadband business and a literature review and the basic thinking underpinning our analysis. Then we explain models used in our analysis and present simulation results. Finally, we derive some policy implications that are applicable not only to the case of Japan, but also to other markets where smartphone use is rapidly proliferating.

2. CHANGES IN THE MOBILE BROADBAND BUSINESS

The MIC (2012) has published several interesting data sets that highlight changes that have occurred in the mobile broadband market. Between FY2009 and FY2011, the average revenue per user from data communications exceeded that of voice communications; at the same time, mobile operators expanded their non-telecommunications sales (e.g. content distribution, billing charges and terminal sales). KDDI and Softbank in 2011 saw non-telecommunications revenues of JPY 251.6 billion and JPY 237.2 billion, respectively; these numbers represent a 50 per cent increase compared to the previous year. These facts indicate that content-related revenue has become more important for mobile operators than traditional voice-related revenue.[3]

The MIC (2013) provided the most recent overview of global and Japanese smartphone markets. The total number of smartphones shipped worldwide has grown from 172 million in 2009 to 680 million in 2012 and the market shares of various operating systems (OSs) changed greatly during that time. In 2009, 41.1 per cent of all smartphones used a Symbian OS and 19.9 per cent used BlackBerry (RIM); in 2012, however, the market share of Android smartphones reached 66.4 per cent while 19.1 per cent were iOS. In Japan, smartphone shipments have also grown from 17.4 million in 2009 to 32.2 million in 2012. In terms of OSs there, in 2009, 50.3 per cent were Symbian OS, 39.9 per cent were Linux, 7.4 per cent were iOS and only 0.7 per cent were Android; in 2012, however, 53.7 per cent were Android and 27.9 per cent were iOS. See Figures 2.1 to 2.4 for more information on global and Japanese market share during 2009 and 2012.

It should be noted that the shares of smartphone makers in Japan strongly reflect the market share of feature phones. Before the advent of mobile broadband services and smartphones, Japanese makers provided high-spec feature phones sold in the domestic market only and they were quite popular there. Those makers have retained their shares in the age of smartphones. Today, things are changing. As one can see in Figure 2.4, the shares of all Japanese makers fell from 2009 to 2012, which appears to indicate that Japanese mobile broadband users are choosing smartphones, more like users in the global market. Also, the reason behind Samsung's lower share may be that the company was late in entering the Japanese

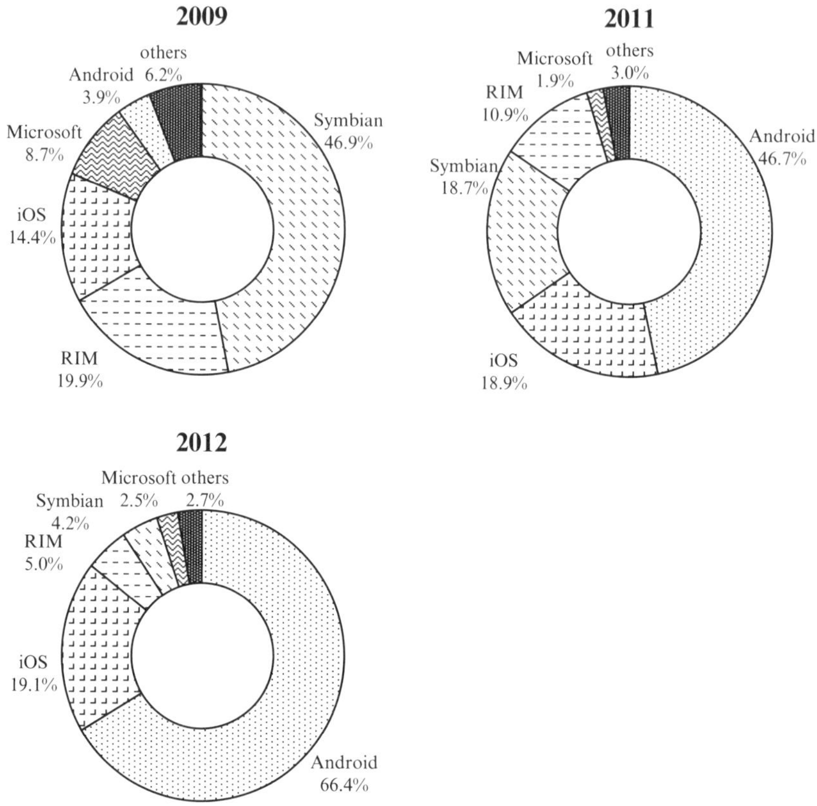

Source: MIC (2013).

Figure 2.1 Smartphone OS market shares worldwide

market. Samsung first entered the Japanese market in 2006 and was not very successful in selling feature phones.

Samsung launched its first smartphone in Japan in October 2010, 15 months later than in the global market, when NTT DOCOMO needed high-spec smartphones to compete with iPhones, which were exclusively sold by Softbank. Today, Samsung is finally succeeding in selling smartphones in the Japanese market, although the success may be temporally a little behind that of the global market.

Overall, it appears that the global smartphone market has consolidated into two giants and a number of other smaller players: the iOS smartphones of Apple and the Android smartphones of Samsung, with smartphones provided by other makers holding smaller market shares.

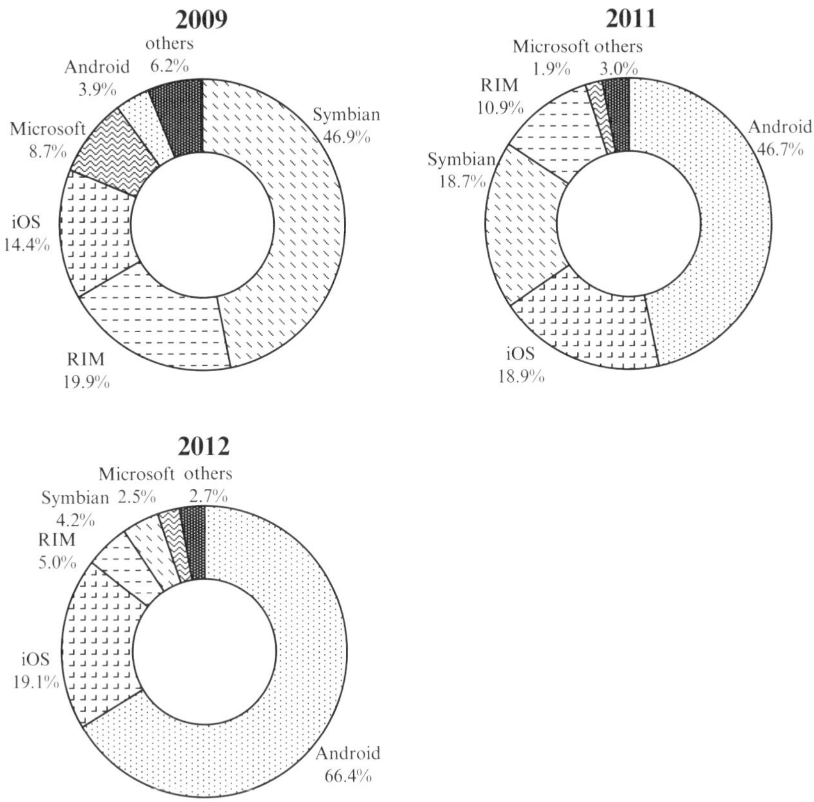

Source: MIC (2013).

Figure 2.2 Smartphone maker market shares worldwide

One finds that the Japanese smartphone market is still influenced by its previous feature phone market, but it appears that Japan is also following the global trend.

Both iTunes and the Google Play Store provide content aggregation services that include such functions as displaying content menus to consumers, providing authorisation, and billing. These functions, as well as mobile terminal platforms, are essential to consumers' enjoyment of the diversity of content on their smartphones.

Apple and Google are major players in content aggregation. However, the MIC (2012) stated that there are key differences between the business models of Apple and Google, and we believe that the ways in which these companies aggregate content constitute an important one. On one

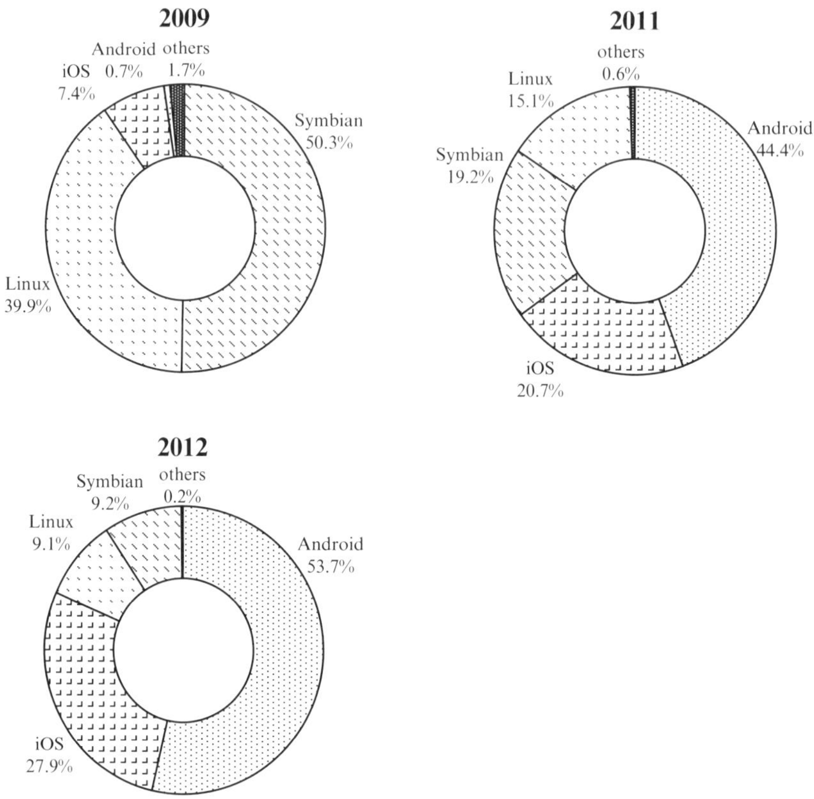

Source: MIC (2013).

Figure 2.3 Smartphone OS market shares in Japan

hand, Apple exclusively provides iOS smartphones and vertically integrates content aggregation, through iTunes, with its smartphone business. Here, one may find vertical integration to be similar to the i-mode of NTT DOCOMO and other legacy business ecosystems. On the other hand, Google develops Android OS as an open-source project; Google also actively urges manufacturers to choose Android OS as their mobile terminal platform and collaborates to develop software development kits (SDK).[4] Moreover, Google accepts third-party aggregators for Android smartphones, aside from its own Google Play Store. As a result, mobile operators, smartphone makers and content companies provide content aggregation simultaneously on Android smartphones.

Much of the digital content accessed through smartphones is available

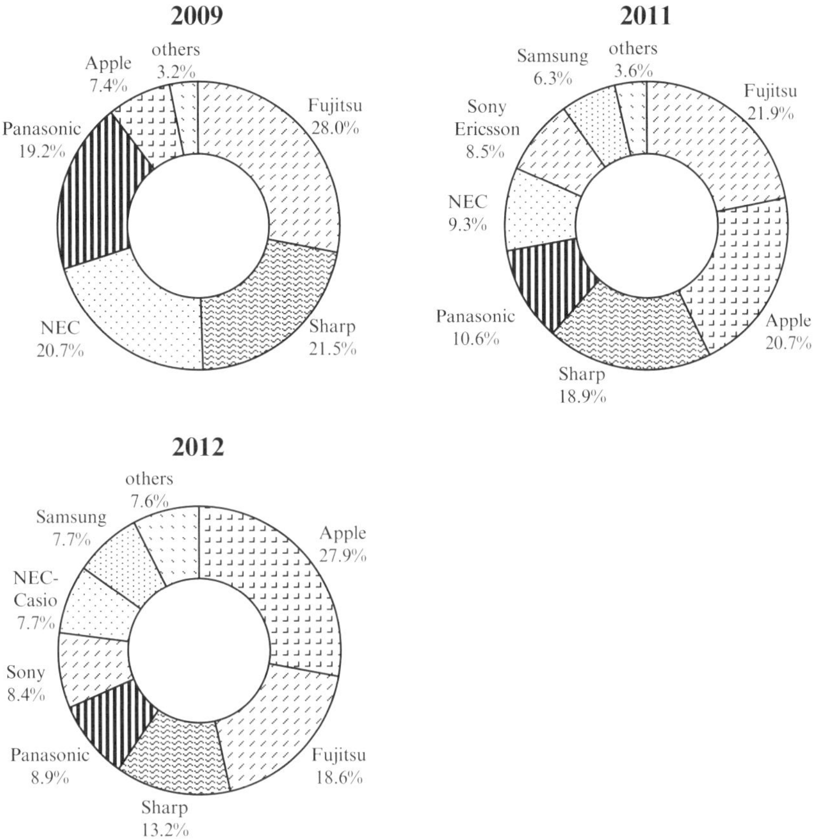

Source: MIC (2013).

Figure 2.4 Smartphone maker market shares in Japan

for free, as their success derives from advertising revenue; this free content naturally attracts many consumers. Looking at the overall trend in Japan, advertising on the Internet is rapidly increasing. Dentsu Inc. (2013) found that Internet advertising expenses in 2012 totalled JPY 662.9 billion, representing a 107.1 per cent increase compared to the previous year despite the fact that there had been a decrease in total advertising expenses across all media. Advertising revenue flows into the smartphone market through various paths, including search portals such as Google.com, social networking sites such as Facebook or Mixi, and online game sites such as Mobage or GREE.

Table 2.1 Business ecosystems of mobile broadband

Ecosystems	Number of content providers	Number of aggregators	Number of makers
i-mode feature phones	N	1 (NTT DOCOMO)	N
iOS smartphones	N	1 (Apple)	1 (Apple)
Android smartphones	N	2 or more (Google and other aggregators)	1 dominant maker (Samsung) and others with smaller shares

Note: N = unspecified large numbers.

Table 2.1 provides an overview of the business ecosystems of the aforementioned mobile broadband businesses. In Apple's business model, content aggregation is vertically integrated with iOS smartphone production. Samsung's business model may 'win and take all' in Android smartphone production, but content aggregation is still open to competition.

3. FOCUS OF ANALYSIS

Consumers take two steps when they begin to use smartphones: they purchase smartphones in the first step, then use content through those smartphones in the second step. In the first step, a consumer selects his or her favourite smartphone while considering the variety of available content. During this step, the consumer has not yet used any content; to 'sell themselves', smartphone makers need to prepare and offer to consumers a menu of available content, either by themselves or by exploiting the services of third-party content aggregators. In the second step, the consumer actually uses the available content, some of which is available free of charge, and some of which incurs a fee.

This study focuses on this first step in consumer behaviour, because we consider it important in analysing consumer choice vis-à-vis smartphones. Today, consumers are increasingly choosing a favourite smartphone, rather than a favourite operator. Consumers who purchase smartphones often change mobile operators – e.g. from DOCOMO to KDDI to use the iPhone, which was not available through DOCOMO. This consumer behaviour is different from that of pre-smartphone days, in which mobile operators controlled the customer experience. Moreover, consumer choice in the first step can significantly influence market competition in the second step, since the revenues of market entrants other than smartphone makers (e.g. mobile operators and content providers) depend solely upon consumer choices made in the first step.

As shown in Table 2.1, in Apple's operations, content aggregation and smartphone manufacturing are exclusively integrated. Android, in comparison, is more open and third-party content aggregators provide the same functions that Google does. The current mobile broadband business and the legacy business model are different in this regard; therefore, in our analysis, we compare two basic models: 1) an Apple-like model, in which aggregation and smartphone manufacturing are vertically integrated; and 2) an Android-like model, in which aggregation and smartphone manufacturing are mutually independent.

The relationship between a smartphone maker and content aggregation is now more closely approximating the relationship between a mobile operator and the telecommunications platform in the i-mode ecosystem: both of them are controlled by a few powerful companies, and they are essential to the content business. What the MIC has examined in its policy discussions is the possibility of strategic behaviour on the part of a few powerful mobile operators – behaviour that could harm content providers' businesses and eventually benefits to consumers by closing platform functions to third-party content providers. Currently a few smartphone makers, rather than mobile operators, are more powerful than ever in the mobile broadband business. Therefore we focus below on the behaviour of smartphone makers in relation to content aggregation, while assuming a monopoly in the smartphone manufacturing market, for the simplicity of analysis.[5] In short, in the next section we examine the profitability of a vertically integrated and monopolistic smartphone maker, as well as that of a monopolistic smartphone maker that separates content aggregation.

According to Eisenmann, Parker and Van Alstyne (2006), the value of platforms – which is to say, where products and services that bring together groups of users come together in a two-sided network – to any given user largely depends on the number of users on the network's other side. Successful platforms enjoy increasing returns to scale because of network effects, a phenomenon whereby two groups of users are attracted to each other. The relationship between smartphones and content closely resembles the hardware software paradigm in the computer industry, which Katz and Shapiro (1985) illustrate as a classic example of a two-sided network. Takachi, Kondo, Shishikura and Takahashi (2007) undertook a choice experiment and found the variety of content to have an indirect network effect on consumer demand for mobile terminals. In consideration of the results of these studies, smartphones (with content aggregation) should be treated as an economic platform, since content and consumers are connected to smartphones and attract each other with indirect network effects.

With respect to two-sided networks, Rochet and Tirole (2003) and Parker and Van Alstyne (2005) each discuss optimal pricing therein; Armstrong (2006), in addition, analysed the influence of users' multi-homing among economic platforms with a two-sided network effect. With respect to vertical integration, Takachi and Jitsuzumi (2011) established a payment model between a mobile operator and content aggregators. Here, we focus on an advertising model, because it is advertising revenue that makes it possible to offer a variety of free content, and it also attracts consumers to smartphones. Economides and Tåg (2012) also examined the relationship between content and consumers by focusing on the efficiency of network neutrality regulations, to which we refer in the following.

4. ECONOMIC MODEL

We use two basic models in our analysis: 1) Type S is a Google-like model, in which a smartphone maker separates itself from content aggregation; and 2) Type V is an Apple-like model, in which a smartphone maker vertically integrates a content aggregator. In the following, we first describe Type S with a single content aggregator (referred to as Type S-1 as necessary); we then describe Type V and Type S with a second competitive content aggregator (referred to as Type S-2 as necessary).

4.1 Demand

Economides and Tåg (2012) assume a generalised Hotelling model and introduce $u_i = \gamma + \alpha\theta_{cp} - tx_i - \rho$ for the utility that a consumer obtains from Internet access. The platform sells the Internet access at a price ρ. Consumers are differentiated in terms of their Internet access preferences. A consumer's 'location (type)' x_i indexes his or her preference for Internet access. A consumer receives an intrinsic value γ from connecting to the Internet, and pays a transportation cost equal to t per unit of distance 'travelled'. Consumer locations are uniformly distributed on the interval of zero to one, with the platform located at $x = 0$. Parameter α is the marginal value that a consumer places on an additional content provider on the Internet and θ_{cp} is the number of content providers that are active. In their model, the demand for content depends on the expected variety of content provided, since more consumers will connect to the network if more content is expected to be available. On the other hand, the provision of content depends on the expected number of consumers, which is consistent with the relationship between consumers' demand for smartphones and the variety of content provided via smartphones. When the expected number of consumers is θ_{cs}, the marginal consumer, x_i – who is indifferent between subscribing to the Internet and not subscribing – is located at $x_i = \theta_{cs} = (\gamma + \alpha\theta_{cp} - \rho)/t$. In our study, we simplify this demand function and assume that consumer demand n is

$$n = A + Bm - p \tag{2.1}$$

where m is the number of varieties of content provided on a smartphone; p is the price consumers pay for a smartphone; and B is the marginal value that consumers receive from the variety of content, thus representing indirect network effect.

4.2 Smartphone Maker

A maker produces n smartphones that can list m kinds of content on their menus. We assume that a smartphone maker must bear the cost of using higher-spec chips, if it wants to increase m. Actually makers now tend to use higher-performance CPU and memory chips, so as to cater to consumers' demand for richer content.[6] Generally speaking, higher-density chips are more expensive at any point in time and so the cost of smartphone production increases when a smartphone maker produces a smartphone with a larger m, since smartphones should be produced so as to be as compact as possible. Therefore the production cost of smartphones can be defined as $T = \varepsilon(m)n$ in which $\varepsilon(m)$ is the average cost of production that increases with a larger m. We assume $\varepsilon(m) = m^a (a \geq 0)$ and $T = m^a n$ for further analysis. A smartphone maker carries digital content by charging to content aggregators r per one piece of content, and we assume that a content aggregator pays rm to a smartphone maker. For example, Google develops the Android OS and offers it to smartphone makers as open-source software; in return, it provides its services (e.g. Gmail or Google Maps) on Android smartphones. This expense incurred by Google in the course of developing Android OS can be interpreted as a form of r. In contrast, smartphone makers often prepare various SDKs (software development kits) to lower the business cost of content aggregators and providers on their terminal platforms.[7] In the model, an SDK offered by a smartphone maker can be considered an incentive to a content aggregator at a negatively priced r. Based on the discussion above, the profit of a smartphone maker is

$$\Pi_S = pn + rm - m^a n. \tag{2.2}$$

4.3 Content Aggregator

A content aggregator produces advertising value Y. We assume that the efficiency of producing advertising value increases as n grows; the efficiency also increases when m, the variety of content, becomes larger since consumers then have more opportunities to contact content. When $\beta(m)$ is the amount of advertising value per smartphone for a given m, the total advertising value is $\beta(m)n$. If we assume a linear function for $\beta(m)$ (e.g. bm), we can write $Y = bmn$ $(b > 0)$. For simplicity, it is assumed that advertising is sold at a unit price, which leads the sales of a content aggregator to be bmn. Q, representing the production cost for advertising value, is clearly a function of r. In addition, the average production cost of advertising value $\varpi(m)$ is assumed to increase when m grows, since the marketing expense for acquiring advertisers is assumed to grow with larger m, and we can write Q as $Q = rm + \varpi(m)m$. We assume that $\varpi(m) = m^c$ $(c > 0)$, for further analysis. The profit of a content aggregator is

$$\pi = bmn - rm - m^{c+1}. \tag{2.3}$$

4.4 Equilibrium: Type S

A content aggregator determines m in maximising its profit subject to r and n, assuming that it has information on how n is determined through m and p. According to (2.1) and (2.3), a content aggregator knows the optimal relationships among r, m and p from the first-order condition (2.4) for maximising its profit and determines m by fulfilling (2.5).

$$\frac{\partial \pi}{\partial m} = b(Bm - p + A) - (c + 1)m^c + Bbm - r = 0 \tag{2.4}$$

$$r = b(Bm - p + A) - (c + 1)m^c + Bbm \tag{2.5}$$

From (1) and (5), (2) is rewritten as

$$\Pi_S = -Bm^{a+1} - (c + 1)m^{c+1} + 2Bbm^2$$
$$-Am^a + Abm - p^2 + Ap + m^a p + Bmp - bmp. \tag{2.6}$$

Since the smartphone maker maximises its profit, p fulfils the first-order condition.

$$\frac{\partial \Pi_S}{\partial p} = m^a + Bm - bm - 2p + A = 0 \tag{2.7}$$

$$p = \frac{1}{2}(m^a + Bm - bm + A) \tag{2.8}$$

From (6) and (8), the first-order condition that m fulfils in equilibrium is obtained as

$$\frac{\partial \Pi_S}{\partial m} = a(-Am^{a-1} + m^{2a-1}) - (B + b)(a + 1)m^a$$

$$-2(c + 1)^2 m^c + (B^2 + b^2 + 6Bb)m + A(B + b) = 0. \tag{2.9}$$

When there is an appropriate m that fulfils (9), we name it m^* as the optimal variety of content in equilibrium. Then m^* gives p^*, r^* and n^* as described below.[8]

$$p^* = \frac{1}{2}(m^{*a} + Bm^* - bm^* + A) \tag{2.10}$$

$$r^* = \frac{1}{2}bm^{*a} - (c + 1)m^{*c} + 1/2 * b^2 m^* + 3/2 * Bbm^* + 1/2 * Ab \tag{2.11}$$

$$n^* = \frac{1}{2}(-m^{*a} + Bm^* + bm^* + A) \tag{2.12}$$

Finally, we obtain consumer surplus, CS, and total surplus, TS_S, from (2.13) and (2.14).

$$CS = \int_0^n ((Bm - n + A)dn) - pn \tag{2.13}$$

$$TS_S = \Pi_S + CS + \pi \tag{2.14}$$

4.4.1 Vertical integration: Type V

Given the lack of a content aggregator, the profit of a vertically integrated smartphone maker is

$$\Pi_V = pn - m^a n + bmn - m^{c+1}. \tag{2.15}$$

Therefore (15) is rewritten from (1) as

$$\Pi_V = -Bm^{a+1} - m^{c+1} + Bbm^2 - Am^a + Abm - p^2$$

$$+ Ap + m^a p + Bmp - bmp. \tag{2.16}$$

When the vertical integrator maximises (2.16) with regard to p, it is equal to (2.8). From (2.1), (2.8) and (2.15), the first-order condition that m fulfils in equilibrium is

$$\frac{\partial \Pi_V}{\partial m} = -Aam^{a-1} + am^{2a-1} - (B+b)(a+1)m^a - 2(c+1)m^c + (B+b)^2 m + A(B+b) = 0. \quad (2.17)$$

When there is an appropriate m* that fulfils (17), p* and n* are obtained as follows.

$$p^* = \frac{1}{2}(m^{*a} + Bm^* - bm^* + A) \quad (2.18)$$

$$n^* = \frac{1}{2}(-m^{*a} + Bm^* + bm^* + A) \quad (2.19)$$

Finally the total surplus of Type V is

$$TS_V = \Pi_V + CS. \quad (2.20)$$

4.5 Duopoly in Content Aggregation: Type S-2

In this section, we describe a situation in which there is competition with regard to content aggregation. For simplicity of analysis, we assume a situation where a second content aggregator enters the market and shares advertising revenue with the first content aggregator. Since content is assumed to be basically homogeneous between content aggregators, they compete in Cournot competition for a larger share in the variety of content. The profit of a content aggregator is

$$\pi_k = bm_k \frac{m_k}{m_k + m_l} n - rm_k - m_k^{c+1} (k,l = 1,2). \quad (2.21)$$

The total marketing expense in acquiring advertisers is smaller for Type S-2 than for either Type V or Type S-1. As shown in (2.3) and (21), a vertical integrator or a content aggregator bears the full marketing expense m^{c+1}, but in Type S-2, two content aggregators bear the marketing expense (m_k^{c+1} and m_l^{c+1}). Therefore the sum of the marketing expense in Type S-2 is always smaller than that in Type V or Type S-1 as long as c is greater than zero. We assume a situation in which there are a large number of advertisers; this market environment makes the marginal cost of searching for additional advertisers gradually increase.

Subject to m_l and r, a content aggregator determines m_k to maximise π_k, and it is again assumed to have information that n is determined by

$m(=m_k+m_l)$ and p. A content aggregator determines m_k, fulfilling (22) given by (2.1) and (2.21).

$$\frac{\partial \pi_k}{\partial m_k} = \frac{1}{(m_k+m_l)^2}$$

$$[b(B(m_k+m_l)-p+A)m_k^2-(c+1)m_k^c(m_k+m_l)^2-r(m_k+m_l)^2$$

$$+(m_k+m_l)(Bbm_k^2+2b(B(m_k+m_l)-p+A)m_k)]=0 \quad (2.22)$$

$$r=-(c+1)m_k^c+\frac{1}{(m_k+m_l)^2}$$

$$[-b(B(m_k+m_l)-p+A)m_k^2$$

$$+(m_k+m_l)(Bbm_k^2+2b(B(m_k+m_l)-p+A)m_l)] \quad (2.23)$$

From (2. 1) and (2.23), (2.2) is rewritten as

$$\Pi_S=-\frac{(c+1)m_k^c}{m_k+m_l}+\frac{1}{(m_k+m_l)^2}[-b(B(m_k+m_l)-p+A)m_k^2$$

$$+(m_k+m_l)(Bbm_k^2+2b(B(m_k+m_l)-p+A)m_k)]. \qquad (2.24)$$

A smartphone maker maximises its profit by fulfilling the first-order condition,

$$\frac{\partial \Pi_S}{\partial p}=(m_k+m_l)^a+B(m_k+m_l)+\frac{bm_k^2-2b(m_k+m_l)m_k}{m_k+m_l}-2p+A \quad (2.25)$$

$$p=\frac{1}{2}\left[(m_k+m_l)^a+B(m_k+m_l)+\frac{bm_k^2-2b(m_k+m_l)m_k}{m_k+m_l}+A\right]. \quad (2.26)$$

We obtain $m_k=m_l$ in equilibrium from (2.26). Subject to (24) and (26), the necessary condition of m_k in equilibrium is

$$\frac{\partial \Pi_s}{\partial m_k}= -2^aAam_k^{a-1}-2^{2a-1}am_k^{2a-1}-2^aB(a+1)m_k^a-2(c+1)^2m_k^c$$

$$+2B^2m_k+\frac{9}{8}b^2m_k+5Bbm_k+2^aam_k^a(2m_k)^{a-1}-\frac{3}{2}abm_k(2m_k)^{a-1}$$

$$+Aa(2m_k)^{a-1}+\frac{1}{2}2^aam_k^{a-1}(2m_k)^a-\frac{3}{4}b(2m_k)^a+AB+\frac{3}{4}Ab=0. \quad (2.27)$$

When there is an appropriate m_k^* that fulfils (2.27), p*, r* and n* are obtained as follows:

$$p^* = \frac{1}{2}\left(2Bm_k^* - \frac{3}{2}bm_k^* + (2m_k^*)^a + A\right) \quad (2.28)$$

$$r^* = -\frac{3}{8}2^a bm_k^{*a} - (c+1)m_k^{*c} + \frac{9}{16}b^2 m_k^* + \frac{5}{4}Bbm_k^* + \frac{3}{8}Ab \quad (2.29)$$

$$n^* = -\frac{2^a m_k^{*a}}{2} + Bm_k^* + \frac{3}{4}bm_k^* + \frac{1}{2}A. \quad (2.30)$$

Finally, the total surplus TS_S of Type S-2 is

$$TS_S = \Pi_S + \pi_1 + \pi_2 + CS. \quad (2.31)$$

5. SIMULATION

For the simulation, we insert into the model arbitrary numbers for the five parameters; the descriptions are summarised in Table 2.2. These numbers fit current market and technology trends. We do so, as it is beyond the scope of our analysis to empirically estimate each of the five parameters. Through the simulation, we analyse how different business ecosystems influence the profits of smartphone makers and the optimal industrial structure, as well as the rationality of policy intervention.

We give concrete numbers for the four parameters (i.e. $a = 1$, $b = 0.5$, $c = 1$, $A = 1.5$). At first, we show the general simulation results. Then, we focus on the value of r to narrow the range of parameter B to be examined, then compare Type S and Type V in terms of their profits (Π_S and Π_V, respectively) and in terms of total surplus (TS_S and TS_V, respectively).

In the following, for analysis we plot scatter diagrams with lines that connect the data, since it is impossible to solve the equations analytically. We basically change the value of B by 0.1, but by 0.01 to find the point where a second content aggregator enters into the market (Type S-1 changes to Type S-2 and vice versa). We change the value of B by 0.01 in figures from Figure 2.8 to Figure 2.11 to examine the simulation results more closely.

5.1 Simulation Results[9]

Figure 2.5 shows the profits of a smartphone maker in Type S and Type V settings. $\Pi_V > \Pi_S$ (of Type S-1) when B is between 0.5 and 1.24. A second content aggregator then enters the market in the area where B is greater than or equal to 1.24, but still, $\Pi_V > \Pi_S$ (of Type S-2) when B is between

Table 2.2 Parameters used in the simulation

Parameters	Assumed range	Descriptions
a	A positive number	Parameter a expresses the increase in production cost of a smartphone relative to the variety of content it can handle. The average cost of smartphone production increases as the variety of content increases (i.e., $a > 0$).
b	A constant positive number	Parameter b represents the marginal advertising value of a single content per smartphone. As we assume that the price of advertising equals 1, parameter b also expresses the revenue from a single content per smartphone.
c	A constant positive number	Parameter c represents the increase in marketing expense (selling, general and administrative expense) as the variety of content grows.
A	A constant positive number	Regardless of the amount of available content, consumer demand depends on some intrinsic value A that derives from the use of smartphones (e.g., value of voice communications, camera and scheduler).
B	A positive number	Parameter B represents the extent to which consumer demand depends on the variety of content available. It is expected that the more accustomed consumers become to using smartphones, the more available content they will want. Improvements in the quality of content may also increase parameter B.

1.24 and 1.6. Finally, Π_S exceeds Π_V when B is between 1.7 and 2.3. In this area, a vertical integrator separates a content aggregator to maximise its profit and offers open access to a third-party content aggregator.

Figure 2.6 shows the total surplus in Type S and Type V. TS_V is greater than TS_S (of Type S-1) when B is less than 1.24. A second content aggregator enters the market when B is greater than or equal to 1.24, but still, TS_V is greater than TS_S (of Type S-2) when B is between 1.3 and 1.5. Finally, TS_S (of Type S-2) exceeds TS_V when B is greater than or equal to 1.5.

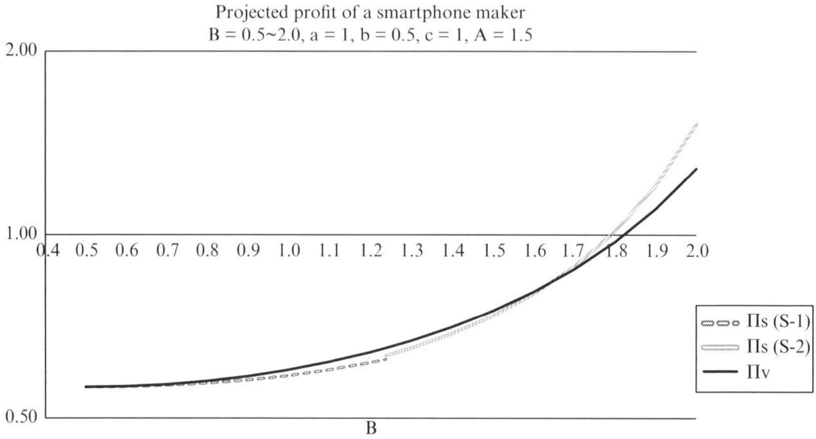

Note: For visibility, we use the logarithm for the y-axis and limit B to ≤ 2.0.

Figure 2.5 Profits of Type S and Type V: Π_S and Π_V.

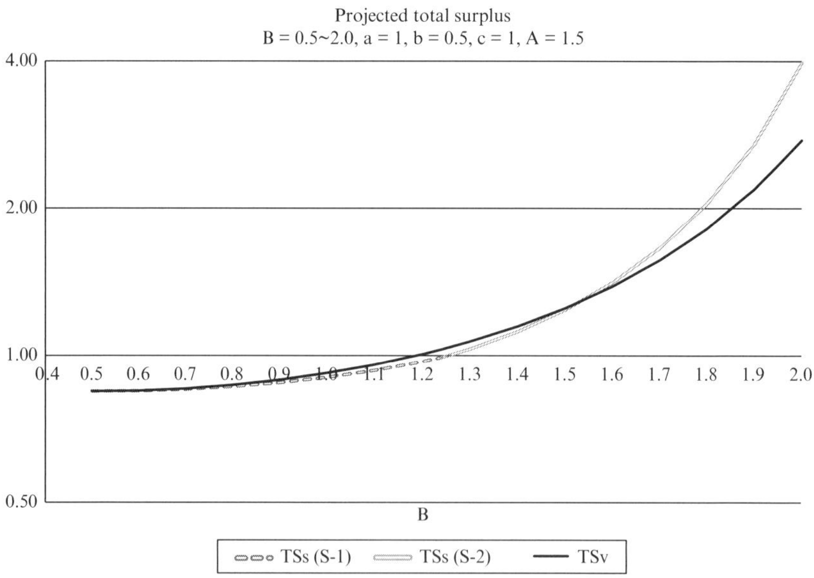

Note: For visibility, we use the logarithm for the y-axis and limit B to ≤ 2.0.

Figure 2.6 Total surplus of Type S and Type V: TS_S and TS_V

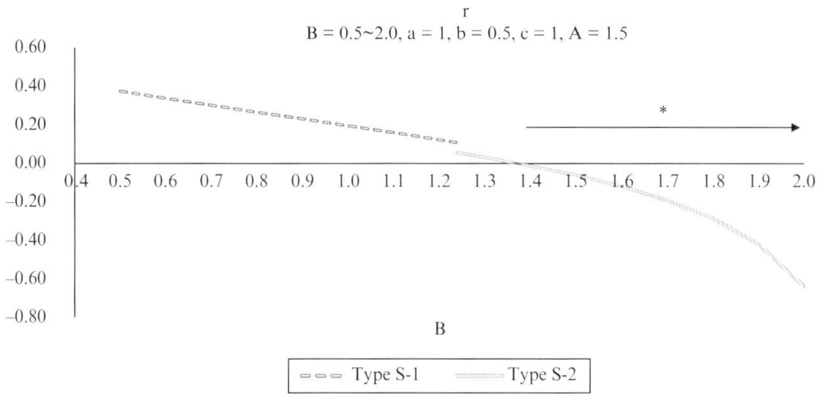

Note: For visibility, we limit B to ≤ 2.0.

Figure 2.7 Value of r

5.2 Value of r

The value of r gradually decreases as parameter B takes larger values. Then r becomes negative when B is greater than or equal to 1.4, which means that a smartphone maker in Type S-2 provides content aggregators with an incentive. This is because a content aggregator cannot generate sufficient revenue when the marketing cost of advertising is lower with a second aggregator, and the smartphone maker must provide content aggregators with an incentive to acquire an optimal variety of content.

In reality, there are several content aggregators in the Android smartphone ecosystem in which Type S-2 appears to fit best, and smartphone makers begin to provide SDKs, where r is considered negative.[10] Therefore in the following, we focus on an area where B is more than or equal to 1.4 (the arrowed area of B marked with * in Figure 2.7).

5.3 Examination of Profit and Total Surplus

Figure 2.8 shows that Π_S exceeds Π_V when B is greater than or equal to 1.65.

Figure 2.9 shows that TS_S exceeds TS_V when B is greater than or equal to 1.54.

Figure 2.10 is created by superimposing Figure 2.9 over Figure 2.8. We find that a smartphone maker's profit maximisation does not always

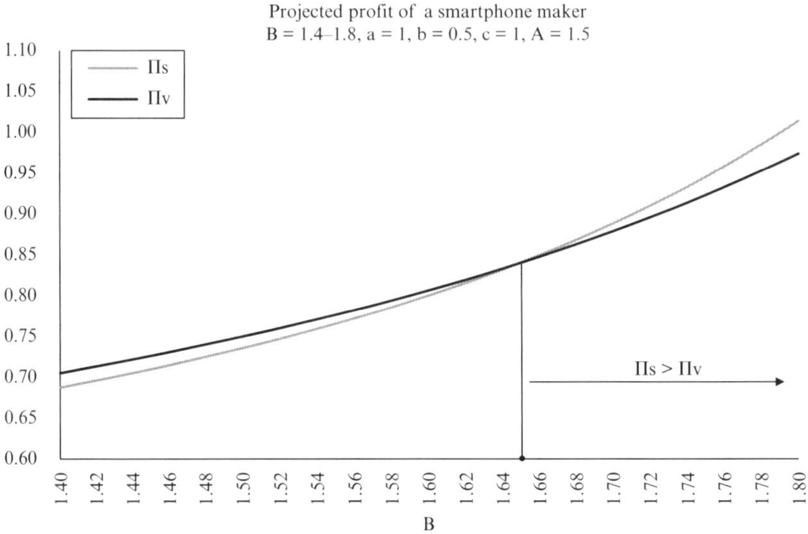

Note: For visibility, we limit B to ≤ 1.80.

Figure 2.8 Π_S *and* Π_V *(in the area marked with * in Figure 2.7)*

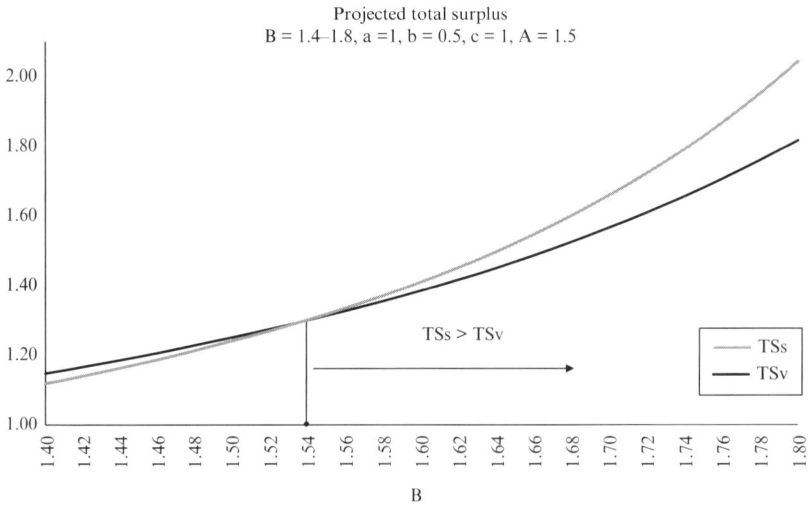

Note: For visibility, we limit B to ≤ 1.80.

Figure 2.9 TS_S *and* TS_V *(in the area marked with * in Figure 2.7)*

end in the maximisation of total surplus. When B is between 1.54 and 1.64 (i.e. the area marked with ** in Figure 2.10), a vertically integrating smartphone maker can maximise profit ($\Pi_V > \Pi_S$); however, it is better to separate content aggregation in order to maximise the total surplus ($TS_S > TS_V$).

5.4 Technological Advancement

In this section, we discuss a situation in which parameter a = 0; this would imply that there is no cost related to the increase in variety of content, on account of technological advancement. It appears that the change in parameter a shifts the demand function rightward, and also increases the variety of content, m.

In fact, we obtain lower Π, CS and TS values in many cases, under the given parameters. This is because the calculation result depends on the magnitude of effect of each parameter on consumer demand and sales. In other words, the advertising value produced by content aggregators is a function of b, n, r and c and it is essential that the value of n be large enough – and that r and c be small enough – to increase advertising revenue.

When parameters other than a are the same here as in previous sections, we find that the slope of $TS_{S,a=0}$ becomes steeper than that of $TS_{S,a=1}$ in all types; this can be considered a result of technological advancement. More specifically, $TS_{S,a=0}$ is smaller than $TS_{S,a=1}$ when parameter B is small; this is because 1) the effect of the increase in n is offset by the increase in p and the decrease in n; 2) the decrease in n reduces the value of advertising; and 3) the growth of m increases the cost of acquiring advertisers. On the other hand, $TS_{S,a=0}$ far exceeds $TS_{S,a=1}$ when parameter B takes a larger value, because the greater magnitude of n absorbs the negative effects of the increase in p, which in turn leads to increases in smartphone makers' profits and consumer surplus. As a result, we believe that the basic effectiveness of our economic model is confirmed when parameter a = 0.

Again, we find that a smartphone maker's profit maximisation does not always end in the maximisation of the total surplus. When B is between 0.88 and 1.15 (i.e. the area marked with ** in Figure 2.11), a smartphone maker can maximise its profit ($\Pi_V > \Pi_S$); but again, it is better to separate content aggregation in order to maximise the total surplus ($TS_S > TS_V$). Compared to Figure 2.10, the area marked with ** moves leftward and covers a broader range of parameter B.

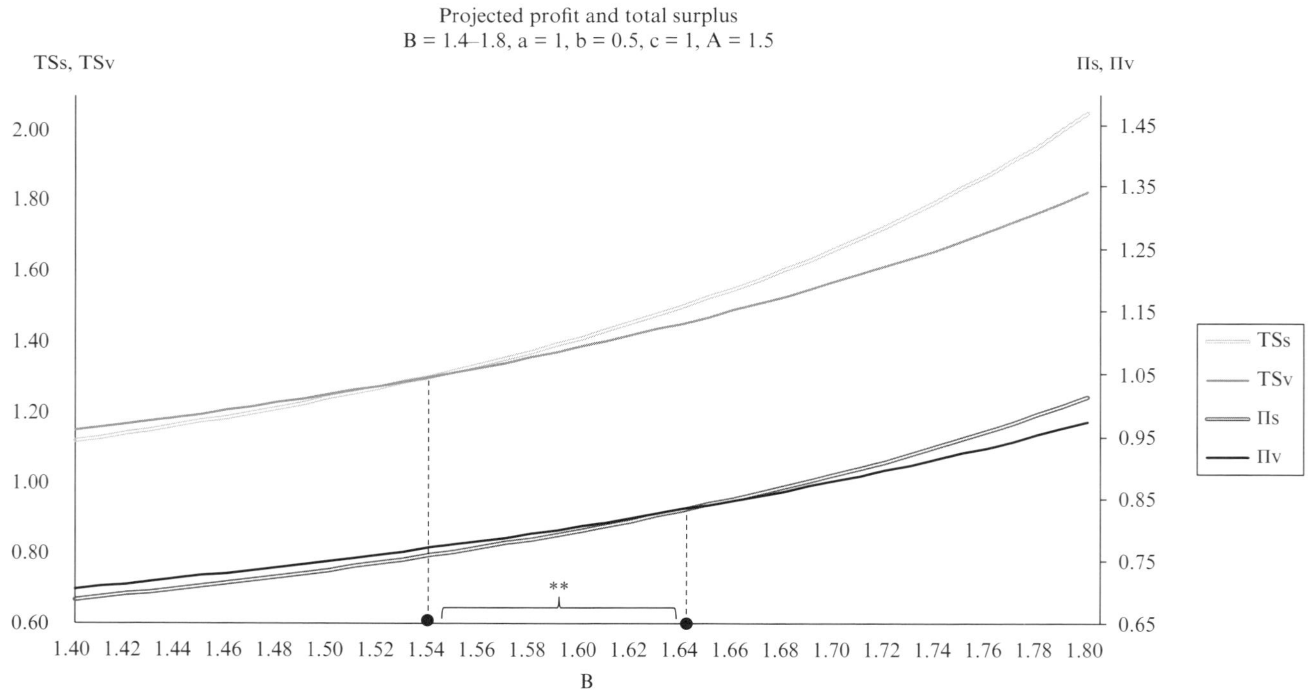

Note: For visibility, we limit B to ≤ 1.80.

Figure 2.10 Π_S, Π_V, TS_S and TS_V

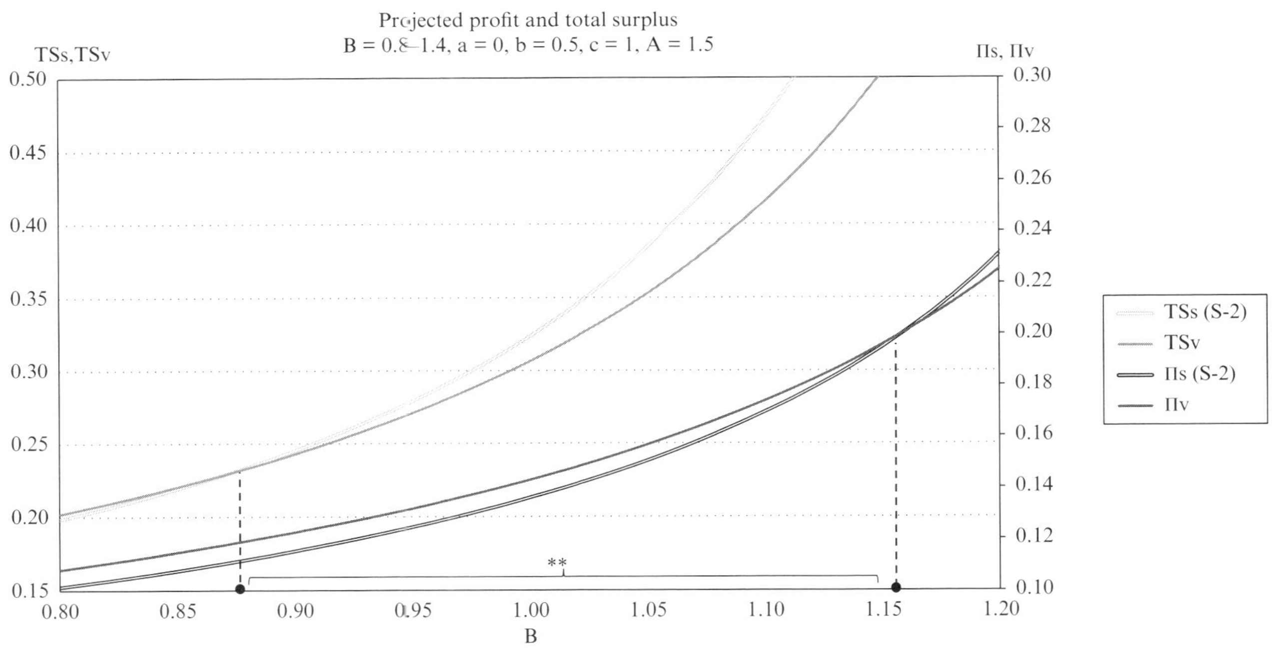

Note: For visibility, we limit B to between 0.80 and 1.20.

Figure 2.11 Π_S, Π_V, TS_S *and* TS_V *(a = 0)*

5.5 Imperfect Information

We assume that content aggregators know about consumer demand trends and act while considering the optimal relationships among r, m and p from the first-order condition (2.4), in order to maximise their profits. As it becomes easier statistically to extrapolate consumer demand by using large-scale customer data – frequently called 'big data' – it is appropriate to assume that a content aggregator can precisely forecast consumer demand for smartphones. It is also possible to obtain such information from smartphone makers.

However, in this subsection, we analyse a case in which a content aggregator does not precisely know consumer demand. For example, for Type S-1, a content aggregator does not know the influence of p on n, and acts while considering the relationships among r, m and n from the first-order condition (32), instead of (4), in maximising its profit.

$$\frac{\partial \pi}{\partial m} = bn - (c + 1)m^c + Bbm - r = 0 \tag{2.32}$$

In this case, as shown in Figures 2.12 and 2.13, TS_V and Π_V are always larger than TS_S and Π_S, respectively; this shows the superiority of vertical integration over separation. Given this result, it is extremely important

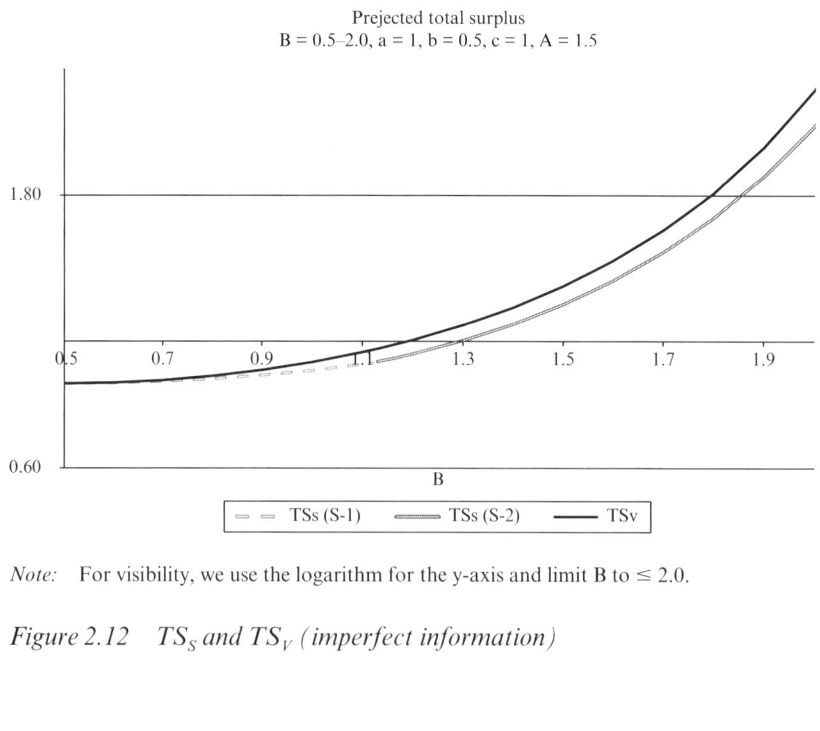

Note: For visibility, we use the logarithm for the y-axis and limit B to ≤ 2.0.

Figure 2.12 TS_S *and* TS_V *(imperfect information)*

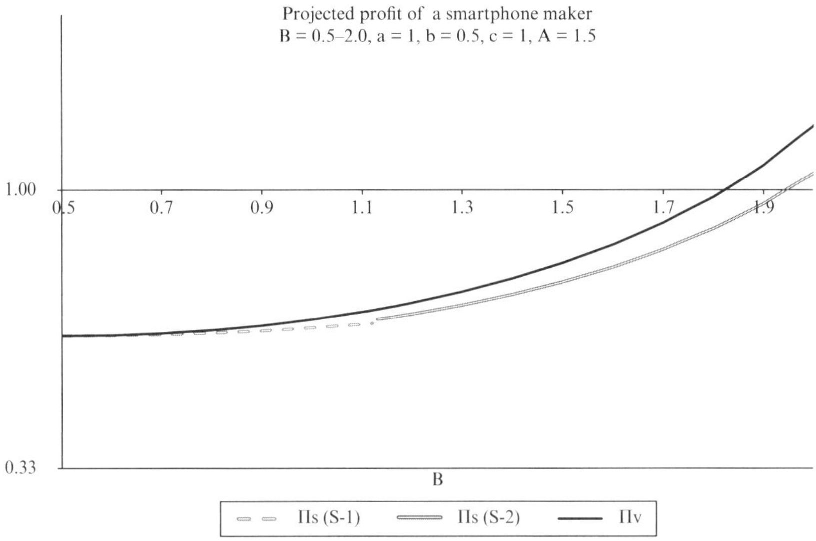

Note: For visibility, we use the logarithm for the y-axis and limit B to ≤ 2.0.

Figure 2.13 Π_S *and* Π_V *(imperfect information)*

that content aggregators have ways of gathering accurate information on consumer demand for smartphones.

6. CONCLUSION

In this chapter, we developed two different economic models to describe the current state of the mobile broadband industry and examine its influence on both enterprise profits and the necessity for policy intervention.

As shown in Figures 2.5 and 2.6, a vertically integrating smartphone maker maximises total surplus, as well as its profit, when consumers place a lower value on the variety of content when purchasing smartphones. As consumers become accustomed to using smartphones, they place more value on the variety of content, which means that the network effect of content, or parameter *B*, gradually becomes greater. In response, a smartphone maker separates content aggregation for profit maximisation and offers a third-party content aggregator on its platform. This move maximises both total surplus and profit when consumers place a higher value on the variety of content.

However, when the strength of the network effect is in the middle area, a smartphone maker that separates content aggregation generates a larger total surplus, while a vertical integrator enjoys a larger profit (Figure 2.10). In this case, it is impossible to expect that a vertically integrating smartphone maker would voluntarily separate content aggregation; as such, government needs to take some policy action that pushes vertical integrators to separate content aggregation and promote competition among content aggregators. If technological advancement eliminates the cost related to the increase in the variety of content, the middle area becomes broader, which means that the necessity for policy action increases.

When one considers the amount of total surplus possibly lost and that the simulation of this study targets only the first step of consumer behaviour (i.e. when consumers purchase smartphones), one might doubt the necessity of strict policy intervention, as proposed above. However, diversity of content aggregation clearly brings various advantages to the second step (i.e. when consumers often pay for the usage of content). For example, as Takachi and Jitsuzumi (2011) discuss, the differentiation of content aggregation will stimulate consumer demand, because consumers can then choose aggregators that best fit their personal preferences.

We also think that vertical integration by smartphone makers tends to be excessive, since smartphone makers tend to focus on the point of sale and pay less attention to the benefits generated thereafter. Excessive vertical integration may give rise to the anxiety of foreclosure, as was the case in earlier policy discussions in Japan. A vertical integrator may discriminate with respect to some content providers' access to smartphones' menu lists, and this would eventually spoil consumer benefits. For these reasons, we propose the separation of content aggregation from vertically integrated smartphone makers, as well as the promotion of competition among content aggregators, especially given that the use of smartphones is rapidly proliferating.

As government will experience greater difficulties in implementing policies once markets reach maturity, it is advised that it take action when markets are in their early stages. With respect to the telecommunications industry, some of the possible regulatory measures have been in action for some time in preventing the abuse of market power by dominant telecom operators. For example, in the Japanese legal system, Article 30 of the Telecommunications Business Law of Japan prohibits dominant carriers from unreasonably disciplining or intervening in the business activities of other telecommunications carriers, or in those of manufacturers or sellers of telecommunications facilities; the MIC and the Japan Fair Trade Commission also state that the article is applicable to content providers. As a measure of structural separation, the Nippon Telegraph and Telephone Corporation Law limits the business operations of two regional monopolies, NTT East and NTT West, to the legally defined 'regional telecommunications business' to prevent the abuse of their market power in adjacent markets.

These existing regulatory measures, however, do not fully cover the ecosystems of the current mobile broadband business. The relevant regulatory authorities need to establish an arena in which to discuss inclusive policy actions, including proactive measures against excessive vertical integration, such as the promotion of open access to smartphone platforms for third-party content aggregators; such matters have already been discussed by the MIC (2009) with regard to NTT DOCOMO's i-mode, as we briefly mentioned in Section 1.

The mobile broadband business continues to be in flux, and our study necessarily involves many assumptions. A smartphone maker is modelled here as a 'pure' smartphone maker, rather than a versatile business system that approximates actual players in the real market. In other words, our model is applicable to a company that places priority on hardware sales. We also assume that content aggregators have accurate information with respect to consumer demand; if content aggregators do not have such information, however, a vertical integrator will have a clear advantage. In future research, we certainly need to refine the assumptions of the model and to examine the second step of consumer behaviour, to cope with broader real-world situations. Empirical analysis should also be undertaken.

ACKNOWLEDGEMENTS

This work was partially supported by JSPS KAKENHI Grant Number 25380304. We would like to express our greatest appreciation to Mr Tetsuya Shimane (Tokyo Institute of Technology) and Dr Manabu Shishikura (Nagasaki University) for their valuable suggestions with regard to our model.

The views and opinions expressed in this chapter are those of the authors and do not necessarily reflect the official policy or position of any agencies.

Finally, we would like to express our gratitude to Haruko Takachi for her continued support and encouragement.

NOTES

1. As there are another 5 million subscribers using the Personal Handy-phone System, long-term evolution or 3G users actually account for 96.3 per cent of all Japanese mobile subscribers. See http://www.tca.or.jp/.

2. Hereafter, the word 'content' is used in reference not only to digital content, but also to any applications or services.
3. Sales of mobile operators in FY2011 were as follows: NTT DOCOMO, JPY 1,061.2 billion; KDDI, JPY 692.4 billion; and Softbank, JPY 605.5 billion.
4. It appears that Google expects to derive revenue from its services. Google designs Android OS in such a way that a consumer can easily use Google's services.
5. This simple treatment may reflect some market reality where Apple exclusively provides iOS terminals, and Samsung has rapidly increased its share of the Android smartphone market; additionally iOS smartphones and Android smartphones are differentiated, to some extent.
6. Examples from March 2013 include Snapdragon of Qualcomm and Tegra of NVIDIA; they have multiple cores by which to increase the overall speed at which one can work with richer content.
7. In March 2013, Apple prepared tools for software development in iOS smartphones (iOS SDK), and Samsung provided tools for the development of applications using a stylus pen (S Pen SDK) and tools for embedding advertising in content (AdHub SDK). Many SDKs are provided free of charge.
8. In the text, including the following Type V and Type S-2, only the necessary condition is explained. Certainly, it is necessary to examine the second-order condition and the parameters, if one is to clarify whether or not Π and π are actually maximised. This chapter focuses on model simulations; however, in Section 5, we check for maximisation by substituting actual numbers into the model.
9. Under the given parameters, Type V is effective when B is between 0.5 and 2.4, and Type S is effective when B is between 0.5 and 2.3. Neither model is effective when B is smaller than 0.5 or larger than 2.5. When one looks at Type S closely, Type S-1 is effective when B is smaller than 1.24. When B is greater than or equal to 1.24, a second content aggregator can enter the market with a non-negative profit – in which case, Type S-2 is applied instead of Type S-1.
10. Samsung provides AdHub SDK to handle advertising in smartphone content (see note 7). In March 2013, it also started to develop Tizen, an open-source OS for smartphones, as well as an SDK for it, in concert with component makers (https://www.tizenassociation.org/). If a smartphone maker bears the cost of the development of an SDK and provides it to content aggregators free of charge, r is considered negative.

REFERENCES

Armstrong, M. (2006), 'Competition in two-sided markets', *The Rand Journal of Economics*, **37** (3), 668–91.

Dentsu Inc. (2013), '2012 advertising expenses in Japan', available at http://www.dentsu.co.jp/news/release/2013/pdf/2013016-0221.pdf (accessed 2 April 2013, in Japanese).

Economides, N. (1998), 'The incentive for non-price discrimination by an input monopolist', *International Journal of Industrial Organization*, **16**, 271–84.

Economides, N. and J. Tåg (2012), 'Network neutrality on the Internet: a two-sided market analysis', *Information Economics Policy*, **24**, 91–104.

Eisenmann, T., G. Parker and M.W. Van Alstyne (2006), 'Strategies for two-sided markets', *Harvard Business Review*, **October**, 92–101.

Farrell, J. and P.J. Weiser (2003), 'Modularity, vertical integration, and open access policies: towards a convergence of antitrust and regulation in the Internet age', *Harvard Journal of Law and Technology*, **17** (1), 86–134.

Katz, M.L. and C. Shapiro (1985), 'Network externalities, competition, and compatibility', *The American Economic Review*, **75** F(3), 424–40.

Ministry of Internal Affairs and Communications (2009), *The Telecommunications Platform As It Should Be*, Tokyo: Ministry of Internal Affairs and Communications (in Japanese).

Ministry of Internal Affairs and Communications (2011), *Evaluation of the State of Competition in the Telecommunications Business Field in FY2010*, Tokyo: Ministry of Internal Affairs and Communications (in Japanese).

Ministry of Internal Affairs and Communications (2012), *2012 White Paper on Information and Communications in Japan*, Tokyo: Ministry of Internal Affairs and Communications (in Japanese).

Ministry of Internal Affairs and Communications (2013), *2013 White Paper on Information and Communications in Japan*, Tokyo: Ministry of Internal Affairs and Communications (in Japanese).

Parker, G.G. and M.W. Van Alstyne (2005), 'Two-sided network effects: a theory of information product design', *Management Science*, **51** (10), 1494–504.

Rey, P. and J. Tirole (2007), 'A primer on foreclosure', in M. Armstrong and R.H. Porter (eds), *Handbook of Industrial Organization*, vol. 3, Amsterdam: North-Holland, pp. 2147–220.

Rochet, J. and J. Tirole (2003), 'Platform competition in two-sided markets', *Journal of the European Economic Association*, **1** (4), 990–1029.

Takachi, K., M. Kondo, M. Shishikura and T. Takahashi (2007), 'A choice analysis of consumers' preference for mobile communications terminals', *Journal of Japan Society of Information and Communications Research*, **24** (3), 39–46 (in Japanese).

Takachi, K. and T. Jitsuzumi (2011), 'An analysis on the economic effects of the market structure related to the platform of mobile content services', *Studies in Regional Science*, **41** (3), 681–94.

Van Schewick, B. (2007), 'Towards an economic framework for network neutrality regulation', *Journal on Telecommunications and High Technology Law*, **5** (2), 329–91.

Whinston, M.D. (1990), 'Tying, foreclosure, and exclusion', *The American Economic Review*, **80** (4), 837–59.

3. Digital power shift from network to smart devices: how will regulators cope with digital power?

Minoru Sugaya

1. INTRODUCTION: WHAT IS 'DIGITAL POWER'?

Digital technologies are core technologies that support the development of information and telecommunication (IT) industries. In this chapter, economic power based on digital technologies is called 'digital power'. Traditionally, economic power in the IT industries has emerged from communication networks. In the age of voice telephony, such networks were owned and operated monopolistically by a telephone carrier. It means that such a carrier had a monopolistic economic power then.

In the advanced countries during the 1980s, a competition policy was introduced in the monopolistic telephone market. To promote competition, regulators started unbundling the network, dividing it into two separate divisions: a network facility division and a service division. This change happened just before the Internet emerged (Crandall and Flamm, 1989).

Soon after commercial Internet services launched, the platform function of the network's service division strengthened. After this, platform enterprises without network facilities, including Google, emerged (Auletta, 2009). This kind of platform division was born between the network division and the content division. The platform division developed an advertising business by taking advantage of its broad business realm, creating a new economic power. In other words, the platform division has become a 'digital power' based on digital technologies.

In addition to wired networks, wireless networks have rapidly developed in the 21st century. The wireless networks have been digitised, and the majority of them have evolved from voice telephony into networks providing non-voice and data services, including email and search engine services. In recent years, the terminal division of the wireless networks includes smart devices – in addition to traditional devices – such as smartphones and smart TVs. Technologically, such smart devices can receive any digital information, which means that these devices are ultimate terminals for the age of media convergence.

While these smart devices are still in the developing period, a new type of 'digital power' is emerging in the devices. This chapter examines the shift of 'digital power' in the ICT industries by reviewing the development of the industry, and illuminates the future development trend in the ICT industry.

2. NETWORK COMPETITIONS

Originally, the IT industry was part of network industries that provided public utilities services, including electricity, gas and water, in which services were offered by a monopoly owing to economies of scale in the network facilities. These industries were called public utilities and their supply of services was monopolistic. Business entities were either government-owned or publicly owned, based on the point of view that this would assure consumers' interests; when business entities were privately owned, regulations on entry and rates charged were imposed by the government.

In the field of public utilities, such as electricity and telecommunication, deregulation has progressed since the 1980s. As a result, the monopolistic market collapsed, allowing new enterprises to enter the market. Traditional network enterprises owned and operated the entire network from the production of services to the provision of services to consumers. On the other hand, new entrants discovered two patterns: 1) they have the same type of network system as the traditional enterprises and compete with them; and 2) the restructured regulatory framework allows a newcomer to use the traditional enterprises' networks. The first pattern is commonly called facility-based competition and the second is called service-based competition.

Competition policies for the network division can be categorised in the following three ways: 1) policy centred on facility-based competition; 2) policy centred on service-based competition, by allowing the monopoly to operate network facilities and carrying out thorough unbundling of facilities; and 3) policy that combines the first pattern and the second by letting facility-based competition and service-based competition coexist to promote the policy. Japanese policy is a typical coexisting policy, while Australia has a government-owned National Broadband Network (NBN), which was established to promote service-based competition. Although it is rare for a country to depend entirely on facility-based competition, US policy after the introduction of the optical network comes close to fitting this pattern.

3. NETWORK TO PLATFORM

Figure 3.1 illustrates the traditional industry structure – the vertical structure of IT industries. The vertical line shows content, platform, network, set-top box and other terminals. In the initial structural drawing, the platform did not exist on the vertical line and the network was divided into a long-distance network and a local network. The emergence of the Internet brought about the current structural drawing.

Contents depicted in this figure do not require any explanation, as they are the contents of information themselves. As for the platform function, it requires further explanation. The platform for telecommunication is a portal site for Internet users to select necessary information, which did not exist during the era of voice telephony. The platform division for telecommunication attracted more attention as an independent division from the network division with the advent of Internet services.

The platform for telecommunication has a rather complicated structure – a structure with various elements, including a general search engine site and an itemised search engine site. In addition, business models used on each structure are divided into free sites and paid sites. Free sites provide contents to users for free, depending on advertising revenues on the sites. Paid sites charge users for the content. On the platform division side, there is no regulation on entry and exit. This is distinct from regulations imposed on the network division, making the platform division rather competitive with easy entry into and exit from the market.

On the other hand, the platform function of broadcasting, which existed before the advent of the Internet, has a simpler structure than that of telecommunication. The platform for broadcasting can be divided into a single-channel model and a multi-channel model. A typical example of the single-channel model can be seen in terrestrial broadcasting. In Japan, broadcasting wireless stations as the network division, with the production and editing departments as the platform division, have been integrated in one company. This type of vertical integration has been called an integration of 'hardware' (infrastructure) and 'software' (content and editing).

In the multi-channel model, such as satellite broadcasting and cable television provision, the platform function can be found in the channel arrangement and customer management. Cable television integrates 'hardware' and 'software' while satellite broadcasting has developed its business on the premise of dividing 'hardware' and 'software'.

As described above, the functions of the platform division vary between telecommunication and broadcasting, but they share the functions of providing content to users and of customer management.

The emergence of the platform division produces a complex industrial

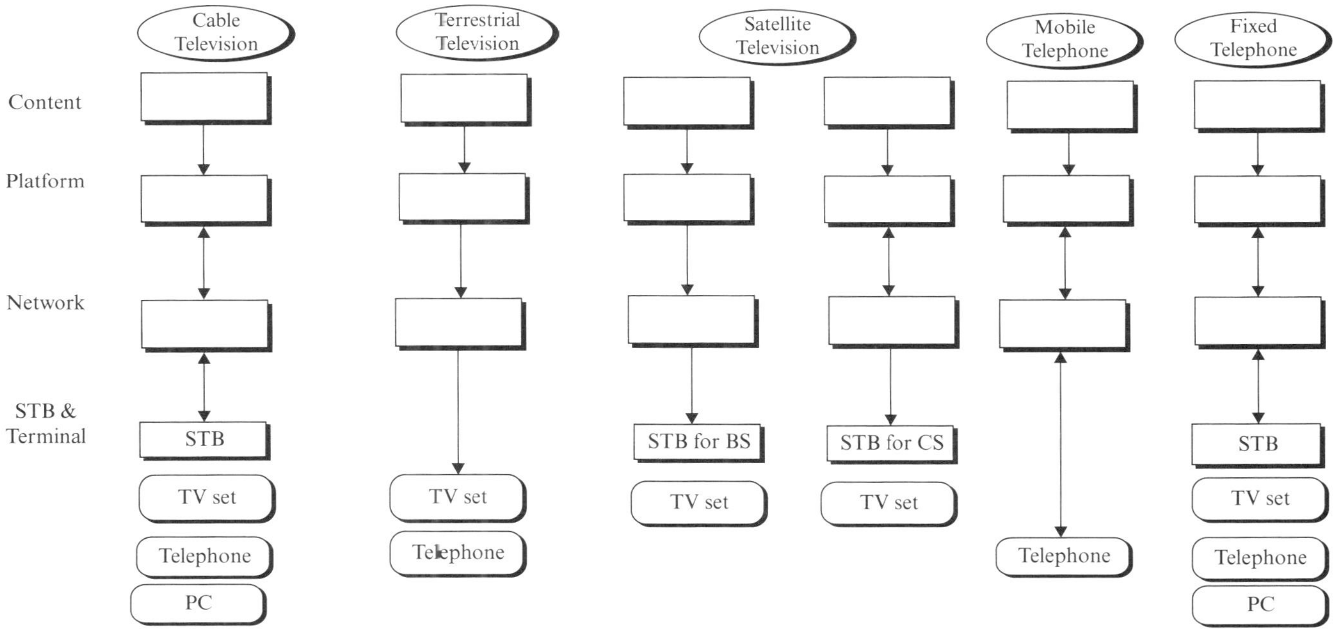

Source: Sugaya (2009), p. 29.

Figure 3.1 *Vertical structure of information and communication industries*

structure as well as a shift of economic power in the IT industry. As described earlier, economies of scale brought about by networks are the source of economic power for the traditional telecommunication business, and the framework for telecommunication policy was based on this power. Furthermore, in the broadcasting business, a limited spectrum constrained the number of enterprises proving broadcasting services, which created economic power.

However, technological innovations have allowed new enterprises to enter into the market, providing competitive services against terrestrial broadcasting. These entrants include enabled wired cable providers and even satellite broadcasters. This change has diminished the economic power of the terrestrial broadcasters.

4. SEPARATION OF PLATFORM NETWORK FUNCTION AND SET-TOP BOX DEVICE FUNCTION

Set-top boxes (STB) and terminals exist closest to users. Users of information services cannot receive information services without terminals, even though there are access points to the network around them. STBs are customer management boxes provided by platform enterprises. Without STBs, users cannot view satellite broadcasting or paid services offered by cable television through TV sets.

Complicated relationships can be seen between the terminals and upstream divisions, and within the vertical integration across these elements in the service provision system. While traditional broadcasting integrated the STB and the platform, no such vertical integration exists between television sets as terminals and the platform.

In contrast, telecommunication had already seen vertical integration between terminals and the network during the monopoly era. In other words, terminals that can be used with the network were owned by a monopolistic telecommunication enterprise. In the 1980s, the liberalisation of terminals occurred, enabling users to use any terminals provided by companies other than the monopoly. However, with the advent of mobile phones in Japan, a vertical integration of the wireless network division and terminal division was reactivated. Currently, competition among mobile phone companies has become fierce based on the premise of vertical integration. In addition, a new business model has emerged, integrating the platform division and the terminal division but not the network division. A typical example of this structure is the iPhone.

Figure 3.2 illustrates the horizontal disaggregation in the IT industry.

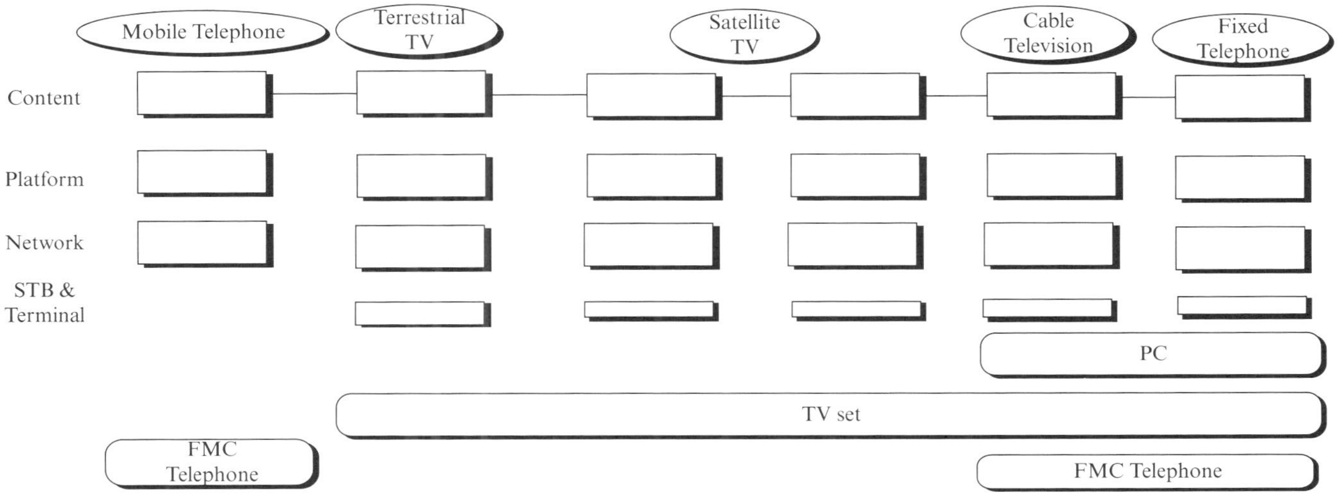

Source: Sugaya (2009), p. 30.

Figure 3.2 Horizontal structure of information and communication industries

In Japan, the idea became concrete by integrating nine laws regulating telecommunication and broadcasting into one law in order to promote the horizontal disaggregation of the industry structure (Sugaya, 2009). This single law was called the Information and Communication Law. However, during the political transition from the Liberal Democratic Party of Japan to the Democratic Party of Japan, the idea of the Information and Communication Law disappeared. Its basic idea is expressed, in part, in a new broadcasting law that integrates four laws related to broadcasting.

In the United States, the idea of horizontal disaggregation was behind the Information Superhighway Plan proposed at the emergence of Internet commercial services in the 1990s. Reed Hundt, then-Chairman of the Federal Communications Commission (FCC, 2005), stated: 'The U.S. Information Superhighway is composed of five lanes: broadcasting, satellite, cable, wireless, wireline telephony, and the FCC is poised to digitize them rapidly.' What is important here is that these five networks do not each occupy distinct lanes in a digitised system, but, rather, are merging into a lane that can be shared by all digital services. This condition will create competitions among lanes, which will expand and develop the IT industry.

In July 2011, terrestrial broadcasting in Japan was digitised. However, three prefectures in the Tohoku area did not implement digitalisation until April 2012 owing to the Great East Japan Earthquake in March 2011. This means that the digital Information Superhighway proposed in the US more than 20 years earlier has been completed in Japan.

However, competition among lanes of the Information Superhighway has just begun, which brings about new competitive relationships but also complementary relationships. A good example of the complementary relationships is Wi-Fi offloaded for mobile phones. Wi-Fi is a wireless LAN service, with the wireless connection between users and the network, which then interfaces with a wired network. That is, Wi-Fi offload is an example of a complementary relationship between the wireless and wired network.

Figure 3.2 indicates that smart devices, including smartphones and smart TV, are terminals connected to the network. To date, smartphones have not yet saturated the market and smart TV has yet to achieve widespread diffusion. When these smart devices achieve market saturation, making it possible to connect with any network illustrated in Figure 3.2, the idea of the Information Superhighway with swappable traffic will be realised.

Two issues have arisen that had not been envisioned during the time when the Information Superhighway concept was proposed. The first issue is fierce competition for profit generated by content, an issue that has already become evident. In existing content trade via telecommunication networks, the content division and network divisions divided user fees. With the advent of the vertically integrated business of the platform and terminal divisions, the profit allocation is organised across the content and platform divisions, leaving the network division unprofitable. The iPhone, with its broad share of the smartphone market, has established a business model of managing all the content trade via its own terminal with its own platform division and managing the collection of information fees by bypassing the network division.

The second issue is the emergence of social media including YouTube, Twitter and Facebook. These media are not the same content providers as the existing media depicted in Figure 3.2. These social media are services used to share information provided by users among other users and they are considered to be part of the platform division, not the content division. As Figure 3.3 shows, social media services are provided by smart devices via three network divisions (cable television, wired telecommunication and wireless communication) and the platform division. Moreover, these social media have brought a new flow of borderless and global information services that had not been seen within the traditional IT industry.

In summary, the economic power that had existed only in the network division has permeated the platform division, while the economic power in the network division has declined. In recent years, the network division of telecommunication has been called a 'dump pipe' that does not yield any profit. This indicates the shift of the economic power; and, indeed, this shift is 'digital power'.

5. MEDIA CONVERGENCE AND SMART DEVICES

As illustrated in Figure 3.2, smart devices are crossover terminals for traditional media – terminals for the post-media convergence era. In Japan, mobile television services started as a business model of providing retransmission service of terrestrial broadcast simultaneously as one-segment broadcasting for free. Moreover, since April 2012, a new type of mobile television service has begun, which NTT DOCOMO developed primarily. This new multimedia broadcasting system, nicknamed 'NOTTV' is a paid broadcasting service utilising the unused VHF spectrum that has become available after digitalising terrestrial broadcasting.

On the other hand, with smart TV, users can enjoy social media services on TV screens. Tablet computers such as the iPad have enabled the use of e-books and created e-textbooks that are coordinated with e-boards. In addition to books, packaged media including movies and newspapers are considered to be part of the content division described in Figure 3.2.

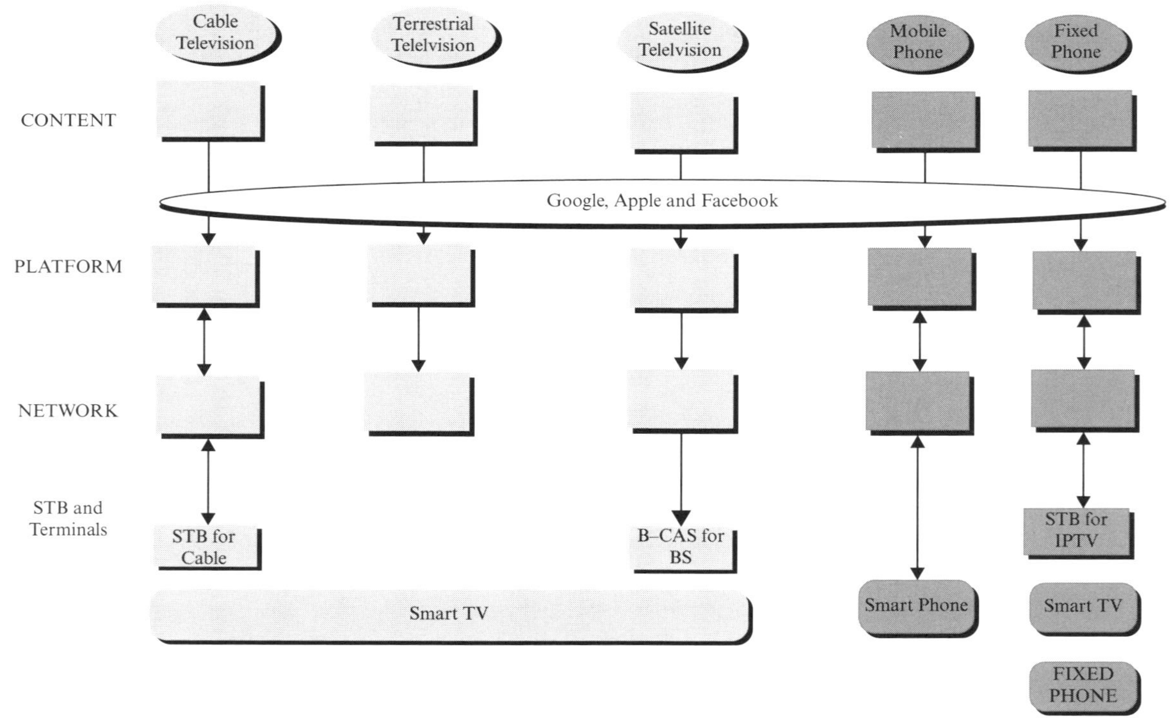

Figure 3.3 Structure of information and communication industries in the post-media convergence age

The traditional media including newspapers and publishing, which relied on physical distribution channels such as parcels, are now providing their own contents via smartphones and tablets.

In the movie industry, the existing distribution channel based on traditional films has been drastically transforming, owing to the digitalisation of movie theatres. The new digitised system connects movie theatres to telecommunication networks. This means that the movie theatres have become part of the terminal division in the industrial structure of the IT industry shown in Figure 3.1. Digital movie theatres already provide live footage of soccer games sent via telecommunication networks, which realises various usages of movie theatres.

In Japan, such packaged media as books, movies and newspapers did not operate in closely regulated markets. However, in the IT industry of the post-media convergence era, these traditional packaged media could be part of the IT industry, including telecommunication and broadcasting, where government regulations have been imposed. While this does not mean that regulations on entry and exit would be imposed on packaged media, this could indicate that contents provided by the packaged media will be provided via telecommunication networks and received by terminals that could also receive telecommunication and broadcast.

It is unlikely that the existing distribution channel for print media will disappear soon, even as these packaged media become part of the IT industry's content division. These traditional media now have new business opportunities – profits from online newspapers, e-publishing and live broadcast at movie theatres, among others. A choice between two media such as print media and e-publishing does not exist here; rather, a strategy for using advantages of both media can be implemented. For instance, both print and online newspapers coexist; in book publishing, releasing an e-book can be used to publicise the same book in hard copy form.

In this way, the diversification in the terminal division produces new business opportunities. The strategy of vertically integrating multiple function, high performance terminals with the platform division or the network division has provided the platform and network enterprises with a choice of strategies for assuring an advantage in the competitive market. Such new opportunities also mean that the terminal division is in the process of shifting to implementing the 'digital power'.

In other words the diversification of platforms, wider bandwidth and advent of smart devices have expanded the distribution channels of traditional media, which had relied on physical channels, into the digital networks that further enlarge the duplicated space of the IT industry and the traditional media industry. The influence of 'digital power' has penetrated the terminal division from the platform division beyond the network division.

The business models of smart devices include both paid and free models, as described earlier. In the realm of paid businesses, we need to consider whether network enterprises or platform enterprises will gain customers, and how the content division will utilise the 'digital power'.

In Japan, both wireless telecommunication enterprises and wired cable television companies started selling devices for home television in April 2013. They are called 'smart TV sticks'. These terminals help transform home televisions into smart televisions by sending information via Wi-Fi to home television receivers and that enables users to enjoy the contents on their televisions via telecommunication networks. It can be said that this is a new business model for the network enterprises to utilise the 'digital power' of smart devices.

6. CONCLUSION

This chapter analysed the transition of the IT industry since the 1980s from the perspective of 'digital power'. For the traditional IT industry, the network division had been the source of economic power. This power derived from economies of scale wielded by the network division. Today, all the networks have been digitised. This chapter calls it the transitional period from the media convergence era to the post-media convergence era. In the era of post-media convergence, the economic power of the network division continues. Government regulations have been imposed on the telecommunication networks and broadcasting networks on the premise of the existence of the economic power.

However, the value of the platform division of telecommunications in the IT industry has attracted attention, as the Internet has diffused. This means the shift of the 'digital power' from the network division to the platform division. The social and economic impacts of social network services have spread globally beyond the US, the origin of the services.

In addition to the platform division, the advent of smart devices including smartphones and smart televisions has also drawn attention. It can be said that with these smart devices all the gaps between telecommunication and broadcasting could be dissolved. This chapter considered this phenomenon as the shift of 'digital power' into the terminal division.

Such traditional media as newspapers and publishing had their businesses developed outside of distribution channels used in the IT industry. Now these media have started providing information services via the digitised network division of the IT industry in the forms of online newspapers and e-publishing, while maintaining the existing distribution channels. In addition, movie distribution has changed drastically with the digitalisation of movie theatres.

In the era of post-media convergence, all information services circulated in the market are digitised, making it possible to receive them with smart devices. This change has brought new ‘digital power’ into the terminal division. It can be assumed that the competition among the network enterprises will give rise to strategies of how to make use of the ‘digital power’ in the terminal division to achieve a competitive advantage. It can also be assumed that the content divisions of firms will have the important strategic challenge of how to create original content and deliver it to other companies’ terminals.

REFERENCES

Auletta, K. (2009), *Googled: The End of the World As We Know It*, New York: Penguin Press.

Crandall, W.R. and K. Flamm (eds) (1989), *Changing the Rules: Technological Change, International Competition, and Regulation in Communications*, Washington, DC: The Brookings Institution.

FCC (2005), *Biography of Reed Hundt*, available at http://transition.fcc.gov/commissioners/previous/hundt/biography.html (accessed 24 February 2015).

Grant, A. (2004), *Australian Telecommunications Regulation*, 3rd edn, Sydney: University of New South Wales Press.

Sugaya, M. (2009), ‘The transformation of telecommunication regulatory structure in Japan: vertical and horizontal perspectives’, *Keio Communication Review*, **31**, 23–36.

PART II

What did the smart revolution achieve for business and society?

4. Empirical analysis of the relationship between social media use and product innovation: focusing on SNS use and social capital

Hiroki Idota, Teruyuki Bunno and Masatsugu Tsuji

1. INTRODUCTION

1.1 Innovation Process and ICT

In accordance with the rapid transformation of the information society, it is becoming ever more difficult to conduct innovation and R&D, since changes in the market and technology have become much faster than previously. Firms are required to mobilise all managerial resources and networks to correspond to these changes and then utilise them to achieve innovation. This innovation process is postulated as follows: Firms perceive outside information on new technologies or shifts in consumer tastes, assimilate them with existing knowledge and resources inside the firm, and transform new information or knowledge into innovation (Zahra and George, 2002; Fosfuri and Tribo, 2008). This process determines the competitive advantage of a firm (Barney, 1991). In this process, the internal capability for innovation plays an essential role, which is defined as an integrated ability of a firm to create innovation. In more detail, internal capability includes the following factors: 1) technological level, such as the number of patents, production and R&D facilities; 2) human resources, such as the number of engineers with higher degrees or skills, the level of craftsmanship and work ethics; and 3) organisational capability, such as communication between workers and top management, speed of decision making and leadership capabilities of top management (Idota et al., 2014; Yam et al., 2011). These internal capabilities are particularly categorised as 'absorptive capability' by Cohen and Levinthal (1990) and Zahra and George (2002), for example.

The above innovation process is generic and applicable to all firms in all ages, but a new factor has come into the picture in the age of information, i.e. ICT (Information and Communication Technology). ICT has been regarded as a tool that not only improves the productivity of firms but also enhances innovation. ICT improves greatly the ability to obtain and share information; it enables the identification of new findings from big data on a real-time basis and facilitates the instant sharing of the information thus obtained by various related people. These features of ICT lead to innovation and it has thus become one of the essential bases for promoting innovation (Dodgson et al., 2006; Lee and Xia, 2006; Idota et al., 2012). These are examples of ICT use inside the firm, while ICT also supports collaboration with entities outside the firm, such as other related firms, universities and local research institutions. There are two categories of information source: transactions and knowledge channels. Regarding the former, new information is transmitted and absorbed through a transaction process, such as a supply chain, the purchasing of new capital goods, or inviting guest engineers, while in terms of the latter, new knowledge comes from university laboratories. In these processes, ICT transmits a much larger volume of information at much faster speeds than face-to-face communication. The information networks thus constructed are referred to as 'open innovation' (Chesbrough, 2003, 2006a, 2006b). In the open innovation process, a strategy for sharing information and resources with other firms and from suppliers to customers is required, since ICT can expedite the sharing of information related to innovation (Tsuji and Miyahara, 2010, 2011).

1.2 Social Media and Innovation

Social media such as blogs (a short form of 'web log') and SNSs (social network service/social network site) have recently been introduced and are utilised by a large number of firms. Kaplan and Haenlein (2010) define social media as 'a group of Internet-based applications built on the ideological and technological foundations of Web 2.0, which allows the creation and exchange of user-generated content'. That is, it can be said that 'social media' is a generic term for technology that enables individual employees to send, respond to and share knowledge and information with each other using the web.

A blog is an online service with a diary creation function. It includes mini blogs such as Twitter. An SNS, on the other hand, is a service that constructs a social network online. Boyd and Ellison (2008, p. 211) define

SNSs as 'web-based services that allow individuals to 1) construct a public or semi-public profile within a bounded system, 2) articulate a list of other users with whom they share a connection, and 3) view and traverse their list of connections and those made by others within the system'. In firms, blogs are used to send and share work-related information. Managers send messages to their subordinates via an SNS and it functions as a bulletin board similar to an administrative report, in-house rules and news of people who have passed away and so on. The team that works together is required to share specific information such as on the daily report, the project, customers, market and technology by use of the SNS.

SNSs thus can activate in-house communications. Employees participate freely in the SNS and various kinds of information are transmitted via the SNS on the basis of their own responsibility. Employees who work in the same field can engage in mutual exchanges of information not only related to their business, such as sales and technology, but also on common themes such as hobbies, preferences, local community, alma maters and so on. Thus employees can form interpersonal relationships via communications with those in other sections without being restricted by their current positions and locations. In recent years, however, the difference between blogs and SNSs has become vague.

A large number of papers examining the roles of social media in business focus mainly on marketing and the sales force; they analyse how social media are utilised for advertising products, or sampling consumer needs and the reputation of products (Schultz et al., 2012; Groza et al., 2012; Agnihotri et al., 2012; Rodriguez et al., 2012). Few papers focus on the relationship between social media and innovation, which is the main interest of this chapter.

1.3 Objectives of the Analysis

Social media thus activate communications among employees, connect them all across organisational sections and promote the sharing of information and the creation of new ideas, which increases not only productivity but also innovation (Lang, 2005). These are the internal uses of social media. A further aim of social media is that they act as a tool for obtaining market trends such as consumer needs, demands, or the reputation of particular products. From these, firms can conceive ideas about new products, reduce the lead time for the adoption of products and reduce product development costs. Firms can use social media to develop new products as well as to leverage sales and marketing.

This chapter confines its analysis only to product innovation, such as the introduction of a new product, and focuses on the role of social media

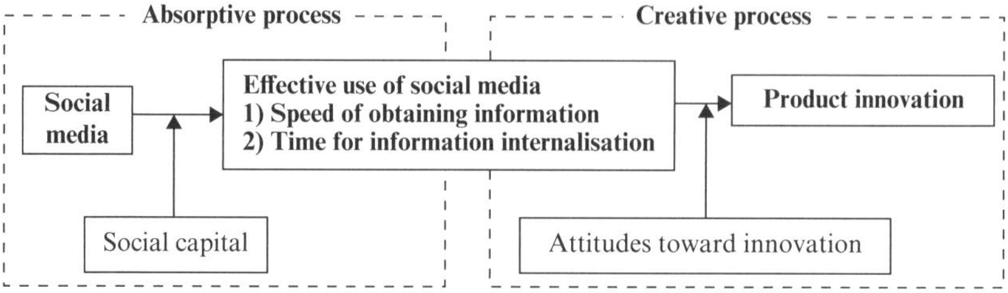

Figure 4.1 Hypothesis on social media and innovation

for this innovation. In particular, social media use inside the firm is specified. Internal capability is required for ICT to be used effectively, since ICT itself is simply a tool and requires an internal organisational infrastructure to function properly. This chapter selects organisational culture, or the collaborative atmosphere among employees, as factors promoting the effects of social media. These factors are referred to as 'social capital', a concept initiated by Putnam (1992). Social media require the cooperation of employees. In other words, embedded social capital, such as trust and reciprocity among employees, is necessary for social media to function well in the knowledge management system. Information is not exchanged openly if such social capital among employees has not been nurtured beforehand.

The hypotheses tested in this study are as follows: 1) how social media inside the firm promote product innovation and how social capital relates to the utilisation and performance of social media; and 2) how the use of social media for product innovation differs between the manufacturing and service industries. The difference in the two industries is owing to the fact that the latter tends to use social media more than the former and is thought to show better performance characteristics. Regarding 1), this paper divides the whole process from the use of social media to the final outcome of product innovation into two sub-processes, namely the 'absorptive process' and the 'creative process', as shown in Figure 4.1. The former focuses on how social media function properly with the support of social capital and in this process the effects of social media, such as in the 'speed of obtaining information' and 'the time required for information to be internalised and put to practical use' are employed as intermediate outcomes for evaluating this sub-process. The latter examines how the effects of social media and the 'attitude towards innovation' enable the fostering of product innovation. The 'attitude towards innovation' consists of factors that constitute internal capability, such as organisational orientation to innovation, namely initiatives of top management, learning processes among employees, the speed of decision making and so on.

2. FRAMEWORK OF ANALYSIS

Here an empirical analysis of social media use for product innovation is presented in order to clarify the characteristics and issues of social media use in this field. In addition, the objective of this analysis is to identify the fundamentals of successful SNS use for innovation.

2.1 Procedure for Estimation

The simple introduction of social media such as blogs or SNSs does not promote their use or achieve the final goal, and some preliminaries or premises are required. The estimation models are constructed in order to clarify these conditions. As mentioned in Section 1.3, the hypotheses to be examined are related to the absorptive process and to the creative process, and in particular to the notions that 1) social capital such as trust and reciprocity among employees is required for the successful use of social media; and 2) attitudes towards innovation must be shared from the top management down to employees for the promotion of innovation. These hypotheses, 1) and 2), are important factors of present organisational culture inside the firm and accordingly they constitute internal capability. The concrete questions are presented in what follows and used for estimation as control variables.

The estimation procedure is divided into two parts. For the absorptive process, how social capital contributes to the performance of social media is examined by taking the speed of obtaining information and time required for newly obtained information to be internalised and put to practical use as the outcomes of social media. In other words, the first procedure is to test the success of social capital inside the firm. The second step is to analyse the creative process, i.e. how product innovation is promoted by attitudes towards innovation and the effective use of social media. Thus the objective of this process is to examine whether firms with a better use of social media tend to achieve better innovation performance.

2.2 Questionnaire and Summary of Statistics

The data for analysis are based on a web survey of 3000 business people conducted by the Research Institute for Socionetwork Strategies, Kansai University, Japan (http://www.kansai-u.ac.jp/riss/en/index.html)

(1) Achievement of product innovation in the most recent three years
(2) Current situation of social media use
This consists of the following questions:
'Degree of social media use compared with three years ago'
'Degree of information exchange on customer awareness and impressions by use of social media'
(3) Social capital
This consists of the following questions:
'Degree of trust in colleagues'
'Degree of atmosphere for easy consultation with colleagues'
(4) Attitudes towards innovation
This presents organisational capability for innovation and consists of the following questions:
'Degree of atmosphere and tendency for an employee's originality and ideas to be linked to the development of products and services'
'Creating new products and services with networks outside the company'
'Human resource support to create new products and services efficiently'
'Managers express understanding when collaborating beyond divisions'
'Speed of decision making'
(5) Effect of social media
This aims to evaluate the organisational efficiency of knowledge management related to social media and consists of the following questions:
'Speed of obtaining information'
'Time required for new knowledge to be internalised and put to practical use'
(6) Characteristics of firms the subject works at, such as the size of the firm in terms of the number of employees and industry

in January 2009. The respondents using social media for business consisted of 1500 managers and 1500 general employees.

The survey questionnaire consists of the above questions, which are used as variables for estimation.

One remark should be given regarding (5). When new information or knowledge is recognised by employees via social media, they interpret and understand it. Then they attempt to use it for the business or job. The second question in (5) asks about the time needed to implement this process.

All questions except (6) require the respondent to answer using the five-point Likert rating scale, since they ask about the qualitative nature of firms. In particular, the survey does not ask about the number and contents of product innovation, since sensitive questions of this nature are hard to respond to. Questions are therefore designed to be as simple as possible.

A summary of the data is shown Tables 4.1 to 4.3. The explained variable of the estimation in this chapter is product innovation and, as stated earlier, the question asks about success in creating new products in the

Table 4.1 Summary statistics

		Freq.	Min	Max	Avg.	Std. Dev.
Innovation	Success in creating new products in the three years	2996	1	5	3.03	0.934
Situation of social media use	Degree of social media use compared with three years ago	3000	1	5	3.44	1.075
	Degree of information exchange on awareness and impressions of customers by use of social media	3000	1	5	3.05	1.208
Social capital	Degree of trust in colleagues	3000	1	5	3.66	0.865
	Degree of atmosphere for easy consultation with colleagues	3000	1	5	3.6	0.959
Attitudes toward innovation	Degree of atmosphere and tendency for an employee's originality and ideas to be linked to the development of products and services	2998	1	5	3.18	0.992
	Creating new products and services with networks outside the firm	2997	1	5	3.05	1.047
	Human resource support to create new products and services efficiently	2996	1	5	3.06	1.048
	Managers express understanding when collaborating beyond divisions	2994	1	5	3.26	1.029
	Speed of decision-making	2994	1	5	2.87	1.115

most recent three years in terms of the five-point scale. The resulting average was 3.03, implying that respondents achieved better than average success (Table 4.1). Table 4.1 also asks about firms' current social media use and attitudes toward innovation. 'Degree of social media use compared with three years ago' shows an average of 3.44 points on the scale.

Table 4.2 Effect of social media

		Freq.	Min	Max	Avg.	Std. Dev.
Effect of blogs	Speed of information gathering per employee	2066	1	5	3.31	0.766
	Time required for information to be internalised and put to practical use	2066	1	5	3.33	0.733
Effect of SNSs	Speed of information gathering per employee	1856	1	5	3.36	0.775
	Time required for information to be internalised and put to practical use	1856	1	5	3.28	0.73

Table 4.3 Size of firm

		Freq.	%
Number of employees	Less than 10	174	5.80%
	10–49	218	7.30%
	50–99	166	5.50%
	100–199	325	10.80%
	200–499	410	13.70%
	500–999	368	12.30%
	Over 1000	1339	44.60%
Industries	Agriculture, forestry and fisheries industry, and mining	12	0.40%
	Construction	136	4.50%
	Manufacturing	798	26.60%
	Sevice industry	1646	54.90%
	Medical treatment and welfare	111	3.70%
	School and educational industry	111	3.90%
	Public office	115	3.80%
	Others	66	2.20%

All social capital factors were larger than the average. In attitudes towards innovation factors, 'Managers express understanding when collaborating beyond divisions' showed the highest value.

The final outcome of social media is the creation of product innovation and there are many factors or channels linking these. This paper postulates that the direct outcome of social media is expressed by 'Speed of obtaining information' and 'Time required for information to be internalised and put to practical use', which are intended as an evaluation of the organisational efficiency of the knowledge management system for innovation. The effects of blogs and SNSs are larger than the average in these questions, implying that respondents recognise these effects (Table 4.2).

Regarding the size of firms shown in Table 4.3, the number of firms with over 200 employees is 2117 (70.6 per cent) and thus the majority of respondents are affiliated to large-sized firms. Regarding industry, 798 firms (26.6 per cent) belong to manufacturing, while 1646 (54.9 per cent) belong to the service industry.

3. ESTIMATION RESULTS AND DISCUSSION

The results of the estimation are presented here and the validity of the hypotheses is also discussed.

3.1 Effect of Introducing SNSs and Social Capital

This chapter mainly analyses SNSs as an example of social media, since the difference in the effects of SNSs and blogs was hardly recognisable from this data set. From here on, this chapter will mainly refer to SNSs. Based on the survey data, an ordered probit analysis is applied to examine the role of SNSs as a promoter of information and knowledge sharing for business processes. In this estimation, explanatory variables referring to social capital are added in order to observe their effects in this process.

3.1.1 Estimation I: absorptive process

First, the effects of introducing SNSs are analysed by taking 'Speed of obtaining information' and 'Time required for information to be internalised and put to practical use' as dependent variables, while explanatory variables include questions regarding social media use such as 'Degree of social media use compared with three years ago' and 'Degree of information exchange on awareness and impressions of customers by use of social media'. As for social capital, 'Degree of trust in colleagues' and 'Degree of atmosphere for easy consultation with colleagues' are selected. In addition, firms' attributes, such as the size of firms, are included in the explanatory variables. Tables 4.4 and 4.5 present the estimation results.

Both estimation results indicate that SNSs raise these two effects, since the coefficients of the two variables of SNS use are positively significant and their marginal effects are relatively large in both tables. In addition, the coefficients of the two variables of social capital are significantly positive; SNSs have better effects when they are robustly supplemented

Table 4.4 Effect of SNS 1: Speed of obtaining information

Effect of SNS 1: Speed of information gathering per employee	Coef.	Std. Err.	z-value	p-value	Marginal Effect
Size of firms	−0.02001	0.014004	−1.43	0.153	−0.00191
Degree of SNS use compared with three years ago	0.191204	0.02777	6.89	0.000***	0.018287
Degree of information exchange on awareness and impressions of customers by use of social media	0.161835	0.024836	6.52	0.000***	0.015478
Degree of trust in colleagues	0.23707	0.039261	6.04	0.000***	0.022674
Degree of atmosphere for easy consultation with colleagues	0.138789	0.035516	3.91	0.000***	0.013274
/cut1	0.127676	0.154127			
/cut2	0.832995	0.145809			
/cut3	2.747827	0.154184			
/cut4	4.120861	0.167007			
Number of obs = 1856					
LR chi2(7) = 380.67					
Prob > chi2 = 0.000					
Pseudo R2 = 0.0911					
Log Likelihood = −1898.44					

Note: ***,** and * indicate significance at the 1%, 5% and 10% levels respectively.

by social capital. In order to raise the efficiency of business processes, it is favourable to introduce SNSs supported concurrently by suitable social capital.

3.1.2 Estimation II: creative process; SNSs and product innovation

These estimations examine whether social media are related to product innovation. Tables 4.6 and 4.7 select the following outcome of introducing social media as a dependent variable; namely, 'Whether a firm has been successful in product innovation in the most recent three years'. As for the selection of explanatory variables in these estimations, the rationale behind this is as follows: For successful product innovation, an organisational atmosphere is required in which employees can freely demonstrate originality and propose new ideas that would result in product innovation. The firm should also establish systems such as arranging for the necessary number of human resources and a suitable decision-making process.

Table 4.5 Effect of SNS 2: Time required for information to be internalised and put to practical use

Effect of SNS 2: Time required for information to be internalised and put to practical use	Coef.	Std. Err.	z-value	p-value	Marginal Effect
Size of firms	0.000662	0.014205	0.05	0.963	0.000487
Degree of SNS use compared with three years ago	0.161069	0.028015	5.75	0.000***	0.011839
Degree of information exchange on awareness and impressions of customers by use of social media	0.165849	0.025231	6.57	0.000***	0.01219
Degree of trust in colleagues	0.211571	0.039637	5.34	0.000***	0.015551
Degree of atmosphere for easy consultation with colleagues	0.134402	0.036006	3.73	0.000***	0.009879
/cut1	−0.03538	0.157197			
/cut2	0.814523	0.146836			
/cut3	2.841237	0.156156			
/cut4	4.181317	0.170331			
Number of obs = 1856					
LR chi2(7) = 1856					
Prob > chi2 = 0.000					
Pseudo R2 = 0.0833					
Log Likelihood = −1813.11					

Note: ***,** and * indicate significance at the 1%, 5% and 10% levels, respectively.

Moreover, collaboration beyond the divisions inside as well as outside the firm is necessary. Therefore for these reasons, managerial attitudes towards innovation, including 'Degree of atmosphere and tendency for an employee's originality and ideas to be linked to the development of products and services', 'Creating new products and services with networks outside the firm', 'Human resource support to create new products and services efficiently', 'Managers express understanding when collaborating beyond divisions' and 'Speed of decision making' are included in the explanatory variables. In addition to these, the above effects of SNS use employed in Tables 4.4 and 4.5 are also included as variables that assist SNSs in contributing to product innovation; i.e. Tables 4.6 and 4.7 include 'Effect of SNS 1: Speed of obtaining information' and 'Effect of SNS 2: Time required for information to be internalised and put to practical use', respectively.

The results of the estimations shown in Tables 4.6 and 4.7 indicate that all variables except the size of the firm are positively significant, implying

Table 4.6 Product innovation and the effect of SNS 1: Speed of obtaining information

Product Innovation: Success in creating new products in the three years	Coef.	Std. Err.	z-value	p-value	Marginal Effect
Size of firms	0.0032568	0.0138772	0.23	0.814	0.0001408
Degree of atmosphere and tendency for an employee's originality and ideas to be linked to the development of products and services	0.1723093	0.0347683	4.96	0.000***	0.0074492
Creating new products and services with networks outside the company	0.1701659	0.0313892	5.42	0.000***	0.0073566
Human resource support to create new products and services efficiently	0.3132976	0.0305232	10.26	0.000***	0.0135444
Managers express understanding when collaborating beyond divisions	0.1887593	0.0321733	5.87	0.000***	0.0081604
Speed of decision-making	0.1925671	0.027118	7.1	0.000***	0.008325
Effect of SNS 1: Speed of obtaining information	0.1903912	0.0357631	5.32	0.000***	0.0082309
/cut1	1.865778	0.1583114			
/cut2	2.922201	0.1596827			
/cut3	4.592972	0.1725937			
/cut4	5.974943	0.1922811			
Number of obs = 1856					
LR chi2(7) = 944.12					
Prob > chi2 = 0.000					
Pseudo R2 = 0.1901					
Log Likelihood = −2011.7215					

Note: ***,** and * indicate significance at the 1%, 5% and 10% levels, respectively.

that they contribute to product innovation. Of these, 'Human resources support to create new products and services efficiently' has the largest marginal effect and this is followed by the two effects of SNS use and 'Speed of decision making'. In other words, to promote product innovation, firms are required to arrange for 1) human resources who are

Table 4.7 Product innovation and the effect of SNS 2: Time required for information to be internalised

Product Innovation: Success in creating new products in the three years	Coef.	Std. Err.	z-value	p-value	Marginal Effect
Size of firms	0.0007605	0.0138796	0.05	0.000***	0.0000329
Degree of atmosphere and tendency for an employee's originality and ideas to be linked to the development of products and services	0.1691925	0.0347912	4.86	0.000***	0.007312
Creating new products and services with networks outside the company	0.1683914	0.0314076	5.36	0.000***	0.0072774
Human resource support to create new products and services efficiently	0.3137148	0.0305215	10.28	0.000***	0.0135579
Managers express understanding when collaborating beyond divisions	0.1931653	0.0320666	6.02	0.000***	0.0083481
Speed of decision-making	0.1881304	0.0271744	6.92	0.000***	0.0081305
Effect of SNS 2: Time required for information to be internalised and put to practical use	0.2063329	0.0383984	5.37	0.000***	0.0089171
/cut1	1.87583	0.1589528			
/cut2	2.932086	0.1603039			
/cut3	4.603526	0.1732376			
/cut4	5.985654	0.1929083			
Number of obs = 1856					
LR chi2(7) = 944.67					
Prob > chi2 = 0.000					
Pseudo R2 = 0.1902					
Log Likelihood = −2011.4495					

Note: ***,** and * indicate significance at the 1%, 5% and 10% levels, respectively.

engaged in innovation; 2) efficient use of SNSs, which reduces the time taken to obtain new information and utilise it freely for innovation; and 3) collaboration beyond work divisions. In sum, SNSs indirectly affect the process of perceiving new information promptly, which can then contribute to the formation of the seeds of new products.

Table 4.8 *Product innovation and the effect of SNS 1: speed of obtaining information; manufacturing industry*

Product Innovation	Coef.	Std. Err.	z-value	p-value	Marginal Effect
Size of firms	−0.0133202	0.0290373	−0.46	0.646	−0.0006937
Degree of atmosphere and tendency for an employee's originality and ideas to be linked to the development of products and services	0.2326792	0.0764257	3.04	0.002***	0.0121171
Creating new products and services with networks outside the company	0.1560769	0.0675202	2.31	0.021**	0.0081279
Human resource support to create new products and services efficiently	0.3839561	0.065541	5.86	0.000***	0.0199951
Managers express understanding when collaborating beyond divisions	0.2184251	0.0656431	3.33	0.001***	0.0113748
Speed of decision making	0.2388998	0.0560419	4.26	0.000***	0.0124411
Effect of SNS 1: Speed of obtaining information	0.1660439	0.0693306	2.39	0.017**	0.008647
/cut1	2.192201	0.3132006			
/cut2	3.332537	0.3227732			
/cut3	4.725625	0.3447421			
/cut4	6.307579	0.3807861			
Number of obs = 471 LR chi2(7) = 314.8 Prob > chi2 = 0.000 Pseudo R2 = 0.2331 Log Likelihood = −517.89481					

Note: ***,** and * indicate significance at the 1%, 5% and 10% levels, respectively.

Table 4.9 Product innovation and the effect of SNS 2: time required for information to be internalised; manufacturing industry

Product Innovation	Coef.	Std. Err.	z-value	p-value	Marginal Effect
Size of firms	−0.0143093	0.0290917	−0.49	0.623	−0.0007496
Degree of atmosphere and tendency for an employee's originality and ideas to be linked to the development of products and services	0.2182658	0.0763926	2.86	0.004***	0.0114345
Creating new products and services with networks outside the company	0.1625038	0.0674099	2.41	0.016**	0.0085133
Human resource support to create new products and services efficiently	0.3776465	0.0659018	5.73	0.000***	0.0197842
Managers express understanding when collaborating beyond divisions	0.23176	0.0650995	3.56	0.000***	0.0121415
Speed of decision making	0.2379786	0.0561941	4.23	0.000***	0.0124673
Effect of SNS 2: Time required for information to be internalised and put to practical use	0.1641845	0.0779442	2.11	0.000**	0.0086013
/cut1	2.163037	0.3164319			
/cut2	3.303342	0.3260898			
/cut3	4.696982	0.348085			
/cut4	6.275623	0.3835531			
Number of obs = 471					
LR chi2(7) = 313.5					
Prob > chi2 = 0.000					
Pseudo R2 = 0.2321					
Log Likelihood = −518.54665					

Note: ***,** and * indicate significance at the 1%, 5% and 10% levels, respectively.

Table 4.10 Product innovation and the effect of SNS 1: speed of obtaining information; service industry

Product Innovation	Coef.	Std. Err.	z-value	p-value	Marginal Effect
Size of firms	0.016453	0.0185838	0.89	0.376	0.0006486
Degree of atmosphere and tendency for an employee's originality and ideas to be linked to the development of products and services	0.1034841	0.0453758	2.28	0.023**	0.0040798
Creating new products and services with networks outside the company	0.1952154	0.041517	4.7	0.000***	0.0076962
Human resource support to create new products and services efficiently	0.3135281	0.0404571	7.75	0.000***	0.0123606
Managers express understanding when collaborating beyond divisions	0.1569121	0.0421707	3.72	0.000***	0.0061861
Speed of decision making	0.2567353	0.0355061	7.23	0.000***	0.0101216
Effect of SNS 1: Speed of obtaining information	0.2236928	0.0485797	4.6	0.000***	0.0088189
/cut1	2.037519	0.2150054			
/cut2	3.15429	0.2169655			
/cut3	4.762882	0.2335412			
/cut4	6.18921	0.2632836			
Number of obs = 1050					
LR chi2(7) = 548.66					
Prob > chi2 = 0.000					
Pseudo R2 = 0.1931					
Log Likelihood = −1146.5194					

Note: ***,** and * indicate significance at the 1%, 5% and 10% levels, respectively.

Table 4.11 Product innovation and the effect of SNS 2: time required for information to be internalised; service industry

Product Innovation	Coef.	Std. Err.	z-value	p-value	Marginal Effect
Size of firms	0.1030485	0.0185557	0.7	0.482	0.0005188
Degree of atmosphere and tendency for an employee's originality and ideas to be linked to the development of products and services	0.1022119	0.0454062	2.25	0.024***	0.0040638
Creating new products and services with networks outside the company	0.1924343	0.0415606	4.63	0.000***	0.0076509
Human resource support to create new products and services efficiently	0.3196859	0.0403759	7.92	0.000***	0.0127103
Managers express understanding when collaborating beyond divisions	0.1622279	0.0420463	3.86	0.000***	0.00645
Speed of decision making	0.2505662	0.0355681	7.04	0.000***	0.0099622
Effect of SNS 2: Time required for information to be internalised and put to practical use	0.224139	0.0512233	4.38	0.000***	0.0089115
/cut1	2.012977	0.2154049			
/cut2	3.12736	0.2169618			
/cut3	4.734606	0.2333958			
/cut4	6.157677	0.2629523			
Number of obs = 1050 LR chi2(7) = 546.6 Prob > chi2 = 0.000 Pseudo R2 = 0.1923 Log Likelihood = −1147.5499					

Note: ***,** and * indicate significance at the 1%, 5% and 10% levels, respectively.

3.2 Effects of SNS in Different Industries

This section examines whether the same results obtained in the previous analysis are applicable to different industries. This chapter divides industry into manufacturing and services. Tables 4.8 and 4.9 show the results for manufacturing, while Tables 4.10 and 4.11 those of the service industry.

In both cases, factors related to attitudes towards innovation and the effects of SNS use are employed. Tables 4.8 and 4.10 employ 'Effect of SNS 1: Speed of obtaining information', while Tables 4.9 and 4.11 use 'Effect of SNS 2: Time required for information to be internalised and put to practical use'.

3.2.1 Estimation III: manufacturing industry

Estimation results are shown in Tables 4.8 and 4.9. Except for the size of firms, all variables are found to be positively significant in the two models, but in both tables, 'Creating new products and services with networks outside company' and 'Effect of SNS 1: Speed of obtaining information' have a relatively smaller marginal effect than other variables, while the marginal effect of the two effects of SNS use are relatively small in the manufacturing industry. These will be compared with those in the service industry.

3.2.2 Estimation IV: service industry

The estimation results are indicated in Tables 4.10 and 4.11, in which all variables are positively significant, implying that they contributed to product innovation. In particular, in both tables, three variables, 'Human resources support to create new products and services efficiently', 'Speed of decision making' and 'Effect of SNS 1: Speed of obtaining information' show a larger marginal effect, while 'Degree of atmosphere and tendency for an employee's originality and ideas to link to the development of products and services' and 'Creating new products and services with networks outside company' have rather smaller marginal effect.

3.3 Result of Estimation

In order to answer the question concerning which industry SNSs have a greater effect on in terms of product innovation – the manufacturing industry or the service industry – let us compare the marginal effects in the estimations, since all variables except the size of the firm are significant in both industries. It is found that all factors in attitudes towards innovation in the manufacturing industry are larger than those in the service industry, whereas in the service industry the two effects of SNS use are larger than those in manufacturing. Thus, the results obtained are summarised as follows:

1. Attitudes towards innovation and the effectiveness of SNS use are important for product innovation in both the manufacturing and the service industries.
2. Attitudes towards innovation are more important for the manufacturing industry.
3. The effectiveness of SNS use is more important for the service industry.

These results appear to be consistent with reality, since the service industry is located closer to consumers or the market and the period of product development is shorter than in the manufacturing industry. The service industry is required to obtain market information quickly, analyse it, develop a new product or service and ship it to the market faster than rivals. It is then necessary to monitor the reaction of consumers and improve the product quickly. Thus the service industry requires more prompt action than manufacturing, which is reflected by the importance of the effects of SNS use.

4. CONCLUSIONS

Social media have recently been introduced by many firms in order to obtain market information, enhance information sharing among employees and other sections and speed up the decision-making process, and all these contribute to greater product innovation. This chapter obtained the following conclusions: 1) SNS use promotes the speed of obtaining new information and shortens the time in which it is internalised and put to practical use; 2) for SNSs to achieve these effects, they must be supported concurrently by social capital; 3) product innovation is enhanced by the effect of SNS use and attitudes towards innovation; and 4) SNSs are generally more effective in the service industry than in the manufacturing industry.

Regarding the first and second point, the analysis clarifies that social capital is a prerequisite for successful SNS use. If employees themselves neither fully understand the target of the firm nor collaborate with each other, an SNS cannot function well even if a manager forces employees to use it. If social capital is not well empowered in the firm, the members cannot send and receive information freely. The top management should make an effort to empower social capital and nurture an organisational culture in which there is mutual trust and reciprocal responsibility inside the firm. In so doing, the following measures are required: 1) learning ICT skills for SNSs; 2) establishing ethical rules regarding SNSs; 3) committing to SNS use by top management; and 4) preparing guidelines for SNS use, such as how to use SNSs in the work context.

This chapter contains limitations that will be addressed in future work. First, the estimation method must be improved to handle the endogeneity problem. While this chapter uses probit analysis, it is not appropriate for the analysis of causal relationships among dependent and independent variables. The relationship shown in Figure 4.1 is causal and the usual probit analysis is insufficient; the instrumental variable method should be used. Second, the number of samples is not sufficiently large to enable a suitable econometric analysis to be conducted. In addition, the web survey is not a proper academic tool for rigorous statistical analysis, since irrelevant or inconsistent answers tend to be contained in the responses. Care is required regarding the methodology of conducting web surveys.

ACKNOWLEDGEMENTS

This study was partly supported by a JSPS grant entitled 'Business Innovation Strategy by Social Media' (No. c-24530435), which is gratefully acknowledged. It was also supported by the Promotion Project for Distinctive Joint Research of the Ministry of Education, Culture, Sports, Science and Technology, 2011. Thanks are also due to the Research Institute for Socionetwork Strategies, Kansai University, Japan for permission to access its database.

REFERENCES

Agnihotri, R., P. Kothandaraman, R. Kashyap and R. Singh (2012), 'Bringing "social" into sales: the impact of salespeople's social media use on service behaviors and value creation', *Journal of Personal Selling and Sales Management*, **32** (3), 333–48.

Barney, J. (1991), 'Firm resources and sustained competitive advantage', *Journal of Management*, **17**, 99–120.

Boyd, D.M. and N.B. Ellison (2008), 'Social network sites: definition, history and scholarship', *Journal of Computer-Mediated Communication*, **13** (1), 210–30.

Chesbrough, H.W. (2003), *Open Innovation: The New Imperative for Creating and Profiting from Technology*, Boston, MA: Harvard Business School Press.

Chesbrough, H.W. (2006a), 'Open innovation: a new paradigm for understanding industrial innovation', in H.W. Chesbrough, W. Vanhaverbeke and J. West (eds), *Open Innovation Researching a New Paradigm*, Oxford, UK: Oxford University Press, pp. 1–12.

Chesbrough, H.W. (2006b), *Open Business Model: How to Thrive in the New Innovation Landscape*, Boston, MA: Harvard Business School Press.

Cohen, W.M. and D.A. Levinthal (1990), 'Absorptive capacity: a new perspective on learning and innovation', *Administrative Science Quarterly*, **35** (1), 128–52.

Dodgson, M., D. Gann and A. Salter (2006), 'The role of technology in the shift towards open innovation: the case of Procter & Gamble', *R&D Management*, **36** (3), 333–46.

Fosfuri, A. and J.A. Tribo (2008), 'Exploring the antecedents of potential absorptive capacity and its impact on innovation performance', *Omega*, **36** (2), 173–87.

Groza, M., R. Peterson, U.Y. Sullivan and V. Krishnan (2012), 'Social media and the sales force: the importance of intra-organizational cooperation and training on performance', *The Marketing Management Journal*, **22** (2), 118–30.

Idota, H., T. Bunno and M. Tsuji (2012), 'Empirical study on ICT usage and internal capability for innovation among Japanese SMEs', *Proceedings of International Telecommunications Society 2012 Biennial Conference*, Bangkok, Thailand, pp. 1–15.

Idota, H., T. Bunno and M. Tsuji (2014), 'An empirical analysis of innovation success factors due to ICT use in Japanese firms', in T. Tsiakis, T. Kargidis and P. Katsaros (eds), *Approaches and Processes for Managing the Economics of Information Systems*, Hershey, PA: IGI, pp. 324–47.

Kaplan, A.M. and M. Haenlein (2010), 'Users of the world, unite! The challenges and opportunities of social media', *Business Horizons*, **53** (1), 59–68.

Lang, E.M. (2005), 'Would you, could you, should you blog?', *Journal of Accountancy*, **199** (6), 36–42.

Lee, G. and W. Xia (2006), 'Organizational size and IT innovation adoption: a meta-analysis', *Information and Management*, **43**, 979–85.

Putnam, R.D., R. Leonardi and R. Nanetti (1992), *Making Democracy Work: Civic Traditions in Modern Italy*, Princeton, NJ: Princeton University Press.

Rodriguez, M., R.M. Peterson and V. Krishnan (2012), 'Social media's influence on business-to-business sales performance', *Journal of Personal Selling and Sales Management*, **32** (3), 365–78.

Schultz, R.J., C.H. Schwepker and D.J. Good (2012), 'An exploratory study of social media in business-to-business selling: salesperson characteristics, activities and performance', *The Marketing Management Journal*, **22** (2), 76–89.

Tsuji, M. and S. Miyahara (2010), 'A comparative analysis of organizational innovation in Japanese SMEs generated by information communication technology', in A. Kuchiki and M. Tsuji (eds), *From Agglomeration to Innovation: Upgrading Industrial Clusters in Emerging Economies*, Basingstoke, UK: Palgrave Macmillan, pp. 231–69.

Tsuji, M. and S. Miyahara (2011), 'Agglomeration and local innovation networks in Japanese SMEs: analysis of the information linkage', in A. Kuchiki and M. Tsuji (eds), *Industrial Clusters, Upgrading and Innovation in East Asia*, Cheltenham, UK: Edward Elgar, pp. 253–94.

Yam, R., W. Lo, E. Tang and A. Lau (2011), 'Analysis of sources of innovation, technological innovation capabilities and performance: an empirical study of Hong Kong manufacturing industries', *Research Policy*, **40**, 391–402.

Zahra, H. and G. George (2002), 'Absorptive capacity: a review, reconceptualization and extension', *Academy of Management Review*, **27** (2), 185–203.

5. Efficiency of transport infrastructure and ICT development

Kyoung-Youn Na and Chang-Ho Yoon

1. INTRODUCTION

Transportation infrastructure enhances the quality and safety of mobility in the economy and promotes economic growth by increasing the returns to private investment and expanding the range of business opportunities. Investment in transportation infrastructure has been regarded as key to economic development (World Bank, 2008). For developed countries, transportation accounts for 7 per cent of GDP on average.

Among the various types of transportation infrastructure, road infrastructure has been examined the most often in empirical literature.[1] The motorway increases production efficiency by reducing the cost of physical distribution of products and raw materials, besides stimulating investment in the vehicle industry and passing on substantial spillover effects to the rest of the economy. Furthermore, the convergence of information and communication technology (ICT) and transportation technology has led to the development of the so-called intelligent transportation system (ITS). As a result, the network effect of motorway infrastructure has been expanding over time.

According to UNECE (2012), ITS performance depends critically on the size and quality of ICT networks in the economy.[2] Empirical observations on recent data reveal that unless the broadband penetration ratio exceeds a certain threshold level, ITS efficiency improves very little. This chapter attempts to examine the effects of ICT development on the magnitude network effects of motorway growth. In particular, the chapter explores the possibility of accelerating the productivity effects that arise from widespread use of ICT in transportation.[3]

The productivity effects of motorway infrastructure among OECD countries are investigated by Na, Han and Yoon (2013). They show that the contribution of motorways to productivity growth in the economy tends to increase as the motorway extends over time. In other words, an accelerating network effect prevails when the motorway grows beyond the threshold level. However, they do not investigate the source of these network effects. Using OECD panel data, this chapter complements the previous study by partially answering why the network effects of motorway infrastructure tend to accelerate when ICT infrastructure grows concomitantly beyond a certain threshold level.

2. INFORMATION AND COMMUNICATION TECHNOLOGY AND INTELLIGENT TRANSPORT SYSTEMS

The purpose of ITS is to increase the efficiency and safety of transportation by applying ICT to transportation infrastructure. ITS consists of four subsystems – travellers, vehicles, traffic centres and roadsides – all of which are connected through wide-area wireless communications, wire-line communications, vehicle-to-vehicle (V2V) communications and dedicated short-range communications (DSRC).[4]

The services provided by ITS include advanced traffic management systems (ATMS), advanced traveller information systems (ATIS), advanced public transportation information systems (APTI), commercial vehicle operations (CVO) and advanced vehicle and highway systems (AVHS).[5] In order to offer these services, ITS uses the sensor technology, satellite GPS, wireless communication technology and database management of transport information centres. In the recent past, wireless communication technologies that facilitate processing and the provision of user information on a real-time basis have become essential, as an increasing number of drivers have been using smartphones.

Figure 5.1 illustrates the structure of an ITS by which traffic can be managed on a real-time basis. ITS collects data from roads and running vehicles, controls the traffic flow by using roadside equipment for fast and safe movement, and provides travellers with traffic information. Normally, information on traffic volume, speed and occupancy is collected through on-road vehicle detectors or CCTV, the collected information is processed in a traffic information centre to make it fit for operators and users, and the information is then provided through variable message sign (VMS), the Internet, or mobile phones. With the use of such real-time information, road users can choose the best route to their destinations. The biggest advantage of real-time traffic information is that it helps disperse traffic volume, and therefore reduces traffic congestion. As drivers obtain traffic information and change routes, high traffic volume on roads is dispersed into alternative routes, alleviating traffic congestion and improving the efficiency of traffic flows.[6]

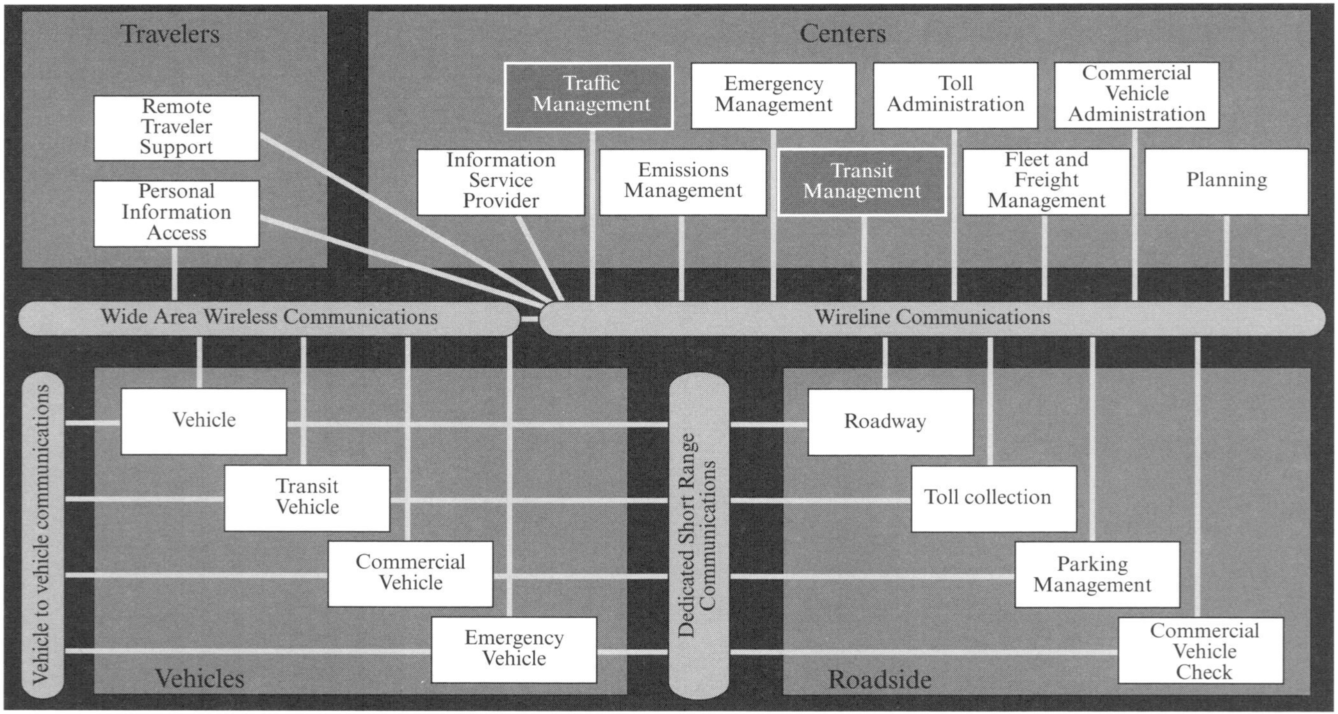

Source: US Department of Transportation (2004).

Figure 5.1 Structure of an intelligent transport system

Table 5.1 Global market for ITS (in million US dollars)

	2011	2012	2013	2014	2015	CAGR (%)
US	5089.62	5884.94	6049.92	6577.11	7141.32	9.57
Canada	221.23	233.77	244.79	253.68	259.92	6.24
Japan	2186.45	2376.65	2576.75	2787.32	3009.28	9.05
EU	4082.45	448.31	4830.01	5233.80	5665.38	9.45
China	574.11	678.74	801.16	944.51	1113.53	18.26
Korea	294.26	326.49	361.74	400.12	441.63	11.47
Other countries	566.33	641.68	726.47	822.27	929.64	13.71
Total	13,014.45	14,590.58	15,590.84	17,018.81	18,560.7	9.99

Source: The Presidential Council on National Competitiveness (2012).

In addition, ITS offers services related to the vehicle industry through communication networks. Drivers and passengers are offered mobile offices that handle traffic information, remote diagnosis of vehicles and Internet services such as financial transactions, news and email. Motorway infrastructure can offer networked services based on sensor technology and become an automated highway system (AHS). For this, a sophisticated communication infrastructure for DSRC or a continuous air interface long and medium ranges (CALM) standard is needed.[7]

Most developed countries have targeted ITS as a national development strategy. The United States passed the Intermodal Surface Transportation Efficiency Act in 1991 and chalked out a plan to invest more than 20 billion US dollars until 2020. The European Union proposes an integrated ITS through a trans-European transport network.[8] According to the Presidential Council on National Competitiveness (2012) of the Republic of Korea, the global market for ITS is likely to grow by more than 10 per cent per annum and reach 18.6 billion US dollars in 2015.

3. EMPIRICAL ANALYSIS

3.1 Background

In most developed countries, investments in both ICT and motorway infrastructure have grown substantially during recent decades. However, an exact comparison of the growth rates of the two different types of infrastructure is unfortunately difficult because the capital stock data on

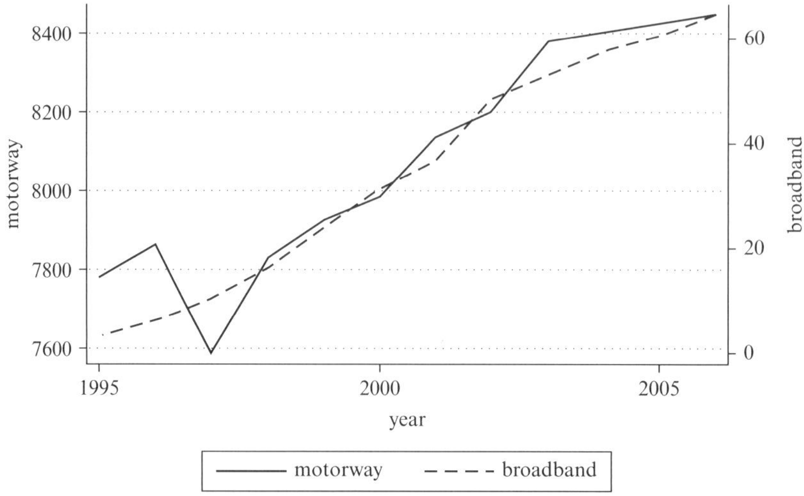

Sources: OECD Factbook (2008) and ITU World Telecommunication/ICT Indicators (2010).

Figure 5.2 Time profile of investment in transportation and ICT infrastructure

motorway infrastructure are not readily available. For expositional convenience, in this chapter, we measure the motorway capital by motorway length per worker (z). The motorway length and number of vehicles for the period 1996–2006 for 21 OECD countries are calculated from the OECD Factbook (2008).[9]

The data on GDP, total capital stock and number of workers are obtained from the OECD Economic Outlook (2012) database and each country's information infrastructure (ICT_{it-1}) is measured by the number of broadband subscribers, as available in the ITU World Telecommunication/ICT Indicators (2010).

Figure 5.2 illustrates the investment trends in transportation and information infrastructure of OECD member countries. It is obvious that since the mid 1990s, both ICT and motorway infrastructure began to grow together very rapidly until the early 2000s. While the ICT infrastructure continues to grow rapidly, the investment growth in motorway infrastructure shows a decline, reflecting fiscal austerity.

The efficiency contribution of each type of infrastructure is measured in terms of resulting change in total factor productivity (TFP), which can be obtained from the OECD Productivity Database. The control variables such as degree of openness of the economy and human capital, which affect the production efficiency of the economy, are obtained from the OECD Stat Extracts and Cohen and Soto (2007); these present the years of schooling for persons 25 years and over.

Figure 5.3 shows that the telecom infrastructure of OECD countries, as measured by the broadband penetration ratio, has steadily increased since the late 1990s. The adjusted correlation coefficient (based on windowed cross-correlation technique) between the growth rates of transportation infrastructure (or motorway capital) and telecom infrastructure remained mostly negative prior to 2003 and began to increase rapidly after 2004, to remain positive. It is interesting to note that the correlation coefficient between the growth rates of TFP and transportation infrastructure show a U-shape and began to increase only after 2003. In the next section, we empirically analyse the causality between these variables.

3.2 Model

We begin with the aggregate production function $Y = AF(L, K)$, where Y stands for the GDP, A represents the TFP and L and K denote labour and capital stock, respectively. It is assumed that the TFP depends on the telecom and transportation infrastructure and is influenced by time- and country-specific socio-economic effects.

The basic model for our estimation is

$$\ln TFP_{it} = \alpha_{0i} + \alpha_i t + f(\ln z_{it-1}) + X_{it}\beta + \mu_t + u_{it},$$

where i and t represent the country and the time period, respectively. In this model, α_{0i} is the country-specific level of the TFP, $\alpha_i t$ the country-specific trend coefficient and z_{it-1} lagged road infrastructure; $f(\cdot)$ denotes a non-linear functional relationship that describes the productivity impacts of motorway capital, while X_{it} denotes a vector of control variables, which include human capital, degree of congestion and openness of the economy.[10] The error term u_{it} represents various macro variables, often containing unobservable factors, and is serially correlated. To control for country-specific trends and integrated errors, we propose a first-difference model as follows:

$$\Delta \ln TFP_{it} = \alpha_i + \Delta f(\ln z_{it-1}) + \Delta X_{it}\beta + \tau_t + \varepsilon_{it}, \quad (5.1)$$

where $\tau_t \equiv \Delta\mu_t$, $\varepsilon_t \equiv \Delta u_t$.

We assume that ε_{it} is independent over time. In other words, given the past sequence of TFP growth rates, the current TFP growth rate depends on one-year lagged transportation and telecom infrastructure, global time

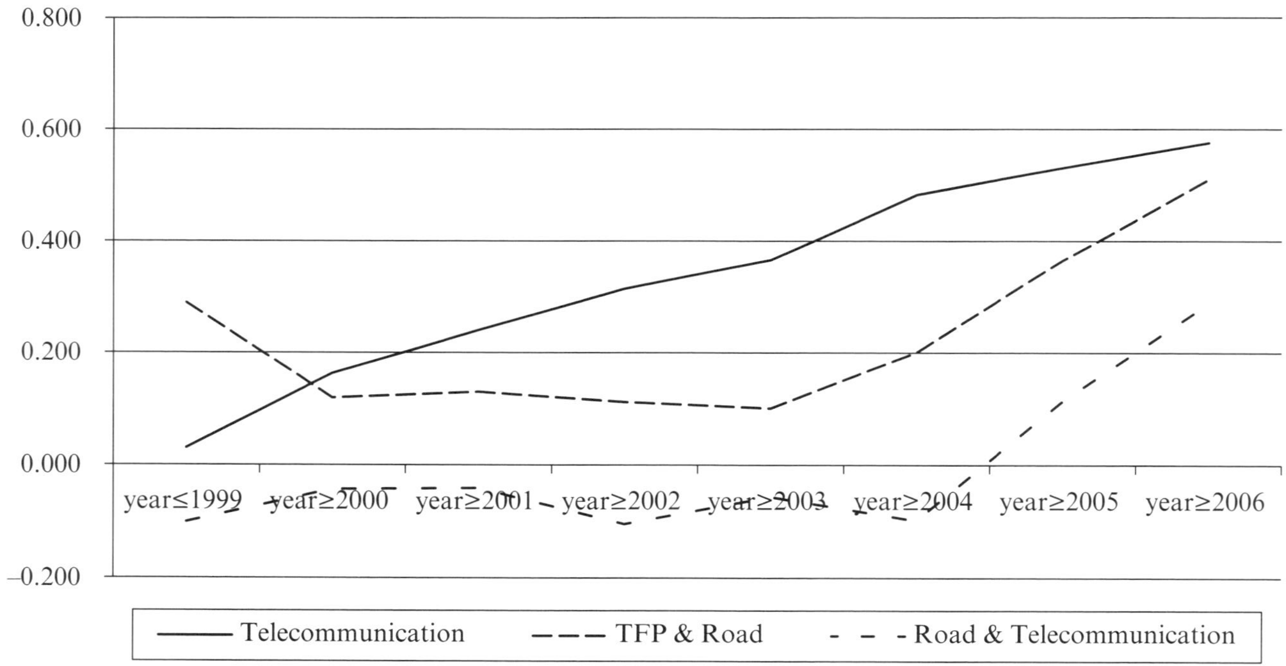

Note: Telecommunication denotes a lagged value of number of broadband subscribers (ICT_{it-1}), while TFP and transportation denote a TFP growth rate of the current period ($\Delta \ln TFP_{it}$) and the growth rate of motorway length per worker in the previous period ($\Delta \ln z_{it-1}$)respectively.

Figure 5.3 Productivity relationship between telecom and transportation infrastructures

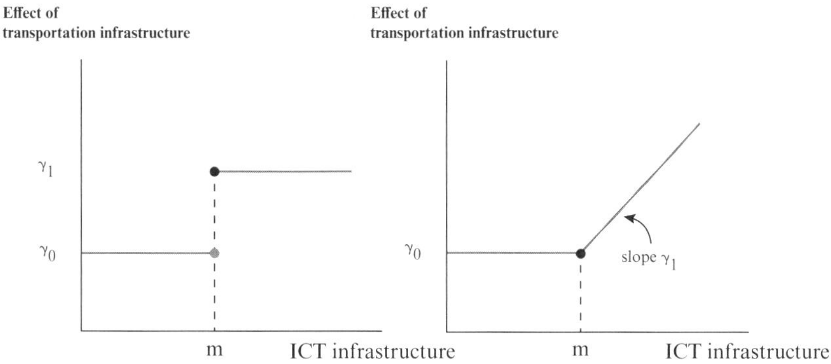

Figure 5.4 Productivity impacts of motorway capital

effects and unpredictable shocks. Finally, the following specification is used to estimate the effect of telecom infrastructure on the motorway infrastructure productivity effect:

$$\Delta \ln TFP_{it} = \alpha_i + (\gamma_0 + \gamma_1 D_{ICT>m})\Delta \ln z_{it-1} + \Delta X_{it}\beta + \tau_t + \varepsilon_{it}. \quad (5.2)$$

The dummy variable $D_{ICT>m}$ takes the value of 1 when the telecom infrastructure or broadband penetration ratio exceeds m. Similarly, an alternative specification can be used to estimate non-linear network effects as follows:

$$\Delta \ln TFP_{it} = \alpha_i + \{\gamma_0 + \gamma_1 Max(0, ICT_{it-1} - m)\}\Delta \ln z_{it-1} + \Delta X_{it}\beta + \tau_t + \varepsilon_{it}. \quad (5.3)$$

The difference between the two specifications stems from the relationship between motorway capital and TFP growth. In place of the discontinuous jump in (2), continuity is preserved in (3). As the ICT infrastructure exceeds a certain threshold level, productivity growth increases linearly with motorway capital. Figure 5.4 compares the two different specifications.

The robustness of the estimation results on the role of ICT and motorway infrastructure can be examined by estimating the effect of infrastructure on labour productivity growth. Given the constant returns to scale (CRS) production function, $Y = AF(L, K)$, we have $\ln y_{it} = \ln A\{f(\ln z_{it-1})\} + \theta \ln k_{it}$, where $y_{it} = Y_{it}/L_{it}$, $k_{it} = K_{it}/L_{it}$, and θ is the share of capital. Then, from Equation (5.3), the following equation can be obtained:[11]

$$\Delta \ln y_{it} = \alpha_i + \{\gamma_0 + \gamma_1 Max(0, ICT_{it-1} - m)\}\Delta \ln z_{it-1} + \Delta X_{it}\beta$$
$$+ \theta \Delta \ln k_{it} + \tau_t + \varepsilon_{it}. \quad (5.4)$$

Table 5.2 Descriptive statistics

Variables	Obs.	Mean	Std. Dev.	Max	Min
$\Delta\ ln\ TFP_{it}$	310	0.011	0.014	0.076	−0.069
$\Delta\ ln\ z_{it-1}$	287	0.033	0.111	1.389	−0.328
ΔS_{it}	320	0.074	0.034	0.169	0.009
$Open_{it}$	357	0.347	0.162	0.924	0.081
$\Delta\ ln\ V_{it}$	320	0.015	0.032	0.177	−0.210
ICT_{it-1}	356	0.245	0.259	0.860	0.000
$\Delta\ ln\ y_{it}$	335	0.017	0.015	0.074	−0.028
$\Delta\ ln\ k_{it}$	302	0.029	0.020	0.119	−0.160

Equation (5.4) indicates that the effect of motorway capital on labour productivity growth depends in general on ICT infrastructure and a set of control variables, $\Delta X_{it} = (\Delta S_{it}, Open_{it}, \Delta \ln V_{it})$. The stock of human capital and degree of congestion are used as control variables for X_{it}. The data on average schooling are used for human capital (S_{it})[12] and congestion is measured by the proxy variable V_{it}, which measures the number of vehicles. Degree of openness ($Open_{it}$) is measured as the ratio of trade volume to GDP.[13] The descriptive statistics of major variables are given in Table 5.2.

3.3 Empirical Results

Table 5.3 presents the fixed-effect estimation results for Equation (2).[14] The threshold level m is adjusted to vary between 0.5 and 0.8. Only when m exceeds 0.7 does the estimated coefficient for the impact of ICT become positive. Statistical significance is obtained when m increases further to 0.8.

This result confirms that the productivity impact of motorway capital depends non-linearly on ICT infrastructure. Models (f) and (g) use two dummy variables to confirm that the productivity effect of motorway capital is positive and statistically significant when $m = 0.8$, below which the estimated coefficient can be negative.

The estimated coefficients for the control variables such as degree of openness turn out to be statistically significant, while the coefficients for the other control variables are not significant (see Na, Han and Yoon, 2013). Although Equation (2) shows that non-linear network effects are pervasive, this does not reflect the growth effect of motorway extension. We use Equation (2) to search for the threshold effects.

Table 5.4 presents the estimation results for Equations (3) and (4). Both

Table 5.3 Estimation results (Equation 5.2)

$\Delta lnTFP_{it}$	Model (a)	Model (b)	Model (c)	Model (d)	Model (e)	Model (f)	Model (g)
ΔS_{it}	0.096	0.096	0.096	0.097	0.091	0.090	0.092
	(0.061)	(0.061)	(0.061)	(0.061)	(0.061)	(0.061)	(0.062)
$Open_{it}$	0.060**	0.061***	0.065***	0.058**	0.059**	0.061***	0.057**
	(0.021)	(0.021)	(0.020)	(0.021)	(0.021)	(0.021)	(0.021)
$\Delta ln V_{it}$	0.021	0.021	0.023	0.020	0.019	0.020	0.019
	(0.034)	(0.033)	(0.033)	(0.034)	(0.034)	(0.033)	(0.034)
Δlnz_{it-1}	0.008*	0.009**	0.010**	0.007*	0.007*	0.008**	0.007*
	(0.004)	(0.004)	(0.004)	(0.004)	(0.004)	(0.004)	(0.004)
$D_{50}\Delta lnz_{it-1}$		−0.008					
		(0.019)					
$D_{60}\Delta lnz_{it-1}$			−0.031				
			(0.026)				
$D_{70}\Delta lnz_{it-1}$				0.027			
				(0.027)			
$D_{80}\Delta lnz_{it-1}$					0.104*	0.101*	0.105*
					(0.051)	(0.050)	(0.050)
$D_{5080}\Delta lnz_{it-1}$						−0.011	
						(0.020)	
$D_{7080}\Delta lnz_{it-1}$							0.021
							(0.030)
Constant	−0.015	−0.016	−0.017	−0.015	−0.014	−0.015	−0.014
	(0.010)	(0.010)	(0.010)	(0.010)	(0.010)	(0.010)	(0.010)
R^2	0.222	0.222	0.227	0.223	0.223	0.224	0.224
Obs.	251	251	251	251	251	251	251

Note: *, ** and *** are significant at the 10%, 5% and 1% levels, respectively.

the fixed-effect and dynamic panel data estimation results are summarised. The AR(1) process is adopted in dynamic panel data estimation.[15]

Since the broadband penetration ratio of OECD countries could reach 60 per cent only after 2005, *m* is adjusted to 0.6.[16] We find that Model (i) is static, while Model (ii) is dynamic. Model (iii) adopts the difference GMM. In any model, the estimated coefficient (γ_1) of the role of ICT infrastructure in the productivity impact of motorway capital is found to be statistically significant and positive. The estimation result allows us to draw meaningful implications for the network effects of infrastructure investment.

In Model (iii), when *m* is less than 0.6, the effect of growth of motorway capital on TFP growth is only 0.006. On the other hand when *m* exceeds 0.6, ICT infrastructure complements motorway capital

Table 5.4 Estimation results (Equations 5.3 and 5.4; m = 0.6)

	$\Delta lnTFP_{it}$			Δlny_{it}		
	Model (i) FE (static)	Model (ii) FE (dynamic)	Model (iii) Diff GMM	Model (iv) FE (static)	Model (v) FE (dynamic)	Model (vi) Diff GMM
$\Delta lnTFP_{it}$		0.050	0.050			
		(0.069)	(0.064)			
ΔS_{it}	0.095	0.091	0.091*	0.033	0.028	0.028
	(0.060)	(0.058)	(0.055)	(0.031)	(0.027)	(0.025)
$Open_{it}$	0.056**	0.054**	0.054***	0.034*	0.022	0.022
	(0.021)	(0.022)	(0.020)	(0.016)	(0.017)	(0.016)
ΔlnV_{it}	0.019	0.018	0.018	0.012	0.008	0.008
	(0.034)	(0.034)	(0.032)	(0.031)	(0.030)	(0.028)
Δlnz_{it-1}	0.006	0.006	0.006*	0.005	0.003	0.003
	(0.004)	(0.004)	(0.004)	(0.003)	(0.004)	(0.004)
$Max(0, ICT_{it-1} - m)$	0.401***	0.384**	0.384***	0.314*	0.267*	0.267**
$\times \Delta lnz_{it-1}$	(0.130)	(0.134)	(0.125)	(0.152)	(0.130)	(0.121)
Δlny_{it-1}					0.190***	0.190***
					(0.063)	(0.059)
Δlnk_{it-1}				0.357***	0.352***	0.352***
				(0.065)	(0.059)	(0.055)
Constant	−0.014	−0.014	−0.014	−0.010	−0.009	−0.004
	(0.010)	(0.010)	(0.009)	(0.007)	(0.007)	(0.006)
R^2	0.226	0.228		0.453	0.481	
Obs.	251	250	231	257	257	238

Note: *, ** and *** are significant at the 10%, 5% and 1% levels, respectively.

and produces a statistically significant additional beneficial effect of $\{0.384(\gamma_1) \times (ICT_{it-1} - 0.6)\}$. For example, in a country where the broadband penetration ratio was 0.7 in the previous period, the marginal effect of growth of motorway capital on TFP growth becomes $0.006(\gamma_0) + 0.0384(\gamma_1 \times 0.1) = 0.039$.

In other words, when the ICT infrastructure exceeds a certain threshold level, it complements motorway capital in TFP growth. Figure 5.5 reports the productivity effect of motorway capital for individual OECD member countries at each time point.

Models (iv) to (vi) present the estimation results for Equation (4). Again, we see that a complementarity between ICT and motorway infrastructure prevails when the ICT infrastructure grows beyond a certain threshold level. In other words, the effect of motorway capital on labour

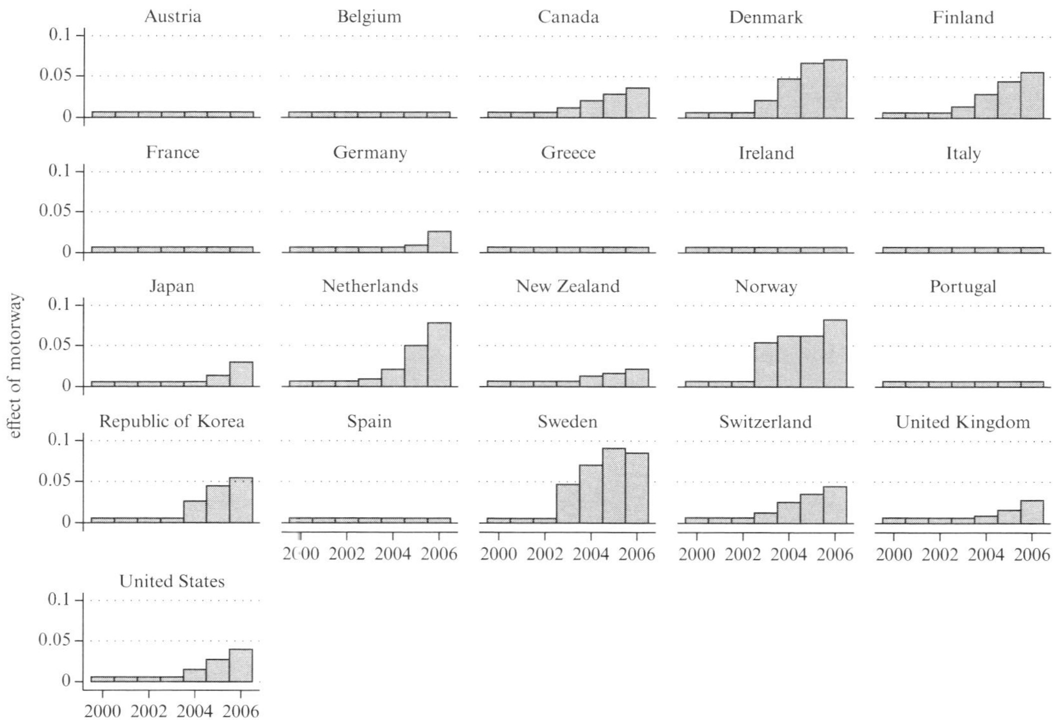

Note: The effect of motorway $\frac{\Delta \ln TFT_{it}}{\Delta \ln Z_{it-1}}$ was calculated as $\gamma_0 + \gamma_1 \times Max\{0,(ICT_{it-1} - m)\}$.

Figure 5.5 Productivity effects of motorway infrastructure among OECD countries [Model (iii)]

productivity depends critically on the broadband penetration ratio of each country. The degree of openness of an economy seems to contribute to TFP growth as well. The estimated range for a share of capital is 0.352–0.357 and statistically significant.

Our estimation results confirm the necessary conditions for productivity impacts of ITS in UNECE (2012). The convergence of ICT and transportation technology engenders the accelerating network effects of motorway infrastructure only when the ICT infrastructure grows beyond a certain threshold level. It is interesting to note that in many OECD member countries, the ICT infrastructure has grown beyond the threshold level. Our data reveal that after 2004, the broadband penetration ratio began to exceed 60 per cent in 13 countries, including the Republic of Korea, the United States and Japan. As shown in Figure 5.5, the productivity effect of motorway infrastructure has been accelerating recently in these countries.

4. CONCLUDING COMMENTS

The productivity effects of infrastructure investment need to be closely examined under fiscal austerity measures. This chapter presented a simplified analysis of complementarity between transportation and ICT infrastructures by showing empirically that the contribution of motorway infrastructure to productivity growth increases over time only if the ICT infrastructure grows beyond a certain threshold level. Although our empirical analysis was carried out for OECD member countries, the pattern of complementarity can be applied to developing countries as well because ICT convergence also takes place in developing countries.

NOTES

1. See, for example, Sim and Yoon (2001); Choi, Kim and Kim (2007); Kunihisa and Kaiyama (1998); Canning and Bennathan (2000); and Calderón and Servén (2003).
2. Intelligent transport systems (ITS) are explained in the next section. UNECE (2012) emphasises that the capacity of an ITS to offer a variety of services does not depend linearly on the size of the information network. Only after the network reaches a certain threshold level does the system perform the expected roles properly.
3. The non-linear network effects of telecommunication infrastructure are examined in Röller and Waverman (2001) for OECD countries. Jung, Na and Yoon (2013) examine the network effects of ICT infrastructure in Korea.
4. An example of DSRC application is electronic toll collection (ETC).
5. See Bonsall (2008).
6. See Korea Transport Institute (2012), p. 274.

7. See Williams (2008) for wireless communication network technologies that are needed for ITS.
8. Presidential Council on National Competitiveness (2012), Republic of Korea, Policy Report.
9. This includes Austria, Belgium, Canada, Denmark, Finland, France, Germany, Greece, Ireland, Italy, Japan, New Zealand, the Netherlands, Norway, Portugal, the Republic of Korea, Spain, Sweden, Switzerland, the United Kingdom and the United States. According to the OECD Factbook (2008), a motorway is defined as follows: *Road, specially designed and built for motor traffic, which does not serve properties bordering on it and which: (a) is provided, except at special points or temporarily, with separate carriageways for the two directions of traffic, separated from each other, either by a dividing strip not intended for traffic or exceptionally by other means; (b) does not cross at level with any road, railway or tramway track, or footpath; (c) is specially sign-posted as a motorway and is reserved for specific categories of road motor vehicles. Entry and exit lanes of motorways are included irrespectively of the location of the sign-posts. Urban motorways are also included.*
10. Fernald (1999) presented a theoretical model for the relationship between capacity and use of roads. If motorway usage exceeds the given road infrastructure capacity, transportation cost increases from congestion. It is natural to use motorways as a control variable.
11. The annual PPP-adjusted GDP (Y), total capital stock (K) and total employment (L) are taken from the OECD Economic Outlook (2012) database. The data for total capital stock for the Republic of Korea are taken from Pyo, Jung and Cho (2007).
12. For human capital, we consider the Mincerean form (see Mincer, 1974; Heckman and Klenow, 1997; Bils and Klenow, 2000), which, in its simplest form, specifies the logarithm of human capital as a linear function of years of schooling.
13. Following Harrison (1996), openness is specified to influence TFP growth, not the TFP level.
14. The Hausman test reveals that the p-value is close to zero.
15. See Arellano and Bond (1991) and Arellano and Bover (1995) for an analysis of an efficient estimator for the dynamic panel data in the presence of endogeneity.
16. The cross-correlation results between motorway and ICT infrastructure turned positive in 2005 when the average broadband penetration of OECD countries reached 0.605. The exogenous level for k is determined taking into account these data.

REFERENCES

Arellano, M. and S. Bond (1991), 'Some tests of specification for panel data: Monte Carlo evidence and an application to employment equations', *Review of Economic Studies*, **58**, 277–97.

Arellano, M. and O. Bover (1995), 'Another look at the instrumental variable estimation of error-components models', *Journal of Econometrics*, **68**, 29–51.

Bils, M. and P. Klenow (2000), 'Does schooling cause growth?', *American Economic Review*, **90**, 1160–83.

Bonsall, P. (2008), 'Modelling response to information systems and other intelligent transport system innovations', in D.A. Hensher and K.J. Button (eds), *Handbook of Transport Modelling*, 2nd edn, Amsterdam: Elsevier, pp. 559–74.

Calderón, C. and L. Servén (2003), 'The output cost of Latin America's infrastructure gap', in W. Easterly and L. Serven (eds), *The Limits of Stabilization: Infrastructure, Public Deficit and Growth in Latin America*, Palo

Alto, CA and Washington, DC: Stanford University Press and World Bank, pp. 95–118.
Canning, D. and E. Bennathan (2000), 'The social rate of return on infrastructure investments', World Bank Working Paper, No. 2390.
Choi, M.S., A.Y. Kim and E.J. Kim (2007), 'The spatial economic spillover effects of road stock using the production function of the manufacturing industry', *Journal of the Korean Regional Development Association*, **19**, 139–150 (in Korean).
Cohen, D. and M. Soto (2007), 'Growth and human capital: good data, good results', *Journal of Economic Growth*, **12**, 51–76.
Fernald, J.G. (1999), 'Roads to prosperity? Assessing the link between public capital and productivity', *American Economic Review*, **89**, 619–38.
Harrison, A. (1996), 'Openness and growth: a time-series, cross-country analysis for developing countries', *Journal of Development Economics*, **48**, 419–47.
Heckman, J. and P. Klenow (1997), *Human Capital Policy*, Chicago, IL: University of Chicago Press.
ITU World Telecommunication/ICT Indicators (2010).
Jung, H.J., K.Y. Na and C.H. Yoon (2013), 'The role of ICT in Korea's economic growth: productivity changes across industries since the 1990s', *Telecommunications Policy*, **37**, 292–310.
Korea Transport Institute (2012), 'ITS (intelligent transport systems)', in *Transport Technologies and Policies*, Chapter 9, pp. 260–85.
Kunihisa, S. and M. Kaiyama (1998), 'The economic effect of highway construction: a comparative analysis for the UK, Germany and Japan', *Review of Urban and Regional Development Studies*, **10**, 28–41.
Mincer, J. (1974), *Schooling, Experience and Earnings*, New York: Columbia University Press.
Na, K.Y., C. Han and C.H. Yoon (2013), 'Network effect of transportation infrastructure: a dynamic panel evidence', *The Annals of Regional Science*, **50**, 265–74.
OECD Economic Outlook (2010–2012).
OECD Factbook (2008).
OECD Productivity Database, available at http://www.oecd.org/std/productivity stats/2352458.pdf (accessed 24 February 2015).
Presidential Council on National Competitiveness (2012), 'Intelligent transport systems (ITS) plan: smart green transport system and new growth engine for development' (in Korean).
Pyo, H., S. Jung and J. Cho (2007), 'Estimates of gross fixed capital formation, net capital stock and capital intensity in Korea: 11 assets by 72 sectors (1970–2005)', *The Analysis of Korean Economy*, **3**, 138–90 (in Korean).
Röller, L. and L. Waverman (2001), 'Telecommunications infrastructure and economic development: a simultaneous approach', *American Economic Review*, **91**, 909–23.
Sim, J.K. and J.H. Yoon (2001), 'Contribution of highway capital to output growth in Korean manufacturing industries', *The Journal of Korean Policy Studies*, **10**, 289–316 (in Korean).
UNECE (2012), 'Intelligent transport systems (ITS) for sustainable mobility', available at http://www.unece.org/trans/publications/its_sustainable_mobility.html (accessed 24 February 2015).
US Department of Transportation (2004), 'Telecommunications and the national ITS architecture', in *Telecommunications Handbook for Transportation*

Professionals: The Basics of Telecommunications, Washington, DC: Federal Highway Administration, Chapter 3.
Williams, B. (2008), 'Wireless communications standards used for ITS', in *Intelligent Transport Systems Standards*, Norwood, MA: Artech House, Chapter 4, pp. 190–233.
World Bank Commission on Growth and Development (2008), 'The growth report strategies for sustained growth and inclusive development', Washington, DC: World Bank.

6. Telecommunication service countermeasures against disasters: Japanese people's willingness to pay for telecommunication services

Akihiro Nakamura

1. INTRODUCTION

On 11 March 2011, the Great East Japan Earthquake hit Japan's Tohoku area, resulting in severe damage to public services. The subsequent blackout around base and telephone stations rendered telecommunication services non-functional for a certain period (MIC, 2011). Since Japan is well known for earthquakes, devising a telecommunication services backup plan in the event of a disaster is currently of the highest importance. In fact, government departments have undertaken several studies on this issue, which include investigating possible countermeasures that could be put in place in the event of an earthquake occurring directly beneath the Tokyo Metropolitan Area (CDMC, 2011). Because of financial/budget conditions, however, these recommended countermeasures are to be undertaken under budgetary constraints. This indicates the necessity of striking a balance between costs and benefits, even for disaster mitigation.

This analysis investigates Japanese people's preferences for taking countermeasures against disasters in the field of telecommunication. The data for this research were obtained from an online survey. Since countermeasure services markets were not explicitly observed, at least under normal circumstances, disaster prevention measures could be considered as non-market goods. Moreover, the price mechanism for such measures would not be functional under disasters. A typical tool for measuring non-market goods is the stated preference (SP) approach, in which researchers survey whether consumers value goods and services by asking them to state their preferences under hypothetical scenarios. In general, researchers employ the contingent valuation method (CVM) to estimate willingness to pay

(WTP) for such goods or services. The resulting estimates can be used to calculate total social benefits.

The CVM approach is used for analysing various policy issues, such as those that concern the environment (Solomon and Johnson, 2009; Feitelson et al., 1996), health care (Gyldmark and Morrison, 2001) and cultural heritage (Ruijgrok, 2006; Kim et al., 2007), among others. A review of the literature on the telecommunication field reveals, although there are many studies on estimating WTP for both market and non-market services (among others, Mitomo and Tajiri, 2010; Lee et al., 2006; Nakamura, 2010; Nakamura, 2011), that there are no empirical studies on estimating WTP for countermeasures in the telecommunication field. The cost for and quantity of countermeasures supplied could be forecast/estimated based on various available technical information; i.e. by combining information on WTP for countermeasures, the question of the extent to which the countermeasures should be carried out could become clearer.

In January 2012, an SP survey was conducted in collaboration with the Information Communications Research Institute of Japan. The intention of the survey was to estimate WTP for measures intended to prevent the suspension of various telecommunication services in the event of a disaster. For measuring the relative importance of or preference for installing a telecommunication services backup system, this analysis estimated preferences for other utility services, such as electricity and water supply. The analysis revealed the following results. First, WTP for measures that prevent the suspension of telecommunication services in the event of a disaster is not insignificant. Second, a higher priority should be placed on mobile (voice) services when formulating appropriate backup plans for communication services during times of disaster. Third, estimation results of victims' WTP reveal that a backup provision within a 24-hour period would be beneficial even when suspension prevention is difficult. Finally, there was a national consensus on the importance of taking the necessary countermeasures so as to ensure that services are functional even in the event of a disaster.

2. A BRIEF REVIEW OF TELECOMMUNICATION SERVICES DURING THE GREAT EAST JAPAN EARTHQUAKE

This section briefly reviews the damage to and recovery of Japan's telecommunication services as a result of the Great East Japan Earthquake. On 11 March 2011, the earthquake hit a large area from Kanto to Tohoku, resulting in severe damage to public services. The tsunami caused by

- Alleviating Congestion in Emergency Situations
  - Ensuring voice call capability
  - Expanding/Improving means of communication other than voice calls
  - Keeping users informed of available means of communication during emergency
  - Designing robust networks to cope with congestion
- Minimizing Disruption to Communications in the Event of Damage to Base and/or Relay Stations
  - Emergency repairs to damaged telecommunications infrastructure
  - Providing communications capability to disaster-affected regions and evacuation centers
- Planned Revision to Technical Standards on Measures against Power Outage
  - Longer operation times for backup power generators and storage batteries
  - Reporting/disclosure of backup power systems and procedures
- Implications of the Recent Disaster for Future Network Infrastructure
  - Improving the disaster resilience of network
  - Setting up systems and structures for responding to disaster
- Implications of the Recent Disaster for Future Internet Usage
  - Maintaining the viability of Internet connections
  - Effective use of the Internet
  - Application of cloud services
  - Building collaborative frameworks among Telecommunications operators to prepare for disaster

(excerpt from MIC 2011)

Figure 6.1 Topics covered by MIC (2011)

this earthquake severely damaged the Tohoku region. The carnage this 9.0 magnitude earthquake caused is as follows: the number of dead was 15,854, the number of missing was 3155, and 419,457 houses were either damaged or destroyed (National Police Agency, 2012: figures taken were for 11 March 2012).

As for the telecommunication field, in April 2011, the Ministry of Internal Affairs and Communications (MIC) established the Study Group on Maintaining Communications Capabilities during Major Natural Disasters and Other Emergency Situations. This study group consisted of telecommunication carriers, vendors, Internet service providers, application service providers, academia and MIC members. The group published its final report in December 2011, which summarises the 11 March devastation and recovery efforts. It also advocates an action plan to protect Japan's communication network against future disasters. These include network congestion countermeasures that should be put in place in the event of emergency situations. Figure 6.1 shows the topics covered in the report.

According to the MIC (2011), there were 240 million communication lines and 132,000 base stations in the Tohoku and Kanto regions. In all, the earthquake damaged 1.9 million communication landlines, most of them in these two regions. As for mobile communications, about 29,000 base stations, again most of them in the Tohoku and Kanto regions, were damaged. By the end of April 2011, all services along these damaged lines, except in the severely damaged areas, had been recovered. The damaged base stations had also been recovered by the end of April, except for those in severely damaged areas.

Although telecommunication infrastructure, especially in the Pacific coast region, was destroyed or carried away by the tsunami, most service disruptions were caused by drained batteries because of disruptions to the power grid, which took several days to repair, and a lack of fuel for electrical generators. In fact, the MIC (2011) showed that the peak number of service interruptions occurred one or two days after the earthquake. Nojima (2011) reported that the electricity supply was recovered a few days after the earthquake in almost all areas except those most severely damaged by the tsunami. Once the electricity supply was recovered, the number of base station and landline communication service disruptions decreased significantly.

Despite the service disruptions, the large number of calls caused congestion in both mobile and landline services. This congestion occurred not only in the damaged Tohoku region but also in metropolitan areas where damage was relatively low. In metropolitan areas, a large number of commuters were stranded because of traffic paralysis; this resulted in the large number of calls during that time. The MIC (2011) reported that the normal volume of landline voice call traffic increased by four to nine times (NTT East case) and that each carrier restricted the phone traffic by as much as 80 per cent to 90 per cent. As for mobile phone calls, 50 to 60 times the normal volume of traffic (the NTT DOCOMO case) occurred. Each carrier restricted the voice traffic by as much as 70 per cent to 95 per cent. On the contrary, au (KDDI) and Soft Bank did not restrict the packet traffic and NTT DOCOMO restricted it only by 30 per cent.

Information on damage to sectors outside of the telecommunication field was publicised by government offices in the Cabinet (CDMC, 2011). In fact, except for regions affected by the tsunami and those contaminated by the damaged nuclear power facilities, the recovery of public services has gained momentum.

3. STATED PREFERENCES: RESEARCH AND DATA

As the MIC (2011) pointed out, in the telecommunication field there are various countermeasures against potential future disasters. This chapter's analysis focuses on network protection. This section briefly explains the data used for investigating Japanese people's preferences for network protection countermeasures against disasters in the telecommunication field.

3.1 Stated Preference Research

In January 2012, an SP survey was conducted. It is difficult to estimate telecommunication service demand or availability under extraordinary circumstances by observing usual usage behaviours. Therefore this study did not explicitly explore usage fees or consumer preferences in subscribing to or using each service. The study's questionnaire explored consumers' acceptance of a fixed surcharge increase to establish a backup system for recovering telecommunication services in the event of a disaster. The questionnaire focused on three types of telecommunication service and two other utility services for comparison: mobile phone voice transmission, mobile phone packet transmission, landline transmission (including voice and data), electrical supply and water supply.

For the aforementioned five services, WTP was measured for establishing backup systems in case of a disaster wherein 'Service would absolutely never fail (non-interruption)' and for establishing backup systems that were 'Guaranteed to be restored within a certain time frame'. For the questionnaire, pre-survey results were considered with specific monetary amounts proposed and two answer choices were provided: 'Accept (need backup)' or 'Reject (no need for backup)'. Three variations were proposed for the option 'Guaranteed to be restored within a certain time frame': 'within 3 h', 'within 24 h (1 day)' and 'within 72 h (3 days)'.

Seven variations in values – JPY 10/month, JPY 100/month, JPY 250/month, JPY 500/month, JPY 800/month, JPY 1200/month and JPY 3000/month – were considered in this survey. Respondents first replied to the JPY 500/month query, and if a respondent chose 'Need backup', the same question was then posed with the JPY 800/month option. If the respondent again chose 'Need backup' for the JPY 800/month option, the same question was then posed with the JPY 1200/month option, and if 'Need backup' was chosen again, the JPY 3000/month query was posed.

The highest monetary setting was the JPY 3000/month option, regardless of the response. Likewise, if the respondent chose 'No need for backup' when answering the JPY 500/month query, the next query was the JPY 250/month option, and if 'No need for backup' was chosen again, the JPY 100/month option was given, and so on down to the JPY 10/month option.

QUERY (EXAMPLE)

Imagine a backup system that guaranteed that the voice transmission function (call function) for your mobile phone would absolutely never fail (or if in recovery mode would guarantee to be restored within 3 hours) in any type of disaster. Would you be willing to accept such a system, if the cost to implement it was JPY 500 per month for each subscriber? Please choose your response.

Mobile phone voice transmission	In case of any disasters service would absolutely never fail (or if in recovery mode would guarantee to be restored within 3 hours)
Surcharge for establishing systems	JPY 500/month (every month)

Please assume that countermeasures are not set up for email or Internet functions on mobile and landline telephones, and therefore, these functions may not be available at the time of disaster.

Answer:
1. Accept (need backup)
2. Reject (no need for backup)

Using this process, data of the WTP distribution within two of the seven amounts were obtained for each respondent. Respondents who chose 'No need for backup' even for JPY 10/month and those who chose 'Need backup' even for JPY 3000/month were considered to have chosen below JPY 10 and above JPY 3000, respectively. A sample questionnaire is shown in the box above.

In this process, a minimum of two and a maximum of four questions were asked for the variations in each service and recovery time. Because of this, there was the possibility that the response accuracy might decrease as the burden on respondents increased when they answered all patterns for each service and recovery time. As a result, respondents were divided into nine segments for each query pattern, as shown in Table 6.1.

This survey primarily analyses mobile phone service, which is expected to be in high demand during a natural disaster. As a result, all respondents were made to respond to the option of non-interruption cases for both mobile phone voice transmission and mobile phone packet transmission. For either of the services, one-third of the respondents answered the recovery time options. On the other hand, for the comparative services other than mobile phone services (landline transmission, electrical supply

Table 6.1 Queries asked of each respondent

		Segment 1	Segment 2	Segment 3	Segment 4	Segment 5	Segment 6	Segment 7	Segment 8	Segment 9	Share
Mobile Voice	No interrupt	•	•	•	•	•	•	•	•	•	All
	Within 3 hours	•	•	•							1/3
	Within 24 hours				•	•	•				1/3
	Within 72 hours							•	•	•	1/3
Mobile Packet	No interrupt	•	•	•	•	•	•	•	•	•	All
	Within 3 hours	•	•	•							1/3
	Within 24 hours				•	•	•				1/3
	Within 72 hours							•	•	•	1/3
Land Line	No interrupt	•			•			•			1/3
	Within 3 hours	•									1/9
	Within 24 hours				•						1/9
	Within 72 hours							•			1/9
Electricity	No interrupt		•			•			•		1/3
	Within 3 hours		•								1/9
	Within 24 hours					•					1/9
	Within 72 hours								•		1/9
Water Supply	No interrupt			•			•			•	1/3
	Within 3 hours			•							1/9
	Within 24 hours						•				1/9
	Within 72 hours									•	1/9

and water supply), one-third of the respondents answered the question for the case of non-interruptions for each of these services. One-ninth of the respondents answered each of the recovery time questions for the comparative services.

3.2 Data

The survey participants were taken from a survey panel organised by an Internet survey company (goo research, Inc.). Respondents were asked a series of Accept–Reject-type questions. First, a set number of web surveys were allocated to the three prefectures most affected by the Great East Japan Earthquake – Iwate, Fukushima and Miyagi. The remaining surveys were distributed equally between eastern (including the three affected prefectures) and western Japan.

Although 2000 samples were obtained, after eliminating samples with inconsistent answers, the effective sample size was 1652. As Nakamura (2013) pointed out, the issue of inconsistent or dishonest respondents in surveys has been discussed in several disciplines. Pasveer and Ellard (1998) pointed out that false information might appear in surveys, even on demographic questions. Rogers and Richarme (2009) identified several indications of dishonest responses including short time completers and patterned responses such as straight-line answers. To eliminate these entries from the data set, they recommended that surveys include cheater-trap questions. In this survey, each respondent answered their WTP for establishing backup systems in case of a disaster wherein 'non-interruption' and at least one same-type question on 'Guaranteed to be restored within a certain time frame' for the same service was posed. The former WTP should be higher than or equal to the latter; consequently, respondents whose order of these two sets of answers did not accord with the expected pattern were treated as 'inconsistent' in this analysis.

Attribute distribution (age and province of residence) of the remaining 1652 samples used for the analysis is shown in Table 6.2 and Figure 6.2. Note that the numbers of female and male respondents were approximately the same.

Before reporting the estimation results of WTP for a backup system for recovering telecommunication services in case of a disaster, the summary statistics of the importance weight for telecommunication services, among other utility services during disasters, is presented. The respondents were asked about the three top priority services they would want restored after disasters. Figure 6.3 shows the aggregated counts for these top priority services and the weighted sum of the total three services, as follows: the top service was weighted as 3 points, the second was weighted as 2 points,

Table 6.2 *Age distribution*

Age	Male	Female	Total
10–19	72	91	163
20–29	108	123	231
30–39	159	130	289
40–49	149	104	253
50–59	120	96	216
60–69	128	126	254
70–79	96	139	235
Over 80	4	9	13
Total	836	818	1654

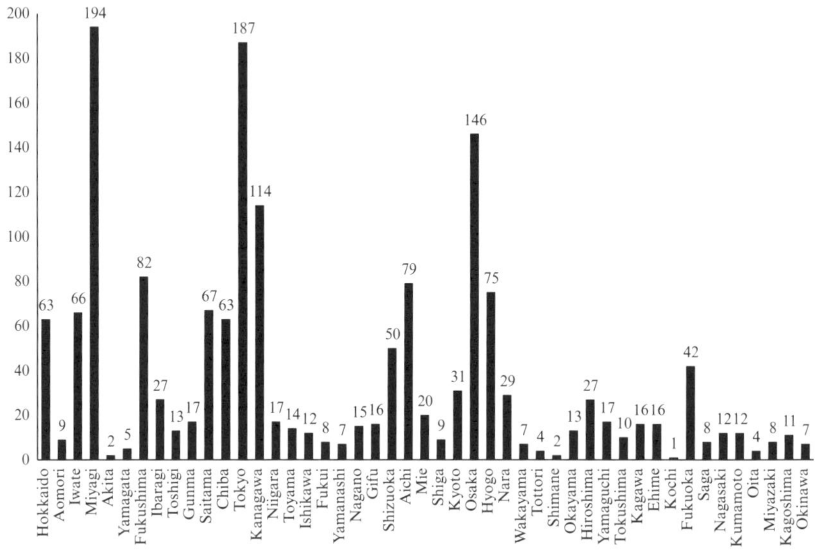

Figure 6.2 *Resident prefecture distribution*

and the third was weighted as 1 point; they were then divided by six for standardisation. Among the various public utility services, respondents assigned higher priority to electricity, water supply and mobile voice communication. Among the top three priority services, electricity supply services were first, followed by water supply and then mobile voice communication services. Considering the multiple-order preference information, television, gas supply, mobile packet (mail/Internet) services

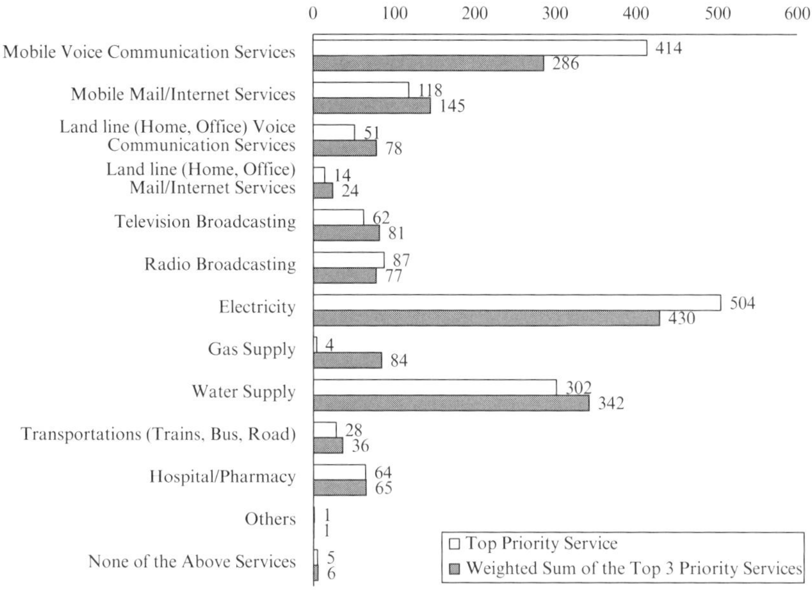

Figure 6.3 Utility services to be restored according to priority

and landline communication each received a higher weighting than did medical services. These results show that, among recovery services that are considered to be critical after a disaster, telecommunication services are a high priority.

Respondents were also asked which telecommunication services should be prioritised to be restored after disasters. Figure 6.4 shows the top priority assigned to telecommunication services and the weighted sum of the top three priority services calculated in the same manner as above. According to the results shown in Figure 6.4, the highest priority telecommunication service is mobile voice communication. If the top priority service could be interpreted as the service most needed just after a disaster, a vast majority of people noted that they would want to use their mobile voice communication service if they found themselves in such a disaster situation. As for SNS (social networking sites/services), the preference for using SNS was not so high; in fact, it was lower than that for news sites.

In addition to priority in restoring services, respondents were also asked about the list of telecommunication services most likely to be used during disasters. Figure 6.5 shows aggregated statistics of the telecommunication services most likely to be used during disasters. It also shows the weighted sum of the top three services most likely to be used. The survey results for

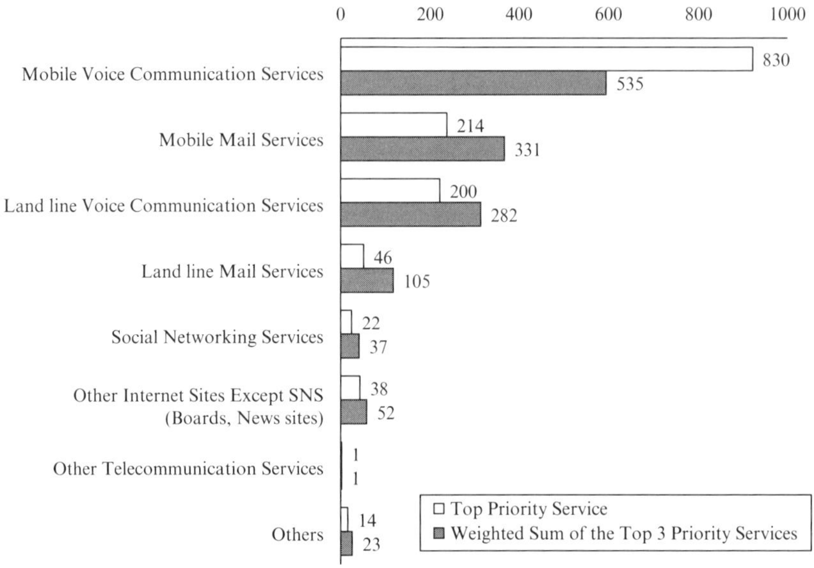

Figure 6.4 Priority of telecommunication services to be restored

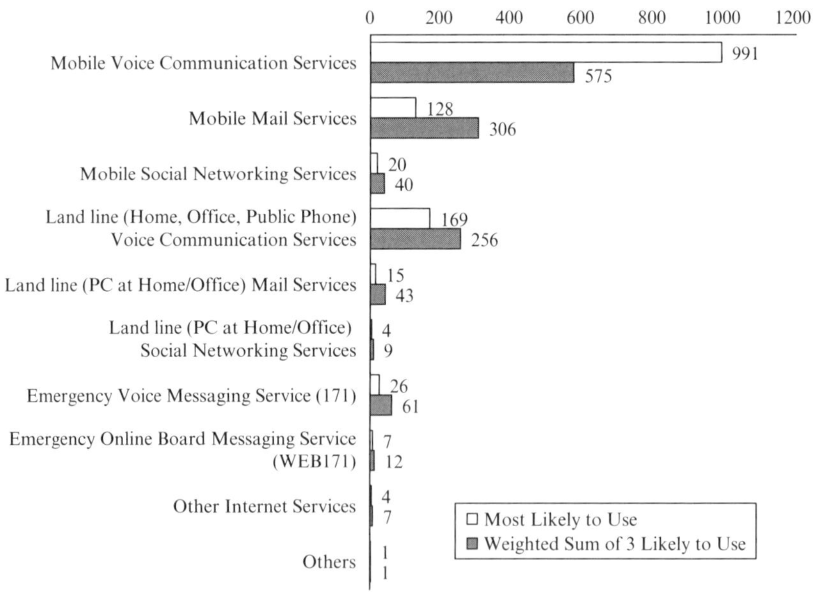

Figure 6.5 Telecommunication services likely to be used

services expected to be used at the time of a disaster were about the same as those for low SNS needs. With regard to the order in which attempts are made, the results show that there is an extremely high need to access disaster relief message services in situations when transmission is possible but the other party cannot be reached, even after trying various methods.

4. ESTIMATED WTP FOR ESTABLISHING BACKUP SYSTEMS

The results of the survey data are reported in this section. The survey data fall within the eight-category range presented in the figures and discussion in the previous section. Each respondent's WTP distribution, taken from the survey data, was measured for each category. Each respondent's WTP was assumed to reflect a normal distribution truncated at zero. The distribution parameters of the mean (μ) and standard deviation (σ) were estimated such that the sample distribution reflecting the eight categories fits as nearly as possible to each one, as shown in Figure 6.6.

4.1 Estimation Results for the Whole Sample Set

First, using the whole sample set, Table 6.3 presents the estimated monthly WTP level (μ in Figure 6.6) for each service and backup system. The parentheses in each cell show the standard deviations for the distribution (σ in Figure 6.6). The lowest column in each cell contains the number of

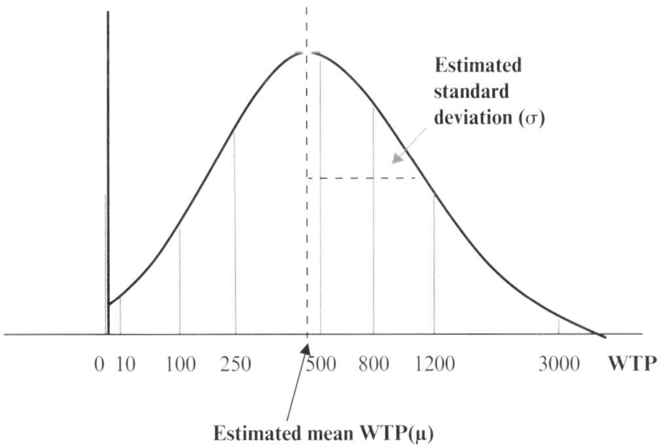

Figure 6.6 Image of estimation parameters

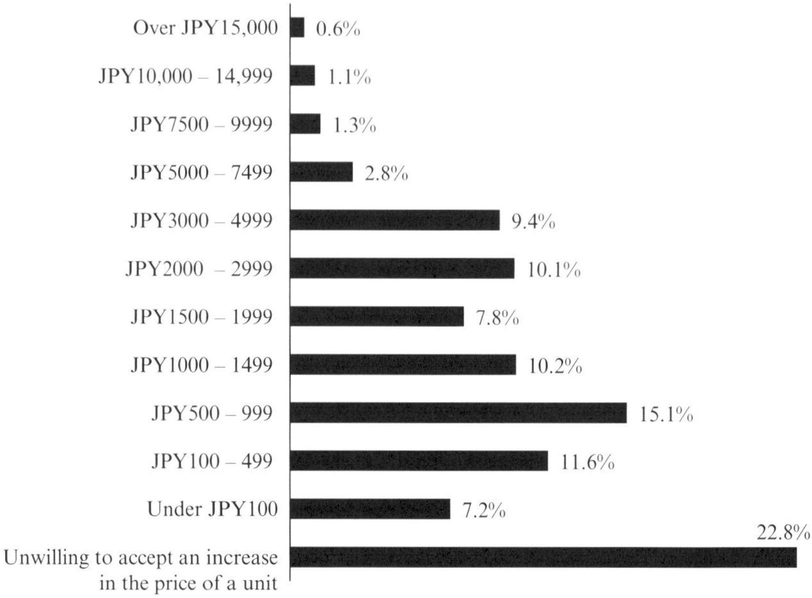

Figure 6.7 Results of question on WTP for increasing battery life

samples used to calculate the value for that cell. While Table 6.3 shows the WTP level, as the directional comparison would be difficult to show, Table 6.4 presents the relative ratios in each direction. The top of Table 6.4 shows a relative ratio for each recovery time with a WTP level of 100 for the 'prevent disruptions' countermeasure for each service (column ratios). The bottom of Table 6.4 contains a relative ratio for each recovery time with a WTP level of 100 for the 'mobile voice service' countermeasure (raw ratios).

As shown in Tables 6.3 and 6.4, in principle, WTP almost consistently decreases as recovery times increase. However, relatively high values are exhibited for landlines, compared to shorter recovery times (restoration within 3 days [72 h]). According to the study's design (refer to Table 6.1), participants responding to the question on landline transmission recovery within 3 days did not respond to questions on shorter recovery times. Consequently, while these cannot be labelled as inconsistent responses, the survey should have been more skilfully designed on this point. When checking for bias in the respondent sample, no major bias in the attribution distribution with other segments was found. When reviewing estimated results other than those for 'landline', the WTP category 'recovery within 3 days' was not weighted markedly less in comparison to the results

Table 6.3 Estimated WTPs for backup systems (using all samples)

	Mobile Voice	Mobile Packet	Land line (both voice and packet)	Electricity Supply	Water Supply
No Interrupt	JPY 385/ month (446) N = 1654	JPY 288/ month (425) N = 1654	JPY 256/ month (412) N = 562	JPY 438/ month (491) N = 547	JPY 442/ month (525) N = 545
Restored within 3 hours	JPY 303/ month (433) N = 546	JPY 243/ month (415) N = 546	JPY 179/ month (370) N = 192	JPY 410/ month (511) N = 178	JPY 404/ month (530) N = 16
Restored within 24 hours	JPY 248/ month (394) N = 564	JPY 182/ month (390) N = 564	JPY 158/ month (345) N = 192	JPY 334/ month (421) N = 190	JPY 355/ month (474) N = 182
Restored within 72 hours (3 days)	JPY 247/ month (417) N = 544	JPY 181/ month (405) N = 544	JPY 209/ month (425) N = 178	JPY 327/ month (437) N = 179	JPY 292/ month (501) N = 187

※ Standard Deviation in parenthesis

for shorter recovery times. While there is no empirical support, it is relatively well known in Japan that 72 h is generally the survival limit when searching for missing persons. Although this is also the reason for including the category 'within 3 days (72 h)' in this survey, it is also possible that respondents reflected on this point and chose the high values.

The results show that for electricity and water supply, the WTP level for establishing uninterrupted service is high and decreases somewhat for mobile voice service, followed by mobile packet transmission and landline service. A review of the difference between 'uninterrupted service' and 'recovery after certain periods of time' shows that for electricity and water supply there is no marked difference in WTP between 'recovery within 3 h' and 'uninterrupted service'. Another way of looking at this is that the tolerance for the loss of these services is 3 h. On the other hand, for communication services there was a definite difference in WTP between 'uninterrupted service' and 'recovery within 3 h'. This may be owing to the need for communication services to confirm the safety of others immediately after a natural disaster.

Regarding the various recovery times, a certain level of difference is observed between the 3 h and 24 h limits. However, one has to be careful when interpreting the results (if not statistically proven) because, as explained above, the responses for questions for each recovery time were not from the same respondent.

4.2. Victims versus Non-victims

In this section, the results of the WTP measured for victims and non-victims of natural disasters are reported. This research required data that realistically reflect disaster conditions. Victims of a natural disaster are likely to respond on the basis of the reality of their own experiences. Where there are differences in the results between disaster victims and non-victims, when devising disaster countermeasures, the data from the actual victims may be considered more important.

In this study, data were collected using questions directly related to the recent Great East Japan Earthquake and other natural disasters. Table 6.5 shows responses to the question on the region the respondent was in at the time of the Great East Japan Earthquake. Much of the damage, however, was caused by the tsunami that followed it. As a result, when distinguishing between natural-disaster victims and non-victims, those who answered 'yes' to either 'affected area (impacted by the tsunami)' or 'affected area (not impacted by the tsunami)' were categorised as natural-disaster victims. As the impact of this earthquake covered a wider area when small earthquakes were included, a note was inserted to define an affected area as 'an area where the surrounding buildings sustained damage'. In addition, a multiple-choice survey question asked about natural-disaster

Table 6.4 Columns and raw ratios of estimated WTPs (using all samples)

	Mobile Voice	Mobile Packet	Land line (both voice and packet)	Electricity Supply	Water Supply
No Interrupt	100%	100%	100%	100%	100%
Within 3 hours	79%	84%	70%	94%	91%
Within 24 hours	64%	63%	62%	76%	80%
Within 72 hours	64%	63%	82%	75%	66%
	Mobile Voice	Mobile Packet	Land line (both voice and packet)	Electricity Supply	Water Supply
No Interrupt	100%	75%	66%	114%	115%
Within 3 hours	100%	80%	59%	135%	133%
Within 24 hours	100%	73%	64%	135%	143%
Within 72 hours	100%	73%	85%	132%	118%

Table 6.5 Area where the respondent was at the time of the Great East Japan Earthquake

	Number of Respondents	Percentage (%)
Disaster area (affected by Tsunami)	50	3.0
Disaster area (non-affected by Tsunami)	313	18.9
Disaster area (no severe damage: ex. Metropolitan area)	445	26.9
Not disaster area	846	51.1
Total	1654	100.0

Table 6.6 Natural disaster experiences other than the Great East Japan Earthquake

	Number of Respondents	Percentage (%)
Ever had difficulty returning because of a natural disaster	198	12.0
Has ever been the victim of a large scale earthquake	312	18.9
Ever been the victim of a large scale typhoon or flood	282	17.0
Ever been the victim of a natural disaster other than those mentioned above	63	3.8
Never been the victim of a large scale natural disaster	1042	63.0
Total	1654	100.0

experiences other than the Great East Japan Earthquake. The results are shown in Table 6.6. In this question, respondents were classified as natural-disaster victims if they responded 'yes' to any of the following: 'Have you ever been the victim of a large earthquake?', 'Have you ever been the victim of a large typhoon or flood?', 'Have you ever been the victim of a major natural disaster other than the aforementioned?'. This survey included 725 victim samples and 929 non-victim samples.

According to Tables 6.7 and 6.8, the overall WTP level is higher for disaster victims. Furthermore, the difference between 'uninterrupted service' and 'recovery within 3 h' is relatively large for disaster victims, while that between 'recovery within 3 days' and 'recovery within 24 h' is relatively

Table 6.7 Estimated WTPs for backup systems (left: using disaster-victim samples; right: using non-victim samples)

VICTIMS	Mobile Voice	Mobile Packet	Land line (both voice and packet)	Electricity Supply	Water Supply	*NON-VICTIMS*	Mobile Voice	Mobile Packet	Land line (both voice and packet)	Electricity Supply	Water Supply
No Interrupt	JPY 404/ month (436) N = 725	JPY 310/ month (419) N = 725	JPY 291/ month (386) N = 255	JPY 491/ month (499) N = 242	JPY 433/ month (571) N = 228	No Interrupt	JPY 370/ month (453) N = 929	JPY 269/ month (428) N = 929	JPY 225/ month (433) N = 307	JPY 398/ month (482) N = 305	JPY 447/ month (493) N = 317
Restored Within 3 hours	JPY 301/ month (428) N = 250	JPY 256/ month (409) N = 250	JPY 191/ month (394) N = 83	JPY 432/ month (526) N = 87	JPY 354/ month (584) N = 80	Restored Within 3 hours	JPY 305/ month (438) N = 296	JPY 232/ month (420) N = 296	JPY 169/ month (351) N = 109	JPY 390/ month (496) N = 91	JPY 443/ month (487) N = 96
Restored Within 24 hours	JPY 301/ month (413) N = 247	JPY 228/ month (398) N = 247	JPY 219/ month (334) N = 99	JPY 429/ month (434) N = 72	JPY 392/ month (547) N = 76	Restored Within 24 hours	JPY 209/ month (371) N = 317	JPY 146/ month (380) N = 317	JPY 92/ month (349) N = 93	JPY 281/ month (404) N = 118	JPY 329/ month (421) N = 106
Restored Within 72 hours (3 days)	JPY 266/ month (381) N = 228	JPY 199/ month (372) N = 228	JPY 302/ month (416) N = 73	JPY 357/ month (434) N = 83	JPY 263/ month (459) N = 72	Restored Within 72 hours (3 days)	JPY 232/ month (445) N = 316	JPY 166/ month (429) N = 316	JPY 142/ month (421) N = 105	JPY 300/ month (437) N = 96	JPY 312/ month (527) N = 115

※Standard Deviation in parenthesis (left table)

※ Standard Deviation in parenthesis (right table)

small. Thus we can assume that victims of natural disasters have a real need for uninterrupted service and place a high monetary value on it, while a similar value level is placed on a system for recovery within 24 h as on the one that takes place within 3 h.

Additionally, in contrast to disaster victims having a relatively low WTP for water supply, non-victims demonstrate a relatively high value. This probably reflects the relative ease of fulfilling material needs through distribution in times of disaster. In fact, commodity goods, such as bottled water, are provided by logistics relatively sooner than utility services. With the exception of water supply, the preference order by both groups generally matches. In addition, although the results are omitted because of space limitations, no marked differences among ages, genders, household incomes and households with/without children were noted in the preference orders. In this sense, a consensus can be reached on which backup services need to be focused on.

5. CONCLUSIONS

The results of the above analysis show that, when formulating appropriate backup plans for communication services during disasters, a higher priority should be placed on mobile (voice) services. The results indicate that, while the ideal scenario is non-interruption of communication services, restoration of services within a 24 h time frame is also useful. The survey results indicate that there is a very high WTP for such a provision. For instance, the WTP for non-interrupted service for mobile voice communications was JPY 385 per month (JPY 4620/year). This amount was calculated by the number of NTT DOCOMO users equivalent to surcharges of JPY 276.6 billion per year. This represents 33 per cent of operating income and 11 per cent of tangible fixed assets for NTT DOCOMO (according to the company's 2011 annual report).

In addition to this point, which was clarified by the results of this study, other telecommunication service preparations for disaster times are also important. The extension of battery life, for example, would potentially prevent the type of communication service interruptions experienced during the recent earthquake, particularly in the loss of electricity services. As discussed in the first half of this chapter, while base stations experienced power loss, mobile communications became unavailable because the power outages made it impossible to charge terminals. In the surveys, respondents were asked about 'Willingness to pay for batteries that could hold a charge for three days'. The results are shown in Figure 6.7. As seen in the figure, while 23 per cent of respondents did not feel the need for

Table 6.8 Raw ratios estimated WTPs (left: using disaster-victim samples; right: using non-victim samples)

VICTIMS	Mobile Voice	Mobile Packet	Land line (both voice and packet)	Electricity Supply	Water Supply	*NON-VICTIMS*	Mobile Voice	Mobile Packet	Land line (both voice and packet)	Electricity Supply	Water Supply
No Interrupt	100%	100%	100%	100%	100%	No Interrupt	100%	100%	100%	100%	100%
Within 3 hours	75%	83%	66%	88%	82%	Within 3 hours	82%	86%	75%	98%	99%
Within 24 hours	75%	74%	75%	87%	91%	Within 24 hours	56%	54%	41%	71%	74%
Within 72 hours	66%	64%	104%	73%	61%	Within 72 hours	63%	62%	63%	75%	70%
VICTIMS	Mobile Voice	Mobile Packet	Land line (both voice and packet)	Electricity Supply	Water Supply	*NON-VICTIMS*	Mobile Voice	Mobile Packet	Land line (both voice and packet)	Electricity Supply	Water Supply
No Interrupt	100%	77%	72%	122%	107%	No Interrupt	100%	73%	61%	108%	121%
Within 3 hours	100%	85%	63%	144%	118%	Within 3 hours	100%	76%	55%	128%	145%
Within 24 hours	100%	76%	73%	143%	130%	Within 24 hours	100%	70%	44%	134%	157%
Within 72 hours	100%	75%	114%	134%	99%	Within 72 hours	100%	72%	61%	129%	134%

Table 6.9 Number of calls made during congestion

Give up after one unsuccessful attempt to connect	0.8%
Attempt to connect 2 or 3 times	27.4%
Attempt to connect 4 or 5 times	31.6%
Attempt to connect 6 or 7 times	7.2%
Attempt to connect 8 – 10 times	17.1%
Attempt to connect 11 – 15 times	2.2%
Attempt to connect 16 – 20 times	1.7%
Attempt to connect 21 – 30 times	2.5%
Attempt to connect 31 times or more	0.0%
Continue attempt to connect (without giving up) until successful	9.6%

Table 6.10 Time waited before attempting to redial during congestion

Attempt to connect immediately	37.9%
Attempt to connect 1 or 2 minutes	14.0%
Attempt to connect 3 – 5 minutes	17.3%
Attempt to connect 6 – 10 minutes	11.3%
Attempt to connect 11 – 20 minutes	4.4%
Attempt to connect 21 – 30 minutes	5.7%
Attempt to connect 31 – 45 minutes	2.2%
Attempt to connect 46 – 60 minutes	2.9%
Attempt to connect 1 – 2 hours	3.6%
Attempt to connect after 2 hours or more	0.9%

extended battery life (since they were satisfied with their current charging function), about 43 per cent were willing to pay over JPY 1000.

This survey also included questions on countermeasures against congestion (the inability to connect because of overcrowding of lines). The answers to the question of whether the respondent had experienced congestion during a natural disaster showed that approximately 60 per cent of respondents had experienced some form of congestion, which indicates some understanding of the need for such a provision. The survey also asked about actions taken during the congestion. The results are shown in Tables 6.9 and 6.10. Table 6.9 shows responses to the question: 'If you were unable to connect with your communication service because of circumstances such as congestion during a natural disaster, how many times do you think you would attempt to connect (redialing, etc.) before giving up?'. In addition, Table 6.10 shows responses to the question: 'If you were experiencing difficulty connecting with your communication

Table 6.11 Acceptable period for forced termination during natural disasters

Forced termination after 10 seconds per call	1.7%
Forced termination after 15 seconds per call	4.5%
Forced termination after 30 seconds per call	21.7%
Forced termination after 45 seconds per call	2.2%
Forced termination after 60 seconds per call	32.2%
Forced termination after 90 seconds per call	6.8%
Forced termination after 2 minutes per call	6.3%
Forced termination after 3 minutes per call	15.7%
Forced termination after 5 minutes per call	5.8%
Forced termination after 10 minutes per call	2.8%
Forced termination after 11 minutes or more per call	0.4%

service because of circumstances such as congestion during a natural disaster, how long do you think you would wait before trying to reconnect (redialing, etc.)?'. Tables 6.9 and 6.10 clearly indicate that during times of congestion, respondents at the very least would attempt to connect two to five times, with 10 per cent persisting until a connection was made. While about 40 per cent of respondents would continue to attempt connecting without waiting in between, over 70 per cent of respondents would attempt to reconnect within a 5 minute wait time. The results indicate that countermeasures are needed for handling congestion during disasters as circumstances that would lead to future congestion are inevitable. Here, the survey posed the following question: 'Supposing that a system was adopted for forcibly terminating calls after a certain amount of time, what do you feel would be an appropriate duration for each call? (Note: From a technological standpoint, connections are more readily available with shorter call durations.)'. The results are shown in Table 6.11, which clearly indicates that over 60 per cent are willing to accept a forced termination of 1 minute, with close to 70 per cent willing to accept a forced termination of 1.5 minutes and 90 per cent willing to accept a forced termination of 3 minutes.

Incidentally, medical and local government facilities were affected by the tsunami caused by the recent earthquake, and medical and residents' information was lost. In light of this, discussions on managing information in a cloud environment took place. Our survey also asked about receptivity to managing diverse personal information in a cloud environment. The results are shown in Figure 6.8, which shows that approximately 25 per cent (425/1654) of respondents 'definitely do not want a cloud management environment' while 50 per cent ([478+366]/1654) are

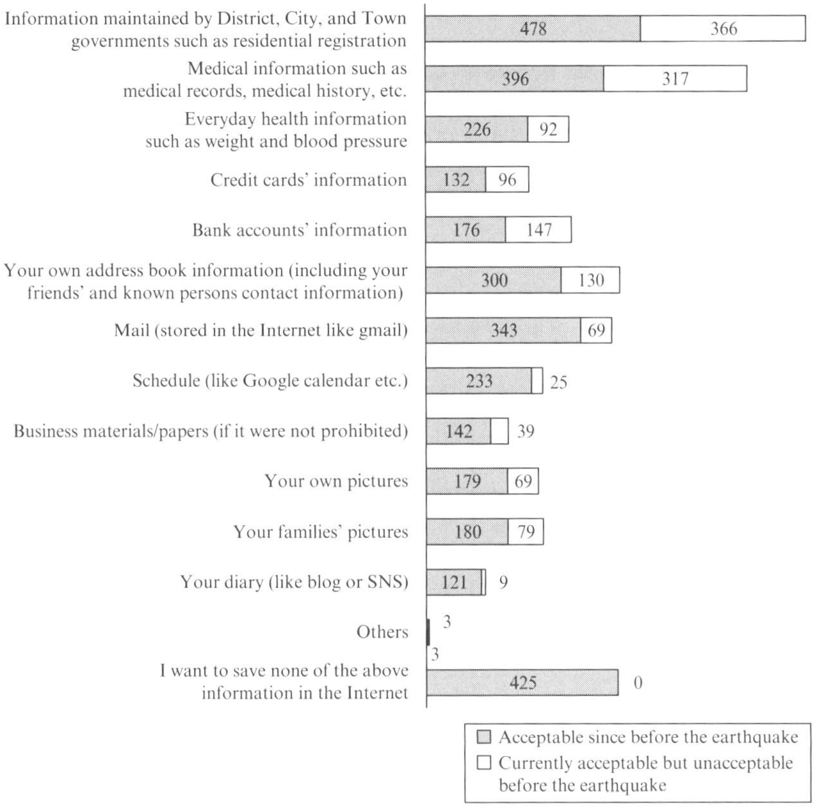

Figure 6.8 Acceptance for managing personal information in a cloud environment

neutral on managing public information, such as resident registrations, in a cloud environment. Moreover about 45 per cent of respondents changed their mind after the earthquake and decided that managing resident registrations and medical information in a cloud environment was acceptable.

After experiencing an earthquake of such magnitude, active discussions on measures for Japan's communication services in times of disasters are taking place among government authorities as well as among business, government and academia. In reality, each communication service provider is implementing various provisions, including power generating measures such as utilising solar panels at base stations and increasing base stations' coverage area. Although this chapter introduced empirical analysis focused on backup measures for networks, discussions on other related topics are ongoing and some countermeasures might be implemented in several fields.

Finally, estimates of WTPs introduced in this chapter are based on collected data using extremely simplified questions. Although some would point out issues such as lack of a behavioural economic perspective, it is the author's hope that the survey results that were collected soon after the unfortunate occurrence of the recent earthquake may contribute to future discussions about countermeasures against such disasters.

ACKNOWLEDGEMENTS

We greatly appreciate all the support provided for recovery from the Great East Japan Earthquake. An earlier version of this chapter was presented at seminars held by InfoCom Research, Inc., at the JSPU annual meeting in 2012. The author is grateful to the participants of these conferences who provided valuable comments on this analysis. In addition, the author thanks InfoCom Research, Inc., for providing the data. All errors are the author's. This research was partly supported by a Grant-in-Aid (No. 2430089) from the Japan Society for the Promotion of Science.

REFERENCES

Central Disaster Management Council at Cabinet Office, Government of Japan (2011), *Report of the Committee for Technical Investigation on Countermeasures for Earthquakes and Tsunamis Based on the Lessons Learned from the 2011 Earthquake off the Pacific Coast of Tohoku.*

Feitelson, E.I., R. Hurd and R.R. Mudge (1996), 'The impact of airport noise on willingness to pay for residences', *Transport Research*, **D1**, 1–14.

Gyldmark, M. and G.C. Morrison (2001), 'Demand for health care in Denmark: results of a national sample survey using contingent valuation', *Social Science and Medicine*, **53**, 1023–36.

Kim, S.S., K.F. Wong and M. Cho (2007), 'Assessing the economic value of a world heritage site and willingness-to-pay determinants: a case of Changdeok Palace', *Tourism Management*, **28**, 317–22.

Lee, J., Y. Kim, J.D. Lee and Y. Park (2006), 'Estimating the extent of potential competition in the Korean mobile telecommunications market: switching costs and number portability', *International Journal of Industrial Organization*, **24**, 107–24.

Ministry of Internal Affairs and Communications (2011), *Report of Study Group on Maintaining Communications Capabilities during Major Natural Disasters and Other Emergency Situations* (in Japanese).

Mitomo, H. and N. Tajiri (2010), 'Provision of universal service and access over IP networks in Japan', *Telecommunications Policy*, **34**, 98–109.

Nakamura, A. (2010), 'Estimating switching costs involved in changing mobile phone carriers in Japan: evaluation of lock-in factors related to Japan's SIM card locks', *Telecommunications Policy*, **34**, 736–46.

Nakamura, A. (2011), 'Estimating switching costs after introducing fixed mobile convergence in Japan', *Information Economics and Policy*, **23**, 59–71.

Nakamura, A. (2013), 'Retaining telecommunication services when universal service is defined by functionality: Japanese consumers' willingness-to-pay', *Telecommunications Policy*, Forthcoming.

National Police Agency (2012), *The Great East Japan Earthquake and Police* (in Japanese).

Nojima, N. (2011), *Lifeline Recovery from the Damages of the Great East Japan Earthquake: Research Subcommittee on Urban Functions Protection Strategy Taking into Account the Interconnectedness of Lifelines under the Earthquake* (in Japanese).

NTT DOCOMO (2012), 'Annual report 2011 (year ended March 31, 2011)', available at http://www.nttdocomo.co.jp/english/corporate/ir/library/annual/ (accessed 25 February 2015).

Pasveer, K.A. and J.H. Ellard (1998), 'The making of a personality inventory: help from the WWW', *Behavior Research Methods, Instruments and Computers*, **30** (2), 309–13.

Rogers, F. and M. Richarme (2009), 'The honesty of online survey respondents: lessons learned and prescriptive remedies', available at http://www.decisionanalyst.com/Downloads/honestyOnlineRespondents.pdf (accessed 25 February 2015).

Ruijgrok, E.C.M. (2006), 'The three economic values of cultural heritage: a case study in the Netherlands', *Journal of Cultural Heritage*, **7**, 206–13.

Solomon, B.D. and N.H. Johnson (2009), 'Valuing climate protection through willingness to pay for biomass ethanol', *Ecological Economics*, **68**, 2137–44.

7. The role of media and ICT to motivate people to take post-quake recovery action: an evidence of the 'Pythagorean effect'[1]

Hitoshi Mitomo, Tokio Otsuka and Mikio Kimura

1. INTRODUCTION

This chapter aims to show empirical evidence of how media and ICT (information and communication technology) motivate people to take after-quake initiatives such as collecting information, preparing for another disaster, participating in volunteer activities and making donations. Assuming that information provided by media and ICT is purely rational and influential because irrelevant information is discarded, the impact of the media information is to be analysed in conjunction with direct and indirect influences of the disaster and the existence of communities.

The Great East Japan Earthquake hit northern Japan on 11 March 2011 (see Figure 7.1 for its seismic intensity). Some 16,000 human lives were lost and 3000 people are still missing as of 23 May 2012 (Japan Reconstruction Agency, 2013). One of the distinct features of the earthquake with respect to the role of ICT and media is that television and other image media provided plentiful high-resolution visual images of the earthquake and the subsequent tsunami to the people. Actually, what modern society could do was to film and record the forces of nature and convey these to areas across Japan except for the severely damaged areas and then to the rest of the world. In addition, social networking services performed an important role providing and exchanging information among people by complementing the role of mass media. In contrast, voice calls on mobile phones, the most popular means of communication, were pretty much useless in terms of establishing communication owing to heavy traffic on telephone lines. On the other hand, broadcasting, most typically television, played a big role in providing reliable information to the mass public (Kimura, 2012).

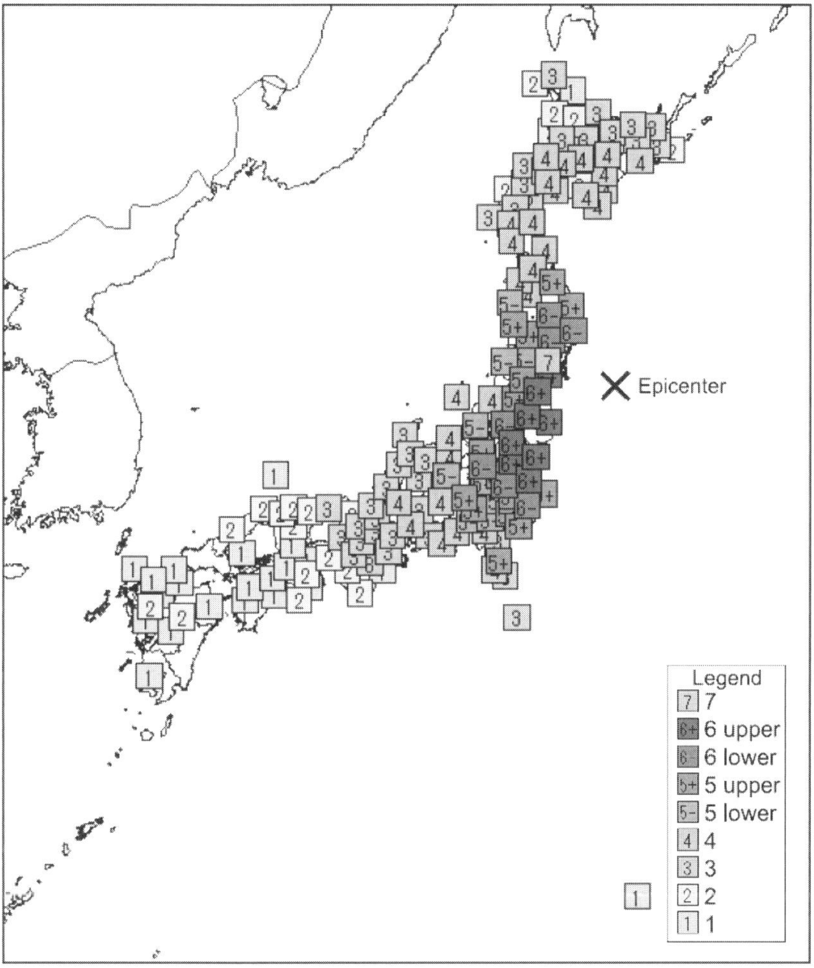

Source: Japan Meteorological Agency (2011), http://www.jma.go.jp/jma/en/2011_Earthquake/chart/2011_Earthquake_Intensity.pdf.

Figure 7.1 Distribution of seismic intensity of the Great East Japan Earthquake

The Ministry of Internal Affairs and Communications, which is in charge of ICT and media policy in Japan, has steered its policies towards recovery from the disaster and strengthened the function of ICT to create a disaster-resistant society (MIC, 2011, 2012). The role of ICT and media is performed not only in areas heavily damaged by a disaster but also in other areas. Namely, they deliver a variety of information to people living outside of the disaster and may motivate them to help people in the damaged areas and/or to prepare for another disaster in the future.

In this chapter, the rationality of media information is defined as the 'Pythagorean effect' as it can be compared to his addiction to 'rational numbers'. People faced with such information are assumed to be more motivated to take immediate action after the disaster. This kind of effect was initially discussed in terms of recognition of words and pictures (see for example, Shepard, 1967; Nelson et al., 1974). Major reactions are to collect information, to take countermeasures against possible earthquake aftermath events and to be altruistic to others. The existence of a similar phenomenon was found in medical education (Shapiro and Rucker, 2004), indicating that physicians feel more emotionally moved when viewing a movie about a patient than when treating a similar patient in an actual situation.

The result suggests that media information has a different contribution to post-quake behaviour as compared to real experience in that it triggers information gathering while the damages from the earthquake provoke countermeasures against possible aftermath events. They both have the same influence on altruism. In addition, real communities motivate people to collect information. In contrast, virtual communities induce altruistic behaviour.

2. THE PYTHAGOREAN EFFECT

Any media information is not genuinely real in the sense that it is provided through some sort of medium and is not acquired by experience. However, its potential is enormous because it can be delivered anytime and anywhere to anyone in the world with almost no delay if he/she has access to it. In most cases, information provided by media and ICT is more rational and influential because irrelevant information is discarded. In that sense, it creates a 'sense of reality' or 'virtual reality'. Since media and ICT are indispensable in our daily lives, we could say that we live in two worlds, i.e., the 'real' and the 'reality'.[2] The latter is not necessarily the genuine truth but rather represents the essence of a thing.

As technology advances, communication through ICT has become more real-time, interactive and effective, and even creates a space to share communication. Video is provided in a higher resolution such as through the high definition TV technology, which contributes to increased reality. This technological advancement will bring about the same awareness and sensation as in the real world. Even without the real experience, people can regard communication through ICT as real and view it with empathy and compassion. In most cases, the reality expresses the essence of the real since communications are usually produced based on the information collected and edited under the intention of the information providers.[3] Media have succeeded in delivering reality more effectively and efficiently and have contributed to creating a sense of the real among their receivers.

Besides an enormous benefit from media information, as the information becomes more realistic, people are apt to fall into a misconception that they have experienced something without experiencing it. While real experience accompanies difficulties, risks, troubles, inconsistencies, etc., created reality has discarded those inconvenient factors. Furthermore, people may tend to believe that the reality provided by media is real. What they believe is not real but the reality created by media.

The relationship between the real and reality could be compared to that between rational and irrational numbers in the era of Pythagorean philosophy. In his doctrines, it is believed that numbers are 'rational'.[4] A disciple of Pythagoras found the existence of an irrational number when investigating the so-called Pythagorean theorem. However, Pythagoras did not accept the existence of irrational numbers since he thought that numbers should be rational and composed of integers and the ratios of integers.

This story suggests a lot in terms of the role of media information. The Pythagorean effect of the digital age can be defined as: an emotional belief that the real (= irrational) is expressed by reality or virtual reality (= rational), in the same way as Pythagoras believed in the world of rational numbers. The effect includes both positive and negative aspects. As the positive aspects, 1) media information can convey the essence of information; and 2) people can take effective and efficient measures in response to the information conveyed. On the other hand, as the negative aspects, 1) people believe that reality or virtual reality is more essential than the real and are satisfied with a virtual world; 2) by accepting the reality, people seem to understand the real without watching it; and 3) people tend to take countermeasures without watching the real.

Media information is rational in that it is more essential and it is easier to understand the essence since complexities are eliminated, necessary information is easy to retrieve and many other factors are discarded. From these advantages of media information, it is natural to assume that it motivates people to take action without experience. It can be assumed that the effect of media information will not be affected by distance and that it will not be decreased even if people live farther away. This means that the impact of media information will be stronger as people's locations get farther away compared to real experience. This chapter examines the positive effect of the rationality of media information in terms of people's contribution to and participation in the post-quake recovery process after the Great East Japan Earthquake.

3. ANALYTICAL APPROACHES

Two approaches can be considered. First, a structure of the cause–effect relationship is hypothesised and given *a priori*. This deterministic approach examines whether a hypothesised structure is statistically significant and represents the real structure. Second, a more heuristic approach is applied, in which an exploratory factor analysis is adopted.

3.1 A Deterministic Approach

To express a comprehensive cause–effect relationship among factors and to express the consequences of the impact of media information on the earthquake in contrast to real damages, structural equation modelling was applied. In the model, two major factors are considered in terms of the contrast between real and virtual, i.e., the source of information and the communities to which people belong. Latent variables adopted are: 1) suffering damage from the earthquake, 2) media information, 3) real communities, 4) virtual communities, 5) collecting information, 6) countermeasures against possible aftermath events and 7) altruistic action. The first two represent the source of information, the next two communities, and the last three people's responses as the consequences of people's decisions. The relationship between the latent variables based on the hypothesis is constructed as given in Figure 7.2. The idea behind the structure is as follows: 1) People know the facts about the consequences of the Great East Japan Earthquake by experiencing it; 2) At the same time, they acquire information through media and ICT; 3) They also collect information in their community from their relatives, friends and acquaintances and are influenced by them; 4) Some people belong to virtual communities such as communities created by social networking services and exchange information through them.

Each potential variable is composed of some relevant observable variables. For example, the variable 'suffering damage from the earthquake' is composed of four observable variables as shown in Figure 7.3. All of the variables are linked appropriately to show causalities among them. The whole structure of the model is represented in Figure 7.4.

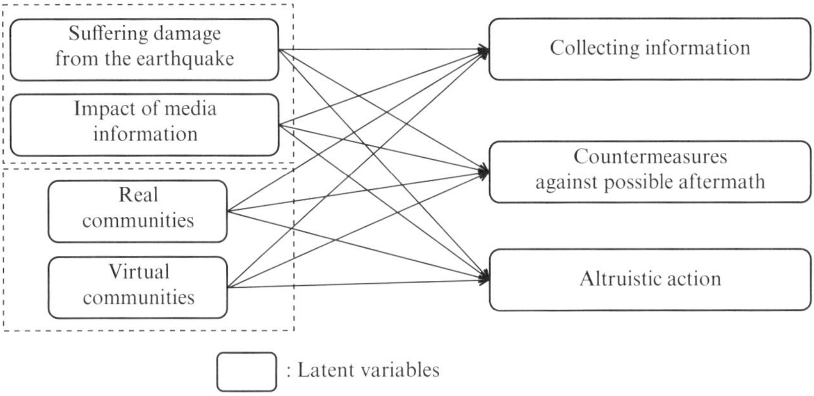

Figure 7.2 The relationship between the latent variables

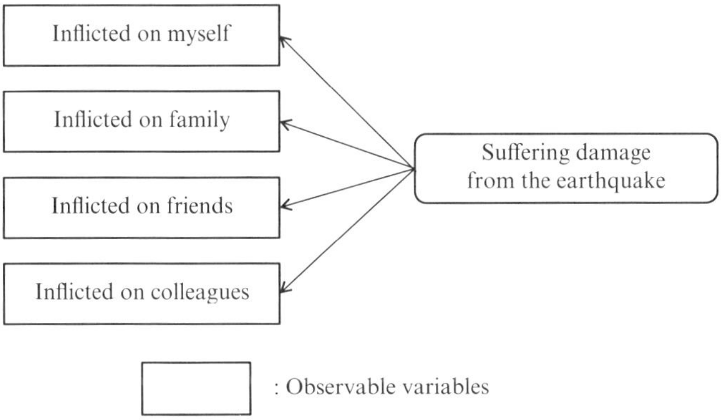

Figure 7.3 Composition of a latent variable 'suffering damage from the earthquake'

3.2 An Exploratory Approach

In addition, a cluster analysis based on the factors derived from an exploratory factor analysis was applied to identify the consequence of exposure to media information. Specifically, their responses upon experiencing the shock of the earthquake and being exposed to the media information were grouped into several behavioural patterns. This analysis aids understanding of how people typically behaved after the earthquake. More concretely, it demonstrates how their reactions were related to the types of response after exposure to the shock of the earthquake and the media information.

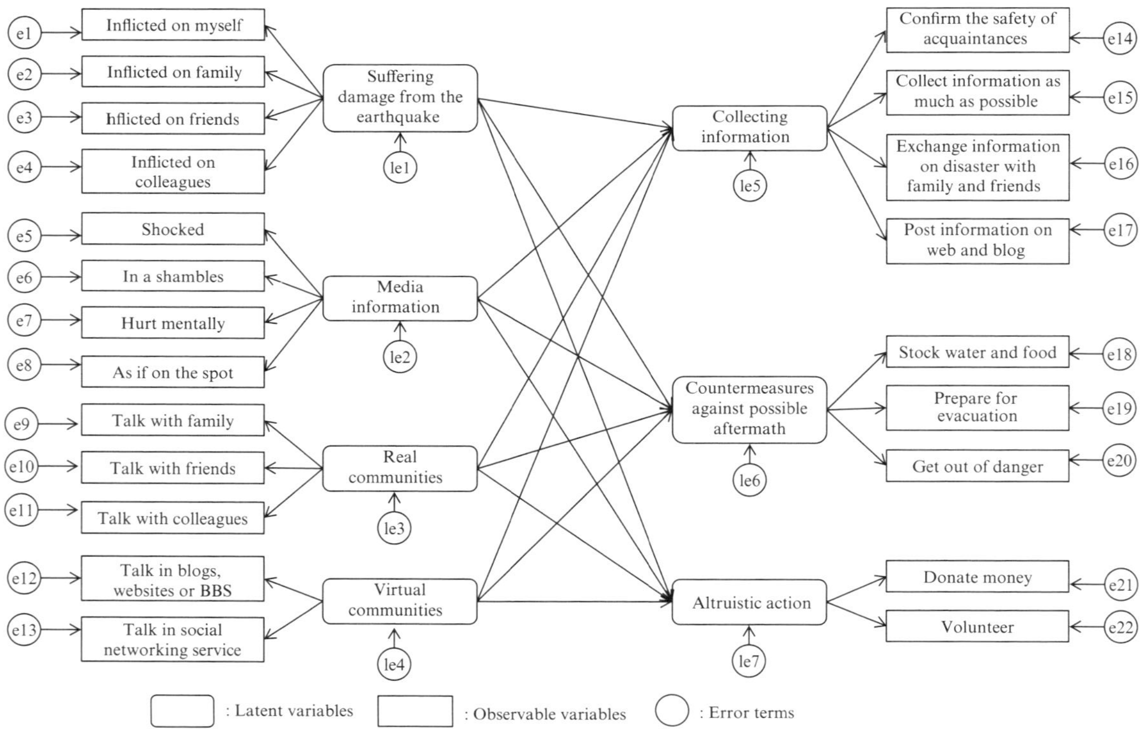

Figure 7.4 The structure of the deterministic model

3.3 The Data

An online questionnaire survey was carried out over sample panels pre-registered in a survey research company. The data were collected in February 2012 from across Japan, excluding the three heavily damaged prefectures, and included more than 2000 samples aged from teens to sixties. The percentage of the samples collected from each region is shown in Table 7.1. The largest number of samples was collected from the Kanto region, which includes Tokyo, while the least number was collected from the Tohoku region, which includes areas heavily damaged by the earthquake. Age and gender compositions are listed in Tables 7.2 and 7.3. The questionnaire survey was intended to collect the same number of samples from each age group. However, there are some variations in the numbers of samples collected.

4. THE RESULTS

4.1 A Deterministic Approach

The structural equation modelling was solved using SPSS/AMOS software. The result of the estimation for television, the most influential media, is shown in Figure 7.5. As the statistic indicator in the bottom of the figure suggests, the estimation result is statistically acceptable.

The result indicates that:

1. damages from the earthquake trigger countermeasures against possible aftermath events, while media information triggers information gathering;
2. the real experience induced altruistic action such as monetary donations and volunteer activities while the same level of altruistic action was induced by media information;
3. real communities motivate people to collect information. In contrast, virtual communities induce altruistic behaviour.

With the model presented in the previous session, in addition to the analysis of the whole data set, several analyses can be carried out by categorising data according to attributes such as the major media for collecting information, age groups and regions. Among others, it is interesting to show how the influences differ among media. The primary statistics show that TV played a dominant role in providing information over other media such as the Internet, radio and newspaper, as shown in Table 7.4.

Table 7.1 Regional distribution of the samples

Region	Number of samples collected	%
Hokkaido	118	5.7
Tohoku	45	2.2
Kanto	865	41.8
Chubu	328	15.9
Kinki	409	19.8
Chugoku	97	4.7
Shikoku	47	2.3
Kyushu	157	7.6
Total	2066	100.0

Table 7.2 Gender distribution

Male	Female
50.3%	49.7%

Table 7.3 Age distribution

Age group	%
12 or less	0.0
12–19	6.7
20–24	6.7
25–29	9.1
30–34	9.2
35–39	11.4
40–44	10.1
45–49	8.3
50–54	10.6
55–59	7.7
60 or over	20.2
Total	100.0

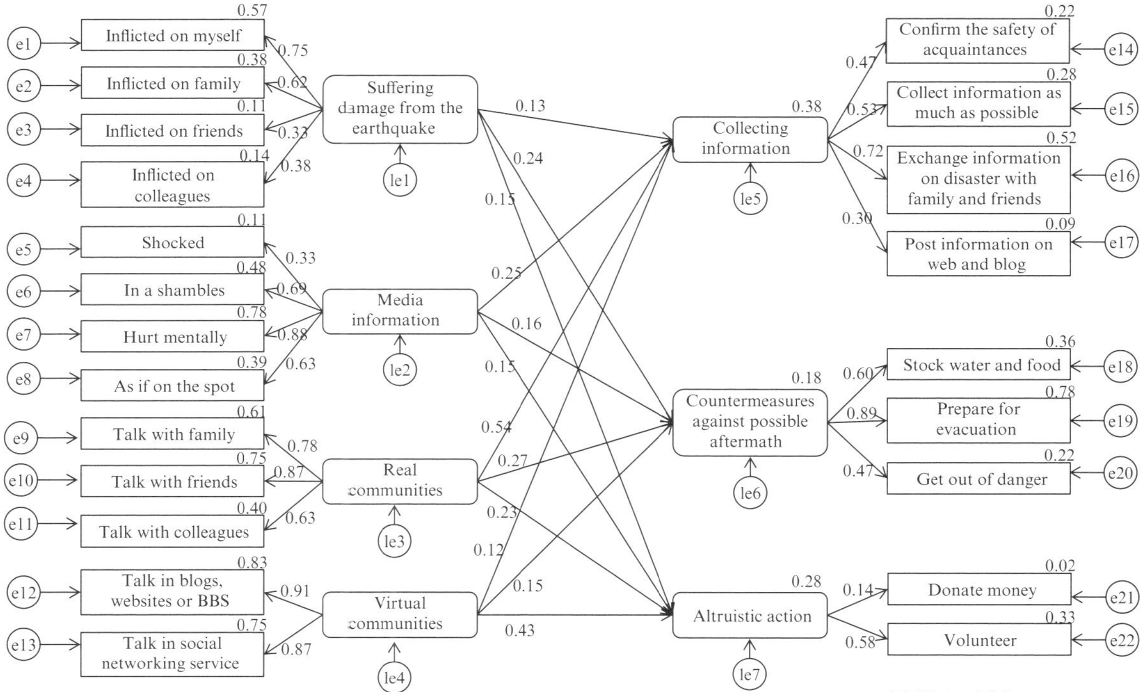

RMSEA = 0.043
P value = 0.000
GFI = 0.845

Notes: Numerical values on arrows: standardised path coefficients.
Numerical values on variables: squared multiple correlation coefficients.

Figure 7.5 The result of the statistical estimation for television

Table 7.4 Media used after the earthquake

		Internet		Total
		No	Yes	
TV	No	71	78	149
	Yes	944	973	1917
Total		1015	1051	2066
Radio		398		

Table 7.5 Comparison between TV and the Internet significant at the 5% level

Cause	Effect	TV	Comparison	Internet
Impact of media information	Countermeasures	0.16	>	0.15
	Altruistic action	0.15	>	0.12
Real communities	Collecting information	0.54	<	0.57
	Countermeasures	0.27	>	0.25
Virtual communities	Collecting information	0.15	<	0.17
	Countermeasures	0.43	<	0.45

Table 7.6 Comparison between TV and radio significant at the 5% level

Cause	Effect	TV	Comparison	Radio
Suffering damage from the earthquake	Countermeasures	0.24	<	0.35
	Altruistic action	0.15	<	0.17
Virtual communities	Collecting information	0.12	<	0.17

Approximately 75 per cent of the respondents regarded TV as the most important source of information. Some may suspect that the role of the Internet is more and more important because it is said that IP-based communication such as emails, blogs and social networking services provided useful information. However, as Table 7.4 indicates, over 90 per cent of Internet users also collected information from TV while only half of the TV viewers accessed the Internet. Consequently, the structural equation modelling approach was not successful in identifying the difference between TV and other media in terms of the creation of motivations. Table 7.5 lists the set of correlations that show a statistical difference at a 5 per cent significance level. From a statistical point of view, six links of the cause–effect relationship showed some difference between TV and the Internet. However, the differences seem very small as far as correlation coefficients are concerned. Table 7.6 shows that radio had a slightly different structure. Experience of the disaster induced more countermeasures against possible aftermath events.

4.2 An Exploratory Approach

This approach examines how people were impressed by media information and how they reacted after the earthquake. An exploratory factor analysis was applied to extract factors regarding the impacts of media information on the earthquake and subsequent behaviours. Then, the respondents were classified into several groups with a cluster analysis by utilising the factors derived. The data used are all the answers from the respondents except those related to 'suffering damage from the earthquake'. A common factor analysis, i.e. the principal factor analysis, was applied with a varimax rotation. The Kaiser rule (i.e. the eigenvalue > 1) was adopted for determining the number of factors.

Five factors were included in the appropriate range of solutions to adopt. The names and implications of the factors were derived from their factor loadings as follows (see Table 7.7):

1. shock of the earthquake – impressed by media information on the earthquake and seeking to collect more information;
2. ties – seeking communication and interaction with families, friends and neighbouring people;
3. virtual ties – seeking communication and interaction online;
4. countermeasures – taking action for preparing for evacuation or joining volunteer activities; and
5. stupors – disengaging from collecting information or taking countermeasures.

Using these factors as principal components, a cluster analysis was applied to categorise respondents who shared a similar feature into a group. Since the data set was large, the k-means algorithm was used instead of hierarchical clustering. The number of clusters was given *a priori* as required by the algorithm. Calculation was successful for the number of clusters from two to five. As the number increases, the iterative process is hard to converge to a solution. As the result of calculation, the maximum number of clusters was five. From the coordinates of the cluster centres in the principal component coordinate space as shown in Table 7.8, the features of people in the clusters can be interpreted as follows:

Table 7.7 The result of factor analysis

	Questions \ Factors	Shock of the earthquake	Ties	Virtual ties	Countermeasures	Stupors
Impression	Shocked	.139	−.017	−.009	−.076	−.074
	Felt the necessity of information	.194	−.024	−.003	−.015	−.224
	In a shambles	.098	−.037	.006	−.012	.309
	Hurt mentally	.212	−.082	.037	.026	.426
	As if on the spot	.082	−.059	.015	.072	.091
	Watched TV repeatedly	.131	−.046	.005	−.020	−.132
	Level-headed	.033	−.023	.021	−.001	−.130
	In deep sorrow	.215	−.032	.009	−.058	.007
	Very anxious	.141	−.014	−.015	−.012	.021
Action	Collected a lot of information	.149	.035	−.013	.016	−.281
	Exchanged information on the disaster	.071	.087	−.043	.080	−.154
	Confirmed safety of acquaintances in the affected area	.015	.004	.013	.078	−.044
	Secured food and water	.009	.021	−.080	.201	−.025
	Prepared for evacuation	−.055	−.015	−.114	.474	.023
	Evacuated	−.056	−.046	.045	.220	.032
	Volunteered	−.015	−.041	.078	.123	−.011
	Posted information online	−.011	−.093	.201	.143	−.047
	Donated	.033	.003	.004	−.002	−.014
Communication	With family members	−.030	.297	−.078	−.013	.033
	With friends	−.101	.574	−.001	−.075	.184
	Acquaintances or colleagues	−.032	.142	.028	−.022	−.001
	Blog, website or BBS	.002	.004	.556	−.158	.002
	Facebook, Twitter or other social networking service	−.010	.016	.247	−.068	−.047

- 1st cluster: community activist;
- 2nd cluster: media-savvy activist;
- 3rd cluster: escapist;
- 4th cluster: resurgent;
- 5th cluster: isolationist.

The number of respondents in each cluster and the percentage share are listed at the bottom of Table 7.8.

Table 7.8 Features of people in the clusters

Factor \ Cluster (group)	Community activist	Media-savvy activist	Escapist	Resurgent	Isolationist
Shock of the earthquake	0.415	0.286	−0.450	−1.792	0.261
Ties	0.665	0.295	0.039	−0.271	−1.300
Virtual ties	−0.636	1.372	−0.174	0.157	−0.331
Countermeasures	0.176	0.315	−0.691	0.591	−0.176
Stupors	0.005	−0.062	0.349	0.064	−0.320
Number of respondents	648	430	408	405	175
Percent constituent	31.4	20.8	19.7	19.6	8.5

5. IMPLICATIONS AND CONCLUSION

This chapter aimed to show empirical evidence of how media and ICT motivate people to take after-quake initiatives such as collecting information, preparing for another disaster, participating in volunteer activities and making donations.

Based on the data collected through an online questionnaire survey, two different approaches were taken for revealing the impact of media information in conjunction with the influence of the experience of the Great East Japan Earthquake and the existence of communities. The idea embedded in the analysis was that the impact of media information was enormous as compared to other disasters we have experienced so far.

The deterministic approach employing structural equation modelling examined the validity of the structural hypothesis based on the idea. The result suggests that media information has a different contribution to post-quake behaviour compared to real experience in that it triggers information gathering while damages from the earthquake provoke countermeasures against possible aftermath events. They both have the same influence on altruism. In addition, real communities motivate people to collect information. In contrast, virtual communities induce altruistic behaviour.

On the other hand, the exploratory approach successfully revealed the behaviour of people after the earthquake. Actually, they are categorised into five groups, each of which shows a typical behavioural response.

There are activists, whether relying on communities or media, who took after-quake initiatives. Those categorised as 'escapist' or 'isolationist' tended to stay away from the fact or ignore it, and did not want to get involved in after-quake activities. Those grouped in the 'resurgent' category were depressed by experiencing the earthquake and/or by being exposed to media information but reanimated. The result seems consistent with a general understanding that reaction can be positive, negative or neutral.

The result in this chapter, however, is solely based on the data collected through an online survey. The existence of sampling bias cannot be denied. In this sense, the result should be taken as general with care. As shown in Section 4, regarding the source of information after the quake, TV played a dominant role and the role of the Internet is still limited although the usefulness of the Internet is becoming more and more important.

Further findings will be provided by additional analyses. Examination of the influence of physical distance on the comparative impacts of experience and media information on people's attitudes towards after-quake initiatives should yield interesting findings. It can be assumed that the impact of the disaster will decrease as the distance from the seismic centre increases, but that the impact of media information will not depend on distance. This means that the impact of media information will relatively increase with increasing distance. The respondents of the questionnaire survey were distributed all over Japan except for the three prefectures close to the seismic centre. For this purpose, another survey is necessary that includes the three prefectures to highlight the impact of closeness to the centre.

Different media will bring about different impacts. Further media-specific analysis will also provide interesting findings. These are left to future research.

NOTES

1. This chapter is based on an earlier version presented at the Biennial ITS Conference in 2012.
2. Though the real and reality are often confused, they are not the same. For example, a person living alone feels loneliness or isolation. However, they may feel similar sensations even if they live with others. This is just the perception; it is not real loneliness or isolation.
3. For example, news information is provided under the control of news editors and issues are sometimes exaggerated to draw attention.
4. Rational numbers are integers or are expressed by the ratio of integers. Nowadays, findings show almost all numbers to be irrational numbers.

REFERENCES

Japan Meteorological Agency (2011), 'Distribution of JMA seismic intensity', available at http://www.jma.go.jp/jma/en/2011_Earthquake/chart/2011_Earthquake_Intensity.pdf (accessed 11 November 2013).

Japan Reconstruction Agency (2013), 'Current status and path toward reconstruction, March 2013', available at http://www.reconstruction.go.jp/english/130528_CurrentStatus_PathToward_FINAL.pdf (accessed 29 June 2013).

Kimura, M. (2011), *Report on the Roles of the Media during the 2011 Great East Japan Earthquake* (in Japanese), Research Institute, National Association of Commercial Broadcasters in Japan.

Ministry of Internal Affairs and Communications (MIC) (2011, 2012), White Paper on Information and Communications in Japan. Japan: Ministry of Internal Affairs and Communications, Economic Research Office, ICT Strategic Policy Division, Global ICT Strategy Bureau.

Mitomo, H., T. Otsuka and M. Kimura (2012), *The Role of Media and ICT to Motivate People to Take Post-Quake Recovery Action: An Evidence of the 'Pythagorean Effect'*, Biennial ITS Conference ITS 2012, Bangkok.

Nelson, T.O., J. Metzler and D.A. Reed (1974), 'Role of details in the long-term recognition of pictures and verbal descriptions', *Journal of Experimental Psychology*, **102** (1), 184–6.

Shapiro, J. and L. Rucker (2004), 'The Don Quixote effect: why going to the movies can help develop empathy and altruism in medical students and residents', *Families, Systems, and Health*, **22** (4), 445–52.

Shepard, R.N. (1967), 'Recognition memory for words, sentences, and pictures', *Journal of Verbal Learning and Verbal Behavior*, 6 (1), 156–63.

PART III

New issues created by the smart revolution

8. Customer service quality and incomplete information in mobile telecommunications: a game theoretical approach to consumer protection

Rafael López, Teodosio Pérez-Amaral, Teresa Garín-Muñoz and Covadonga Gijón

1. INTRODUCTION

According to the concept of consumer sovereignty (Averitt and Lande, 1997) the ultimate goal of a market is twofold: to provide consumers with a wide range of available choice options and to allow consumers to freely choose among those options based on information. The first part is obtained by promoting competition while the second part is ensured through consumer protection. Mobile telecommunications markets have reached maturity in terms of competition and regulators are now turning their attention to consumer protection as has been suggested by the International Telecommunications Union (ITU, 2006) and the Organisation for Economic Co-operation and Development (OECD, 2008).

The two main types of demand-side market failure that prevent mobile telecommunications consumers from freely choosing welfare-maximising options are consumers' behavioural biases (see Xavier and Ypsilanti, 2010) and incomplete information (see Sappington, 2005). Behavioural biases involve issues such as hyperbolic discounting, information overload, default options and the way information is framed in contracts, which may cause consumers to make choices against their best interests. Incomplete information is connected to service quality: consumers cannot observe it when signing a contract and therefore firms may lack incentives to provide high quality. This is particularly true if we refer solely to customer service quality (i.e. call centres and customer care) where providing low quality seems to be spread among mobile operators worldwide (Sappington, 2005; El País, 2012; NBC News, 2012; El Tiempo, 2013). The main purpose of this chapter is to offer theoretical support to this empirical observation. We rationalise this outcome by using simple game-theoretic models where low quality is obtained in equilibrium. We also show that the low-quality equilibrium is inefficient (i.e. overall welfare would improve by providing higher quality) and that the inefficiency may not be solved through repeated interaction or competition. This result is important in terms of policy implications because it suggests that competition may not solve the inefficiencies and therefore justifies regulation via consumer protection (see Smith, 2000; Cherry, 2010).

Our basic model involves the following two-stage game: In the first stage, the firms (mobile operators) providing the good (phone contract) simultaneously choose a price–quality pair. The price refers to the whole mobile phone service while the quality refers solely to the post-sale customer service associated with the phone contract (which is unobservable for the consumer upon purchase).[1] In the second stage, a single (representative) consumer makes his purchase decision after observing all the prices, but remains ignorant about the qualities of the service (which are only revealed at the end of the second stage). The consumer prefers higher quality over lower quality, but the former is more costly to firms than the latter. We consider variations of the model according to the number of participating firms (one, two, several) and whether the game is played just once (one-shot) or several times (finitely and infinitely repeated).

The crux of the problem is incomplete information: mobile phone contracts are what are known as experience goods (Nelson, 1970), that is, goods that possess some features that are not observable by the consumer upon purchase and that can only be learned through experience. Customer service in mobile telephony is such a feature: the mobile operator knows exactly the quality it provides but it is impossible for the consumer to observe its quality before actually using it, therefore the consumer has incomplete information about the service.

Although our work is motivated by empirical evidence in the mobile telecommunications industry, it applies to the general setting of an experience good with an unobservable quality and is thus related to the literature on vertical differentiation, adverse selection and signalling games and reputation.

In their vertical differentiation models, Shaked and Sutton (1982) and Motta (1993) find that under complete information (i.e. when quality is observable for the consumer) firms choose to differentiate their products in order to relax price competition. This no longer holds under incomplete information: in our model we find equilibria where both firms choose the same low quality (i.e. no differentiation), although they may or may not relax price competition. In a similar vein to ours, Bester (1998) finds that when allowing for vertical as well as horizontal differentiation, incomplete information drives firms to no differentiation (i.e. they choose the same quality and locate at the same place) although he focuses on high-quality equilibria.

Our low-quality equilibria bear a resemblance to the 'lemons' problem found by Akerlof (1970) in his adverse selection model for the used-car market, where only the bad cars (lemons) are traded. Despite this resemblance, the major difference is that in our model, quality is an endogenous variable (firms choose between high and low quality), while in his model it is exogenous (a used-car owner either has a lemon or does not).

Repeated purchases play an important role because they may allow for firms to signal their quality and build a reputation. Milgrom and Roberts (1986) show that when quality is exogenous, firms may use prices to signal their true quality and support separating equilibria (i.e. coexistence of different quality levels). In our model, separating equilibria are not possible, because when quality is endogenous, firms are driven to choose the same quality level. For finitely repeated purchases we show that the unique equilibrium outcome implies the low-quality choice (i.e. there is no possibility of building a reputation). However, with an infinite horizon there is equilibrium multiplicity: although the low-quality choice by all firms continues to be an equilibrium, we also show that there are price signalling equilibria that support the high-quality choice (i.e. firms are able to build a reputation). This result goes along the same line of thought as the quality premium and reputation models by Shapiro (1983) and Klein and Leffler (1981). Nevertheless, the quality level obtained in equilibrium relies heavily on the consumers' expectations, and in particular we show that under pessimistic beliefs only the low-quality level may prevail.

The rest of the chapter is organised as follows. In Section 2 we present our benchmark monopoly model, which is a generalisation of Tirole (1988, p. 96). In Section 3 we analyse the duopoly case, which also leads to an inefficient equilibrium outcome. In Section 4 we show that the low-quality equilibrium continues to hold for a large number of firms. Section 5 offers concluding remarks. Proofs and examples are in the Appendix.

2. THE MODEL (MONOPOLY)

We will consider a single representative consumer with utility function

$$U(p,q) = \begin{cases} \theta q - p & \text{if buying quality } q \text{ at price } p \\ 0 & \text{otherwise} \end{cases}$$

where p is the price of the good, q is the quality level and θ is a taste parameter reflecting how much the individual values quality. This corresponds to the standard utility function used in models of vertical differentiation (i.e. when goods may differ only in their quality). For simplicity we assume that q can take only two values: h ('high') or l ('low'), with $h > l \geq 0$. In Tirole's model these levels take the particular values $h = 1$ and $l = 0$. As we will show later on, when $l > 0$, our generalisation allows for equilibria that are not possible in that model.

There is only one firm producing the good, and the per-unit cost of providing the qualities are, respectively, $c_h > c_l > 0$ (i.e. high quality has a higher cost).

We assume that there are no fixed costs and therefore the per-unit profit of the firm is given by

$$\pi(p,q) = \begin{cases} p - c_q & \text{if selling quality } q \text{ at price } p \\ 0 & \text{if not selling} \end{cases}$$

Note that the social surplus of producing quality q is

$$W_q = U(p,q) + \pi(p,q) = \begin{cases} \theta q - c_q & \text{if trading takes place} \\ 0 & \text{otherwise} \end{cases}$$

Therefore the high quality level is socially efficient whenever $\theta h - c_h > \max\{0, \theta l - c_l\}$. For our specific case of mobile telecommunications, recall that quality refers to 'customer service quality' instead of the broader 'quality of service' (QoS), which also includes all aspects of connection (coverage, loudness, data transmission speed, etc.). If we restrict to customer service quality it seems very natural to assume that providing high-quality is socially better than providing low-quality, which in turn dominates not providing the service at all.[2] Therefore we will assume

$$\theta h - c_h > \theta l - c_l > 0$$

It is worth noticing that this assumption does not hold under Tirole's model since when $l = 0$ no production dominates providing low-quality service.

The time structure of the game is as follows: In the first stage the firm (player 1) chooses a price–quality pair (p,q). In the second stage, having observed p but ignoring q, the consumer (player 2) decides whether to buy or not. Let G be this one-shot game and denote by $G^T(\delta)$, the game G repeated during T periods with common discount factor δ. We turn now to study the equilibria of these games.

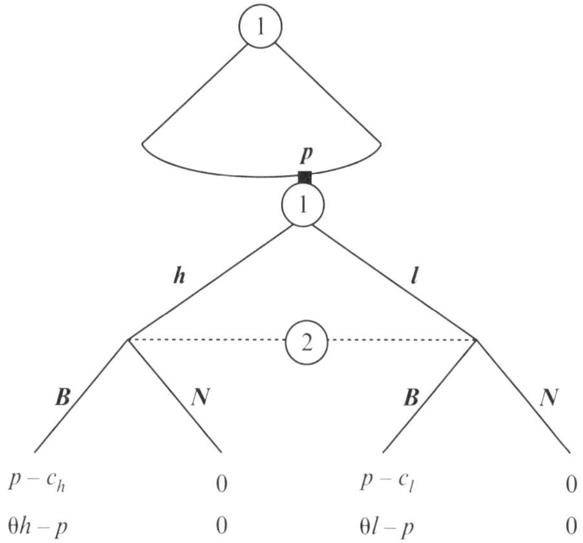

Figure 8.1 *One-shot game decision tree*

2.1 One-shot Game

Notice that since player 2 does not know the level q when she decides her action and since player 1 does not know player 2's action when deciding the quality level q, the induced game is strategically equivalent to a game where player 1 chooses the price p in the first stage and in the second stage there is a simultaneous game where player 1 chooses quality q and player 2 chooses whether to buy or not, as illustrated in Figure 8.1.

Assumption: We will make the standard assumption that the consumer buys whenever indifferent.

Solving the game backwards we obtain a unique subgame-perfect Nash equilibrium (SPNE): for any p chosen in the first stage, the second stage game in strategic form is:

	B	N
h	$p - c_h, \theta h - p$	0,0
l	$p - c_l, \theta l - p$	0,0

Notice that player 1 will play l because it is a dominant strategy. By eliminating strategy h from the game we observe that player 2 chooses to buy as long as

$$\theta l - p \geq 0 \tag{8.1}$$

(the assumption allows for the inequality not being strict). Knowing this, in the first stage the firm will choose the highest price p for which (8.1) holds, i.e. $p = \theta l$. Therefore in the unique SPNE the firm chooses quality l and the payoffs are $U^* = 0$ and $\pi^* = \theta l - c_l > 0$. The result is inefficient since by assumption

$$\theta h - c_h > \theta l - c_l = U^* + \pi^*$$

The SPNE is characterised by the strategy profile $s^* = (s_1^*, s_2^*)$, where

$$s_1^* = (p,q) = (\theta l, l) \quad \text{and} \quad s_2^* = B \Leftrightarrow p \leq \theta l$$

2.2 Finitely Repeated Interaction

Let's consider now game $G^T(\delta)$ with finite T. Since the stage game G has a unique SPNE; solving by backward induction we obtain a unique SPNE for $G^T(\delta)$ that consists of playing s^* unconditionally at each period. Therefore the firm chooses the low quality in every period and per-period payoffs are the ones corresponding to the one-shot game. Thus, we have shown that under a finite horizon and a single firm the inefficient low-quality outcome is not only an equilibrium but is in fact the unique equilibrium.

2.3 Infinitely Repeated Interaction

Let's focus now on game $G^\infty(\delta)$. Once again, playing unconditionally in every period the strategy profile s^* constitutes an SPNE of game $G^\infty(\delta)$ for any value of δ. However, this is no longer the unique SPNE of the repeated game. In fact there are infinitely many equilibria: Applying the 'Folk' theorems (see Fudenberg and Tirole, 1991) it is possible to find an SPNE supporting at each period any outcome yielding feasible per-period payoffs for both players higher than those of s^*, provided δ is sufficiently large (i.e. players are sufficiently patient). In the Appendix we illustrate this result by constructing multiple SPNE of $G^\infty(\delta)$ where in all of them the high quality is provided at each period (they differ on how the social surplus is distributed between the firm and the consumer, and in the minimum value of the discount factor that supports them).

We have thus shown that for the monopoly case and under the three scenarios (one-shot, finitely repeated and infinitely repeated) there exists an inefficient equilibrium where the low quality is produced. In the first two scenarios this is in fact the unique equilibrium while in the last it is the unique equilibrium that holds for any value of the discount factor. However, if δ is sufficiently large, there are also efficient equilibria where the high quality is produced. Notice that in all of the inefficient low-quality equilibria the utility obtained by the consumer is zero while the social surplus is entirely captured by the firm. In these equilibria, the inefficiency obviously calls for regulation and the fact that the consumer does not obtain any gains from trade calls for consumer protection. In the following sections we study the implications of allowing for multiple firms providing the good.

3. DUOPOLY

Consider now the extension of the previous model allowing for two firms. Let's label the firms players 1 and 2, the consumer being player 3.

3.1 One-shot Game

The time structure of the one-shot game G is as follows: In the first stage both firms simultaneously choose a price–quality pair. In the second stage, the consumer observes both prices (but does not observe the chosen qualities) and then decides whether to buy or not, and from which firm. Contrary to the monopoly case, here we cannot find an equivalent game that may be solved by backward induction. In other words, there is no proper subgame of G and therefore subgame perfection is of no use. Of course, the consumer's decision must be based on the observed prices as well as on the expectations or beliefs they have about the chosen qualities. Let q_i^e be the quality the consumer expects from firm i. Since prices are the only information available at the time of the consumer's decision, the consumer can use them to form his expectations, therefore we define beliefs as functions of prices. Following Bester (1998) we can restrict to degenerate expectations since the firm's quality choices are non-random (i.e. we are looking for pure-strategy equilibria), then for any price pair (p_1,p_2), we have $\forall i = 1,2$

$$q_i^e(p_1,p_2) \in \{h,l\} \tag{8.2}$$

Let's define a *beliefs system* $q^e(\cdot)$ as the function that assigns to each possible observed price pair (p_1,p_2) the qualities the consumer expects from each of the two firms, i.e. $q^e(\cdot) = (q_1^e(\cdot),q_2^e(\cdot))$ where $q_i^e(\cdot)$ satisfies (8.2) for $i = 1,2$. The consumer's decision is straightforward given his beliefs system: he just chooses to buy from the firm that, according to his beliefs, offers him the highest utility (as long as this utility is non-negative in which case he would prefer not to buy from any firm). Again, we will assume that the consumer buys whenever there is indifference between buying or not. We also assume that if there is indifference between buying from the two firms he buys from each with equal probability. Formally, given that $q^e(\cdot) = (q_1^e(\cdot), q_2^e(\cdot))$, the consumer's optimal strategy is

$$s_3^*(p_1,p_2,q^e(\cdot)) = \begin{cases} \text{buy from } i & \text{if} \quad \theta q_i^e(p_1,p_2) - p_i \geq \max_{j \neq i}\{\theta q_j^e(p_1,p_2) - p_j, 0\} \\ \text{not buy} & \text{if} \quad 0 > \max\{\theta q_1^e(p_1,p_2) - p_1, \theta q_2^e(p_1,p_2) - p_2\} \end{cases} \tag{8.3}$$

As solution concept for the game G we will use the *rational-expectations equilibrium*, which explicitly states the consumer's beliefs and requires that they are confirmed in equilibrium and that all players respond optimally to them. Clearly, in any rational-expectations equilibrium, the consumer must play the strategy given by (8.3). Therefore we formally define it in terms of the firms' strategies:

Definition: *A beliefs system $q^{e*}(\cdot) = (q_1^{e*}(\cdot), q_2^{e*}(\cdot))$ and a strategy profile $s^* = (s_1^*, s_2^*)$, with $s_i^* = (p_i^*, q_i^*)$ for $i = 1,2$, constitute a rational-expectations equilibrium if, $\forall i = 1,2$*

1. *s_i^* is optimal given s_{-i}^* and beliefs $q^{e*}(\cdot)$*
2. *$q_i^{e*}(p_1^*, p_2^*) = q_i^*$*

The first condition ensures that s^* is a Nash equilibrium for given beliefs q^{e*}, while the second condition guarantees rational expectations (i.e. beliefs are confirmed at equilibrium). Additionally, we will restrict to equilibria where the consumer buys the good with equal probability from any of the firms (otherwise at least one firm would leave the market).

We now show that in equilibrium both firms choose low qualities and set prices equal to marginal cost (and the consumer buys from any of them with equal probability) as stated in the following proposition:

Proposition 1: *Let a strategy profile $s^* = (s_1^*, s_2^*)$ and a beliefs system $q^{e*}(\cdot) = (q_1^{e*}(\cdot), q_2^{e*}(\cdot))$ constitute a rational-expectations equilibrium. Then*

$$q_1^* = q_2^* = l$$

$$p_1^* = p_2^* = c_l$$

Proof: The formal proof is in the Appendix (for the general case with $n \geq 2$ firms). But the intuition is very simple: we first show that, as in the monopoly case, choosing l is a dominant strategy for each firm. In fact it is strictly dominant unless the firm does not sell. This means that the only reason a firm should choose high quality is because it is certain it will not sell, which then implies that high quality would never be provided in equilibrium. Anticipating this, the consumer should have beliefs assigning low quality to both firms. But then we are in the same situation as in a Bertrand duopoly with complete information and, therefore, the unique equilibrium prices are

$$p_1^* = p_2^* = c_l$$

Notice that although there is a unique equilibrium outcome – firms provide low quality and set prices equal to marginal cost and the consumer buys from any of them with equal probability – there are multiple equilibria (all yielding the same outcome). The reason for this is that our equilibrium concept requires that along the equilibrium path, beliefs should be pessimistic (in the sense that the consumer should expect low quality from both firms) but it poses no restriction on how beliefs are formed off the equilibrium path. In fact any of the equilibrium beliefs are admissible as part of a rational-expectations equilibrium, thus giving rise to multiple equilibria. Some equilibrium refinements have been proposed (e.g. perfect Bayesian, sequential, trembling-hand perfect and proper equilibrium) in order to discard untenable beliefs off the equilibrium path, but all of those refinements will yield the same outcome since, in this case, the equilibrium outcome is unique. However, one may wonder what kind of beliefs could be accepted as tenable off the equilibrium path. The answer to this is embedded in the fact that low quality is a dominant strategy for any possible price pair: Therefore, the consumer should expect low quality upon observing any price pair (and not just equilibrium prices). This leads us to select a unique rational-expectations equilibrium, namely, the one involving pessimistic beliefs for any price pair as stated in the following definition:

Definition: *In game G we say that the consumer has* ***pessimistic beliefs*** *if for any given price vector* $p = (p_1, p_2)$ *he expects both firms producing low quality, i.e.* $\forall i = 1,2$

$$q_i^e(p_1, p_2) = l$$

The corresponding payoffs in this unique equilibrium are

$$U^* = \theta l - c_l > 0 \quad \text{and} \quad \pi_1^* = \pi_2^* = 0$$

As in the monopoly case, this equilibrium is inefficient because of the low quality provided. However, in this case, the price competition between both firms drives the equilibrium price down to marginal cost, so that all the gains from trade go to the consumer. As we shall see, this continues to hold if the game is repeated a finite number of times. However, with infinitely repeated interaction, firms may recapture the consumer's surplus by means of tacit collusion.

3.2 Finitely Repeated Interaction

Consider now game $G^T(\delta)$ with finite T and where G is the duopoly game of the previous section. For the finitely repeated game there is a unique equilibrium: the one-shot equilibrium under pessimistic beliefs repeated every period. To see this, let's consider the last period. The consumer observes the period-T prices and makes his purchase decision depending on those prices and his period-T beliefs about qualities. Period-T beliefs are now much more complicated than the one-shot game beliefs. This is so because the consumer now has more information from which to form his beliefs: He knows current period-T prices, he knows the prices from all previous periods and, if he bought from firm i in period $t \leq T$, he also knows q_i^t – the quality chosen by firm i in period t. Therefore period-T beliefs may depend on all of that information. But no matter how complicated those beliefs may be, exactly the same argument used in the one-shot game shows that period-T equilibrium beliefs should also be pessimistic and therefore firms should choose low quality and set T-period prices equal to marginal cost. The key reason for this is that period-T choices have no effect in future actions simply because there is no future beyond period T. Therefore all players should behave in period T exactly as they do in the one-shot game. Since there is a unique equilibrium in the one-shot game (the one with pessimistic beliefs), that must be the predicted equilibrium in period T. Having obtained a unique equilibrium for the last period, the usual unwinding argument follows to obtain the same unique equilibrium played unconditionally in each of the preceding periods. Note that the beliefs supporting this equilibrium are constantly pessimistic in the sense that, for each period t, the consumer expects low quality no matter what the past history of the game has been up to that period. We therefore extend the pessimistic beliefs definition to the case of repeated games and state the result just obtained in the following definition and proposition.

Definition: *In game $G^T(\delta)$ the consumer has* ***pessimistic beliefs*** *if, for every period $t \leq T$, he expects low quality from every firm, independently of what the history of the game has been up to that period.*

Proposition 2: *For game $G^T(\delta)$ the unique rational-expectations equilibrium with tenable beliefs implies*

1. *The consumer has pessimistic beliefs.*
2. *In every period t, firms choose low quality and set prices equal to marginal cost, i.e.* $\forall t \leq T, \forall i = 1,2$

$$(p_i^{*t}, q_i^{*t}) = (c_l, l)$$

The equilibrium inter-temporal payoffs are

$$U^* = \frac{1 - \delta^T}{1 - \delta}(\theta l - c_l) > 0 \quad \text{and} \quad \pi_1^* = \pi_2^* = 0$$

The result is inefficient and, as in the one-shot game, all the gains from trade go to the consumer. We turn now to study the infinitely repeated game.

3.3 Infinitely Repeated Interaction

Infinitely repeated interaction allows for a myriad of equilibria arising in game $G^\infty(\delta)$. In particular, efficient equilibria may be obtained because firms are able to build and maintain a reputation by providing high quality. Notice that in the finitely repeated scenario no high-quality reputation could be formed because both firms have incentives to provide low quality in the last period and then, proceeding with a backward reasoning, firms provide low quality in each preceding period. The infinitely repeated game has no last period and therefore allows for building a reputation. Nevertheless, here we are interested in finding low-quality equilibria that call for consumer protection, since that is the empirical evidence we are aiming at rationalising. The duopoly cases we have analysed (one-shot and repeated interaction) have produced inefficient low-quality equilibria where the consumer obtains all the gains from trade: Although there is inefficiency, there is not too much consumer protection needed so far! In the infinitely repeated version of the game, playing unconditionally in every period the unique rational-expectations equilibrium of the one-shot game is also a rational-expectations equilibrium of game $G^\infty(\delta)$: But all the gains from trade continue to go to the consumer. However, an infinite horizon allows for tacit collusion where the firms continue to provide low quality and they capture all the gains from trade. For our purposes, this is the interesting case because it is an inefficient equilibrium that calls for consumer protection. The following proposition provides such an equilibrium. The idea behind it is the standard trigger strategies used in infinitely repeated games of complete information: By assuming pessimistic beliefs, we may extend that idea to this incomplete information game.

Proposition 3: *Consider each firm* $i = 1,2$ *following the strategy*

$$\sigma_i^* = \begin{cases} t = 1 : s_i^1 = (p_i^1, q_i^1) = (\theta l, l) \\ t > 1 : \begin{cases} s_i^t = (p_i^t, q_i^t) = (\theta l, l) & \text{if } \forall \tau < t, (s_1^\tau, s_2^\tau) = (s_1^1, s_2^1) \\ s_i^t = (p_i^t, q_i^t) = (c_l, l) & \text{otherwise} \end{cases} \end{cases}$$

Then the profile $\sigma^* = (\sigma_1^*, \sigma_2^*)$ constitutes a rational-expectations equilibrium under pessimistic beliefs for $\delta \geq 1/2$ (i.e. provided firms are sufficiently patient).

Proof: The formal proof is in the Appendix (for the general oligopoly case) but the idea is a very simple 'carrot–stick' argument: Firms start by tacitly colluding in the first period (they set, independently, the low quality monopoly price θl). In the subsequent periods they continue to collude as long as both firms have colluded in all the previous periods. This is the 'carrot' incentive. In case a firm deviates in some period t, it triggers an infinite punishment phase (the 'stick') starting in period $t + 1$, because the other firm will permanently set its price equal to marginal cost. When considering a unilateral deviation in some period t a firm must weigh the immediate benefits (by setting its price slightly below θl it captures all the period-t market) against future losses (future profits are zero since the other firm will set the price equal to marginal cost forever after). The more a firm values the future (i.e. for higher values of δ) the less likely the unilateral deviation will be profitable. The key that allows us to use the same argument as in a complete information repeated game (i.e. the trigger strategies) is that under pessimistic beliefs there is no need to consider deviations that involve producing high quality.

This equilibrium is inefficient because low quality is provided. But now all the gains from trade go to the firms because of the tacit collusion; the corresponding inter-temporal payoffs are

$$U^* = 0 \quad \text{and} \quad \pi_1^* = \pi_2^* = \frac{1}{2}\left(\frac{1}{1-\delta}\right)(\theta l - c_l) > 0$$

We have thus obtained an inefficient equilibrium for the infinitely repeated game that may call for consumer protection.

The duopoly results we have obtained for the three scenarios rely heavily on the consumer having pessimistic beliefs. However, it is interesting to note that while pessimistic beliefs are an assumption for the equilibrium in the infinitely repeated game, they are a result for the other two: In the one-shot game and the finitely repeated game equilibria, the consumer must have pessimistic beliefs.

Also notice that only in the infinite horizon scenario do we obtain a low-quality equilibrium (provided firms are patient enough), where the consumer loses all his surplus, thus demanding consumer protection that could only be achieved by regulation. A mobile phone contract is indeed an indefinite repeated game and therefore the infinite horizon should be the appropriate setting to model the game. In the next section we show that these results continue to hold when allowing for several firms.

4. OLIGOPOLY

Consider now game $G^T(\delta)$ with an arbitrary large number of firms $n \geq 2$. By using the same arguments as in the duopoly case we obtain the following results, which are counterparts to those obtained for the duopoly case (in fact the formal proofs in the Appendix are for this general oligopoly case). We also restate the definitions for this general case.

4.1 One-shot Game

Definition: *We say that the consumer has* ***pessimistic beliefs*** *if for any given price vector p he expects all firms producing low-quality, i.e.* $\forall i = 1,2,\cdots,n$; $\forall p = (p_1,p_2,\cdots,p_n)$

$$q_i^e(p) = l$$

Definition: *A beliefs system* $q^{e*}(\cdot) = (q_1^{e*}(\cdot),q_2^{e*}(\cdot),\cdots,q_n^{e*}(\cdot))$ *and a strategy profile* $s^* = (s_1^*,s_2^*,\cdots,s_n^*)$, *with* $s_i^* = (p_i^*,q_i^*)$ *for every i, constitute a rational-expectations equilibrium if,* $\forall i = 1,2,\cdots,n$

1. s_i^* *is optimal given* s_{-i}^* *and beliefs* $q^{e*}(\cdot)$
2. $q_i^{e*}(p_1^*,p_2^*,\cdots,p_n^*) = q_i^*$

Proposition 4: *Let a strategy profile* $s^* = (s_1^*,s_2^*,\cdots,s_n^*)$ *and beliefs* $q^{e*}(\cdot) = (q_1^{e*}(\cdot),q_2^{e*}(\cdot),\cdots,q_n^{e*}(\cdot))$ *constitute a rational-expectations equilibrium. Then,* $\forall i = 1,2,\cdots,n$

$$q_i^* = l \quad \text{and} \quad p_i^* = c_l$$

Proof: See the Appendix.
The corresponding payoffs in this unique equilibrium implying pessimistic beliefs are

$$U^* = \theta l - c_l > 0 \quad \text{and} \quad \pi_1^* = \pi_2^* = \cdots = \pi_n^* = 0$$

4.2 Finitely Repeated Interaction

Definition: *In game $G^T(\delta)$ the consumer has* ***pessimistic beliefs*** *if, for every period $t \leq T$, he expects low quality from every firm, independently of what the history of the game has been up to that period.*

Proposition 5: *For game $G^T(\delta)$ the unique rational-expectations equilibrium with tenable beliefs implies*

1. *The consumer has pessimistic beliefs.*
2. *In every period t, firms choose low quality and set prices equal to marginal cost, i.e. $\forall t \leq T, \forall i = 1,2,\cdots,n$*

$$(p_i^{*t}, q_i^{*t}) = (c_l, l)$$

Proof: Follows the same argument as in the duopoly case.
The equilibrium inter-temporal payoffs are

$$U^* = \frac{1-\delta^T}{1-\delta}(\theta l - c_l) > 0 \quad \text{and} \quad \pi_1^* = \pi_2^* = \cdots = \pi_n^* = 0$$

4.3 Infinitely Repeated Interaction

Proposition 6: *Consider each firm $i = 1,2,\cdots,n$ following the strategy*

$$\sigma_i^* = \begin{cases} t = 1: s_i^1 = (p_i^1, q_i^1) = (\theta l, l) \\ t > 1: \begin{cases} s_i^t = (p_i^t, q_i^t) = (\theta l, l) & \text{if } \forall \tau < t, (s_i^\tau, s_{-i}^\tau) = (s_i^1, s_{-i}^1) \\ s_i^t = (p_i^t, q_i^t) = (c_l, l) & \text{otherwise} \end{cases} \end{cases}$$

Then the profile $\sigma^* = (\sigma_1^*, \sigma_2^*, \cdots, \sigma_n^*)$ constitutes a rational-expectations equilibrium under pessimistic beliefs for

$$\delta \geq \frac{n-1}{n}$$

Proof: See the Appendix.
This equilibrium is inefficient because low quality is provided and all the gains from trade go to the firms because of the tacit collusion; the corresponding inter-temporal payoffs are

$$U^* = 0 \quad \text{and} \quad \pi_1^* = \pi_2^* = \cdots = \pi_n^* = \frac{1}{n}\left(\frac{1}{1-\delta}\right)(\theta l - c_l) > 0$$

Remark: We have seen that all the results of the duopoly case apply to the general oligopoly case. However, notice an important difference: The minimum discount factor needed to support tacit collusion in the infinitely repeated game increases with the number of firms:

$$\delta \geq \frac{n-1}{n}$$

This means that the higher the number of firms, the more patient they need to be in order for tacit collusion to be an equilibrium. This is so because with a higher number of firms the immediate relative profits of deviating are higher and the future relative losses are smaller, therefore the deviation is more likely to be profitable. This implies that the smaller the number of competing firms, the more consumer protection is likely to be needed.

5. CONCLUSIONS

There is growing evidence that low-quality customer service prevails in the mobile telecommunications industry. We provide theoretical support to this observation by using simple game theoretical models where inefficient low-quality service levels are part of an equilibrium strategy for the firms. We do so for three different scenarios of the game: one-shot, finitely repeated and infinitely repeated. We find that any inefficiency is owing to the demand-side market failure generated by incomplete information and that such inefficiency may not be solved through repeated interaction or competition. In the infinitely repeated scenario there are also high-quality equilibria. Nevertheless, the quality level obtained in equilibrium relies on the consumer's expectations. In particular we show that under pessimistic beliefs only the low-quality level may prevail.

In terms of policy implications, these results are important because they suggest that competition will not necessarily solve the inefficiency, which therefore justifies regulation via consumer protection (for example, by means of imposing minimum quality standards; see Leland, 1979).

Our results rely heavily on two particular assumptions. On the one hand, we have assumed that providing high quality is socially better than providing low quality (which in turn implies the inefficiency of the obtained equilibria). This assumption seems very natural if we restrict it to customer service quality (i.e. call centres and customer care) but it may no longer hold if we allow for the broader concept of QoS – which also includes all aspects of connection such as coverage, loudness, data transmission speed, etc. – since the high costs of providing high quality under network congestion are likely to exceed consumer valuation. In this latter case, low-quality equilibria would be efficient and therefore consumer protection would be harder to justify. On the other hand, we have assumed consumer homogeneity (i.e. all consumers have the same valuation for quality). Allowing for consumer heterogeneity is the natural next step in our research agenda. One would expect to find the existence of separating equilibria that in turn could explain, for example, why business consumers (who have a higher valuation for quality) obtain better customer service than private consumers. But notice that consumer protection continues to be needed as long as low-quality equilibria are inefficient. In this sense our present work should be considered as a first step in that direction.

Consumer protection regulations are partly in place in some industrialised countries. However, more specific rules may be needed in terms of, say, training call centre operators, quality standards of call centres, possibilities for accessing supervisors in case of a disagreement and other regulations that are beyond the scope of this chapter.

Certain countries have government services that mediate in unsettled customer complaints (e.g. Spain's Customer Service Office for Telecommunications Users, http://www.usuariosteleco.es/). However, in many cases they are unable to impose fines or penalties on those who fail to apply the remedies mandated by the mediator, which substantially reduces their effectiveness.

Although our work is motivated by empirical evidence in the mobile telecommunications industry, it applies to the general setting of an experience good with an unobservable quality and is thus related to the literature on vertical differentiation, adverse selection and signalling games and reputation.

NOTES

1. To avoid confusion between these two services – the phone service and the customer service – we will refer to the former as the 'good' and the latter as the 'service'.

Therefore the price refers to the whole good while the quality refers only to the customer service.

2. If we allow for the broader concept of QoS this assumption might not hold: for example, under network congestion, high costs of providing high quality may exceed consumer valuation.

REFERENCES

Akerlof, G. (1970), 'The market for lemons: qualitative uncertainty and the market mechanism', *Quarterly Journal of Economics*, **84**, 488–500.

Averitt, W. and R.H. Lande (1997), 'Consumer sovereignty: a unified theory of antitrust and consumer protection law', *Antitrust Law Journal*, **65**, 713–56.

Bester, H. (1998), 'Quality uncertainty mitigates product differentiation', *Rand Journal of Economics*, **29**, 828–44.

Cherry, B.A. (2010), 'Consumer sovereignty: new boundaries for telecommunications and broadband access', *Telecommunications Policy*, **34** (1–2), 11–22.

El País (2012), 'El laberinto de las reclamaciones', available at http://sociedad.elpais.com/sociedad/2012/08/23/actualidad/1345746537_436783.html (accessed 12 February 2014).

El Tiempo (2013), 'Calidad del servicio de la telefonía celular es desesperante: Santos', available at http://www.eltiempo.com/tecnologia/santos-se-queja-de-mala-calidad-de-la-senal-de-celular_12788737-4 (accessed 12 February 2014).

Fudenberg, D. and J. Tirole (1991), *Game Theory*, Cambridge, MA: MIT Press.

ITU (2006), 'ICT Quality of Service Regulation: Practices and Proposals', Background paper presented at the ITU Global Seminar on Quality of Service and Consumer Protection, Geneva, Switzerland.

Klein, B. and K. Leffler (1981), 'The role of market forces in assuring contractual performance', *Journal of Political Economy*, **89**, 615–41.

Leland, H.E. (1979), 'Quacks, lemons, and licensing: a theory of minimum quality standards', *Journal of Political Economy*, **87** (6), 1328–46.

Milgrom, P. and J. Roberts (1986), 'Price and advertising signals of product quality', *Journal of Political Economy*, **94** (4), 796–821.

Motta, M. (1993), 'Endogenous quality choice: price vs. quantity competition', *Journal of Industrial Economics*, **41**, 113–31.

NBC News (2012), 'Verizon best, AT&T worst in consumer reports carrier survey', available at http://www.nbcnews.com/technology/verizon-best-t-worst-consumer-reports-carrier-survey-1C7319906 (accessed 12 February 2014).

Nelson, P. (1970), 'Information and consumer behavior', *Journal of Political Economy*, **78** (2), 311–29.

OECD (2008), 'Enhancing Competition in Telecommunications: Protecting and Empowering Consumers', Ministerial Background Report, OECD Ministerial Meeting on the Future of the Internet Economy, Seoul, Korea.

Sappington, D. (2005), 'Regulating service quality: a survey', *Journal of Regulatory Economics*, **27** (2), 123–54.

Shaked, A. and J. Sutton (1982), 'Relaxing price competition through product differentiation', *Review of Economic Studies*, **49**, 3–13.

Shapiro, C. (1983), 'Premiums for high quality products as returns to reputations', *The Quarterly Journal of Economics*, **98** (4), 659–80.

Smith, R.L. (2000), 'When competition is not enough: consumer protection', *Australian Economic Papers*, **39** (4), 408–25.

Tirole, J. (1988), *The Theory of Industrial Organization*, Cambridge, MA: MIT Press.

Xavier, P. and D. Ypsilanti (2010), 'Behavioral economics and telecommunications policy', in A. Gentzoglanis and A. Henten (eds), *Regulation and the Evolution of the Global Telecommunications Industry*, Cheltenham, UK and Northampton, MA: Edward Elgar Publishing, pp. 83–106.

APPENDIX

Example: High-quality Equilibria for the Infinitely Repeated Monopoly Game

Consider the stage game G with only one firm and let W_h and W_l be the social surpluses associated with providing high and low quality respectively, i.e.

$$W_h = \theta h - c_h \quad \text{and} \quad W_l = \theta l - c_l$$

So that the variation in social surplus Δ is given by

$$\Delta = W_h - W_l = \theta(h - l) - (c_h - c_l) > 0$$

Letting $0 < x < \Delta$, we will construct an SPNE where in each period the consumer obtains utility $x > 0$ and the firm gets profits $\pi_x = \pi_l + \Delta - x > \pi_l$. This is a 'cooperation' situation in the sense that if both players stick to the strategy profile, they will receive, in each period t, a higher payoff than the one obtained in the unique one-shot equilibrium. As is usually the case with infinitely repeated games, we do so by constructing a trigger strategy profile where both players earn the cooperation payoff as long as they stick to the strategy profile. In order to enforce this result, there must be a credible punishment in case some player deviates from the cooperation profile and we do so by permanently punishing them (if there is deviation) with the unique one-period equilibrium (which is credible because it is an SPNE). That is exactly what the following profile does:

Consider the strategy profile $\sigma = (\sigma_1, \sigma_2)$ where

$$\sigma_1 = \begin{cases} t = 1: s_1^1 = (p^1, q^1) = (\theta h - x, h) \\ t > 1: \begin{cases} s_1^t = (p^t, q^t) = (\theta h - x, h) & \text{if } \forall \tau < t, (s_1^\tau, s_2^\tau) = (s_1^1, s_2^1) \\ s_1^t = (p^t, q^t) = (\theta l, l) & \text{otherwise} \end{cases} \end{cases}$$

$$\sigma_2 = \begin{cases} t = 1: s_2^1 = (\text{buy} \Leftrightarrow p \leq \theta h - x) \\ t > 1: \begin{cases} s_2^t = (\text{buy} \Leftrightarrow p \leq \theta h - x) & \text{if } \forall \tau < t, (s_1^\tau, s_2^\tau) = (s_1^1, s_2^1) \\ s_2^t = (\text{buy} \Leftrightarrow p \leq \theta l) & \text{otherwise} \end{cases} \end{cases}$$

If both players follow their strategy, they obtain the following intertemporal payoffs

$$\pi_1(\sigma_1,\sigma_2) = \frac{1}{1-\delta}(\pi_l + \Delta - x) = \frac{1}{1-\delta}(\theta h - x - c_h) \qquad (8.4)$$

$$\pi_2(\sigma_1,\sigma_2) = \frac{1}{1-\delta}x$$

Clearly the consumer has no profitable deviation in any period. However, at any period t the firm could obtain some extra profits in that period by providing low quality but sacrificing future payoffs by triggering the punishment: If it deviates at $t = 1$, by maintaining the price but cutting quality, it will obtain the following:

$$\pi_1(\sigma_1',\sigma_2) = (\theta h - x - c_l) + \frac{\delta}{1-\delta}(\theta l - c_l) \qquad (8.5)$$

The deviation is not profitable as long as (8.4) is higher than (8.5), i.e.

$$\frac{1}{1-\delta}(\theta h - x - c_h) > (\theta h - x - c_l) + \frac{\delta}{1-\delta}(\theta l - c_l)$$

Solving for δ in the above inequality we obtain

$$\delta \geq \frac{c_h - c_l}{\theta(h-l)-x} > 0 \qquad (8.6)$$

Hence, as long as the discount factor δ satisfies (8.6), the profile $\sigma = (\sigma_1,\sigma_2)$ constitutes an SPNE of $G^{\infty}(\delta)$ yielding the stated per-period payoffs.

Proof of Propositions 1 and 4

Let a strategy profile $s^* = (s_1^*, s_2^*, \cdots, s_n^*)$ and beliefs $q^{e*}(\cdot) = (q_1^{e*}(\cdot), q_2^{e*}(\cdot), \cdots, q_n^{e*}(\cdot))$ constitute a rational-expectations equilibrium. Then, $\forall i = 1,2,\cdots,n$

$$q_i^* = l \quad \text{and} \quad p_i^* = c_l$$

Proof: The following three lemmas establish the proof.

Lemma 1: *Choosing quality l is a dominant strategy for each firm i.*

Proof: Let q^e be any beliefs system for the consumer, $s_{-i} = (p_{-i}, q_{-i})$ be any strategy profile for all other firms except i, and p_i be any price chosen by firm i. We will show that

$$\pi_i((p_i,l),s_{-i},q^e) \geq \pi_i((p_i,h),s_{-i},q^e)$$

i.e. choosing low quality dominates high quality. Having observed prices (p_i, p_{-i}), the consumer makes his purchase decision according to his beliefs q^e. In terms of firm i, this purchase decision falls into basically three case types:

1. The consumer does not buy from firm i, in which case choosing high or low quality yields the same payoff (since the true quality chosen by firm i does not affect the consumer's purchase decision):

$$\pi_i((p_i,l),s_{-i},q^e) = \pi_i((p_i,h),s_{-i},q^e) = 0$$

2. The consumer buys only from firm i, in which case choosing low quality yields a strictly higher payoff (since the true quality chosen by firm i does not affect the consumer's purchase decision):

$$\pi_i((p_i,l),s_{-i},q^e) = p_i - c_l > p_i - c_h = \pi_i((p_i,h),s_{-i},q^e)$$

3. The consumer is equally likely to buy from firm i and m more firms, in which case choosing low quality yields a strictly higher payoff (since the true quality chosen by firm i does not affect the consumer's purchase decision):

$$\pi_i((p_i,l),s_{-i},q^e) = \frac{1}{m+1}(p_i - c_l) > \frac{1}{m+1}(p_i \quad c_h) = \pi_i((p_i,h),s_{-i},q^e)$$

Therefore choosing l is a dominant strategy.

Note that, from the three cases in the previous lemma, the only case where a firm should choose high quality is if it were certain that it would not sell. But then, why bother to choose high quality if it knows it is not going to sell? Therefore we can conclude that the firm will never optimally choose high quality. This proves the following lemma:

Lemma 2: *Let (s^*, q^{e*}) be a rational-expectations equilibrium, then $\forall i = 1,2,\cdots,n$, $q_i^* = l$.*

Finally we show that in equilibrium all prices must equal marginal cost. This should not be surprising since now the game resembles a Bertrand game: Players should anticipate low quality to be produced in equilibrium so that incomplete information vanishes resulting in a price competition game. Lemma 3 proves this.

Lemma 3: *Let* (s^*, q^{e*}) *be a rational-expectations equilibrium, then* $\forall i = 1,2,\cdots,n,\ p_i^* = c_l$.

Proof: By Lemma 2; the consumer should anticipate all firms choosing low quality. Therefore the consumer's optimal decision should be to buy from the firm offering the lowest price p_i^* as long as

$$\theta l - p_i^* \geq 0$$

All firms should then set the same price in equilibrium (otherwise some firms would not sell). Clearly, it is not optimal for firms to set prices below marginal cost (since they would get negative profits). If all prices are above marginal cost any firm would find it profitable to unilaterally deviate by setting a slightly smaller price (just as in a regular Bertrand game), therefore, in equilibrium, we must have $\forall i = 1,2,\cdots,n$

$$p_i^* = c_l$$

Proof of Propositions 3 and 6

In game $G^\infty(\delta)$ for the oligopoly case with $n \geq 2$ firms, consider the following strategy for each firm i

$$\sigma_i^* = \begin{cases} t = 1: s_i^1 = (p_i^1, q_i^1) = (\theta l, l) \\ t > 1: \begin{cases} s_i^t = (p_i^t, q_i^t) = (\theta l, l) & \text{if } \forall \tau < t, (s_i^\tau, s_{-i}^\tau) = (s_i^1, s_{-i}^1) \\ s_i^t = (p_i^t, q_i^t) = (c_l, l) & \text{otherwise} \end{cases} \end{cases}$$

Then $\sigma^* = (\sigma_1^*, \sigma_2^*, \cdots, \sigma_n^*)$ constitutes a rational-expectations equilibrium under pessimistic beliefs for discount factor

$$\delta \geq \frac{n-1}{n}$$

Proof: If all firms follow the strategy the inter-temporal payoffs are $\forall i = 1,2,\cdots,n$

$$\pi_i(\sigma_i^*,\sigma_{-i}^*) = \frac{1}{n}\left(\frac{1}{1-\delta}\right)(\theta l - c_l) \tag{8.7}$$

Suppose all firms except i stick to strategy profile σ_{-i}^* and let σ_{-i}' be the best possible unilateral deviation for firm i: Since the future is discounted by factor δ, that deviation should occur at time $t = 1$. There is no need to consider deviations that involve producing high quality because, since the consumer has pessimistic beliefs, those kind of deviations could never be the best possible deviation. So let's consider only deviations in the chosen price. All the other firms are setting their prices equal to θl: By slightly lowering its price, firm i would get all the market in period 1 (thus obtaining immediate profits of at most $\theta l - c_l$) but it would trigger the punishment phase starting in period $t = 2$ and therefore would earn zero profits in all subsequent periods (since, starting in $t = 2$; all other firms would set prices equal to marginal cost on a permanent basis). Therefore the intertemporal payoff of the best possible deviation is at most

$$\pi_i(\sigma_i',\sigma_{-i}^*) = \theta l - c_l \tag{8.8}$$

This deviation is not profitable as long as (8.7) is higher than (8.8), i.e.

$$\frac{1}{n}\left(\frac{1}{1-\delta}\right)(\theta l - c_l) \geq \theta l - c_l$$

Solving for δ in the inequality above we obtain

$$\delta \geq \frac{n-1}{n}$$

9. Industry-specific patterns inherent in personal information leakage incidents in Japan: analyses of an incident database

Naoko Sakurai

1. INTRODUCTION

Significant progress in the development of the Internet has contributed to current society's strong dependence on information. Various types of technology, such as mobile phones and smartphones, pervade Japanese society. It might not be an exaggeration to state that our social lives depend on information systems. Although the rate at which the Internet had penetrated Japanese society in December 1997 was only 9.2 per cent, the rate ultimately grew to 57.8 per cent by December 2002. It then continued to rise until it reached 79.1 per cent (a 0.9 per cent increase, in comparison with the previous fiscal year) in December 2011. At that time, the number of Internet users in Japan equalled 96.1 million or approximately 80 per cent of the population.[1]

In the past, entities were solely able to collect personal data based on customers' willingness to provide personal information. Now without their consent, personal data are collected from information searches or product purchases on the Internet, from location information when they activate global positioning systems (GPS), from transportation information when they employ integrated circuit (IC) cards, from their activities on Twitter or Facebook and so on.

Efforts to achieve full utilisation of this enormous amount of personal data (referred to as 'Big Data') as the source of new insights into marketing began in 2012.[2] As greater amounts of data related to personal behaviour accumulate, greater amounts of information related to users' private lives – that they likely would prefer to keep private – are being transformed into Big Data. This collection of personal data is not solely maintained in Japan. Data collected on the Internet are collected across borders; therefore it is easy to assume that large amounts of data are instantly transmitted across international borders, and that these activities threaten users' privacy.

The Privacy Protection Principles were first adopted by the Organisation for Economic Co-operation and Development (OECD) as the 'OECD's Eight Principles' in 1980. This set of principles became the foundation for basic concepts related to personal information protection across the globe. In 1995, 'The Data Protection Directive' (EU Directive 95/46/EC) requested that member states transpose the directive into internal laws. It also requested developing countries to achieve adequate levels of personal data protection. In January 2012, the EU announced 'a proposal for a Regulation of European Parliament and of the Council on the protection of individuals with regard to processing of personal data and on the free movement of such data (General Data Protection Regulation)'.[3] Because rapid technological developments up to that point had been significant, the existing Privacy Protection Principles were considered inadequate. This proposed amendment suggests the overall strengthening of data protection systems. For example, it expands the concept of the protection of personal information data and proposes that members adopt 'rules' that will be applied directly to member states instead of 'directives' that require legislation. In the case of violations, companies will be required to pay surcharges of a maximum of 1 million EUR or surcharges of 2 per cent of their annual global sales as penalties.

In an effort to comply with the EU Directive's requests that developing countries achieve adequate levels of personal data protection, Japan passed the Act for the Protection of Personal Information in 2005. However, because Japan maintains looser regulations in comparison with other EU countries, it has not yet received accreditation at a sufficient protection level that would permit Japan to engage in the transfer of personal data with EU countries. Currently, entities in Japan bear no responsibility to liaison and report when personal information is leaked. Penalties are loosely applied. A maximum amount of 300,000 yen is imposed for violations. In reality, most entities have adopted technical as well as organisational information security measures. They have also engaged in the formulation of rules related to information security policies and the implementation of information security training, yet personal information leakage incidents continue to occur. The publication of all facts related to personal information leakage incidents is considered important in the 'Basic Policy on the Protection of Personal Information', so entities usually publicise incidents on their websites; however, it is difficult to determine whether all entities publish their leakage incidents, because they are not required to do so. Alternatively, although some entities consistently publicise personal information leakage incidents, it seems that their publication efforts are mere formalities. Thus, concerns exist that information security measures have lost their impact because of the loose application of penalties.

2. PURPOSE OF THE STUDY

This chapter aims to conduct an empirical examination of actual conditions related to personal information leakage incidents that have occurred in Japan since 2000. This study analyses data from personal information leakage incidents from a variety of angles. In comparison with the author's previous studies[4] that analysed 1511 samples collected between 2000 and 2005, the author intends to examine the types of secular changes related to personal information leakage incidents that occurred until 2011, as well as the social changes described above. Specifically, the author sampled more than 5200 incidents that occurred between January 2000 and December 2011 and assembled a database. The author focused on the differences among the types of personal information leakage incidents that occurred in various industries, such as the causes of those incidents, the number of victims of leaked personal information per incident, as well as the number of incidents published by one company.

The following analytical methods were applied:

1. To understand the entire scope of personal information leakage incidents, a primary data analysis using cross-tabulation was performed based on years, industries and causes.
2. To extract regularity and similarity based on various categories of incident, a correspondence analysis was conducted. Correspondence analysis is an analytical method that places attributes with similar patterns in a cross-tabulation. It is often used for the analysis of brand images in marketing fields, as it can reveal a visual pattern of incidents inherent to each industry on a map that is easy to understand.

Based on these methods, the author aimed to perform an empirical extraction of the following three patterns:

1. Incidents inherent to each industry; the relationships that exist between industry sectors and causes, and between industry sectors and the number of victims of leaked personal information.
2. Causes of serious incidents; the relationships that exist between causes and the number of victims of leaked personal information.

3. Recurrences of incidents; the relationships that exist between industry sectors and the number of incidents published by one company.

3. SUMMARY OF THE PERSONAL INFORMATION LEAKAGE DATABASE

The database was constructed by classifying 5219 articles collected from newspapers, the Internet and entities' websites between 1 January 2000 and 31 December 2011. Although the author collected as much data as possible from published articles, the database does not include all incidents related to personal information leakage incidents that occurred in Japan. In addition, many of the articles were related to industries that stood at the forefront of media attention, and articles related to industries that actively published incidents were easily collected. Therefore the existence of bias in the database is undeniable; however, the results of the data analysis will, in all likelihood, indicate the actual state of personal information leakage incidents, because no comparable research or analysis has been conducted using a database that contains 12 years of data.

Causes of incidents were classified into nine categories (see Table 9.1).

The types of information were divided into the following three categories based on the 'Personal Information Protection Management Systems Requirements', a guideline created by Japan's Ministry of Economy, Trade and Industry:

1. Basic information: names, dates of birth, addresses, telephone numbers and so on.
2. Credit information: information that might influence individual interests, such as financial assets, academic backgrounds, academic results, physical descriptions and so on.
3. Sensitive information: information that is prohibited from being collected, such as medical charts, medical histories, criminal histories, human rights information and so on.

Industrial sectors were classified into five categories: 1) public services, 2) educational institutions, 3) medical institutions, 4) companies and 5) others. Moreover companies were classified into 11 service sector industries: 1) fishery, agriculture and forestry, 2) mining, 3) construction, 4) manufacturing, 5) electricity and gas, 6) warehouse and transportation, 7) information and communications, 8) commercial, 9) financial and insurance, 10) real estate and 11) service businesses.

Table 9.1 Causes of incidents

Classification		Explanation
Malicious and Intentional	Theft	Stolen lists of data, PCs, etc.
	Leak that occurred over the network	Stolen by unauthorised access, unauthorised disclosure on the Internet, etc.
	Paper-based leak	Lists taken without permission, inappropriate provision, misappropriation of printed information, etc.
	Inappropriate process	Incorrect disposal, intentional discards, etc.
Accidental and/or Negligent	Lost	Lost documents, PCs, mobile phones, etc.
	Stolen	Stolen documents, PCs, mobile phones, etc.
	Leak that occurred over the network	Sent emails by 'cc' instead of 'bcc', etc.
	Paper-based leak	Provision of lists to the wrong persons, faxes sent to incorrect addresses, etc.
	Inappropriate process	Incorrect disposal, mistaken discards, etc.

4. PRIMARY ANALYSIS BY THE USE OF CROSS-TABULATION

4.1 Yearly Number of Personal Information Leakage Incidents

Figure 9.1 shows the number of incidents divided by industrial sectors for each year. Beginning in 2004, media coverage focused on an increase in the number of personal information leakage incidents because the Act for the Protection of Personal Information was scheduled to be enacted in April 2005. The media paid significant attention to the public's interest in personal information. Some believe that the highest number of incidents was reported in 2005 because of the media's focus on this issue. In addition, companies published violations of the Personal Information Protection Law based on retroactive examinations of their possession of personal data. The number of incidents decreased after 2006 because the media lost interest in reporting small incidents; however, since 2009, the number of incidents has again been increasing. In reality, the actual number of incidents does not appear to have declined.

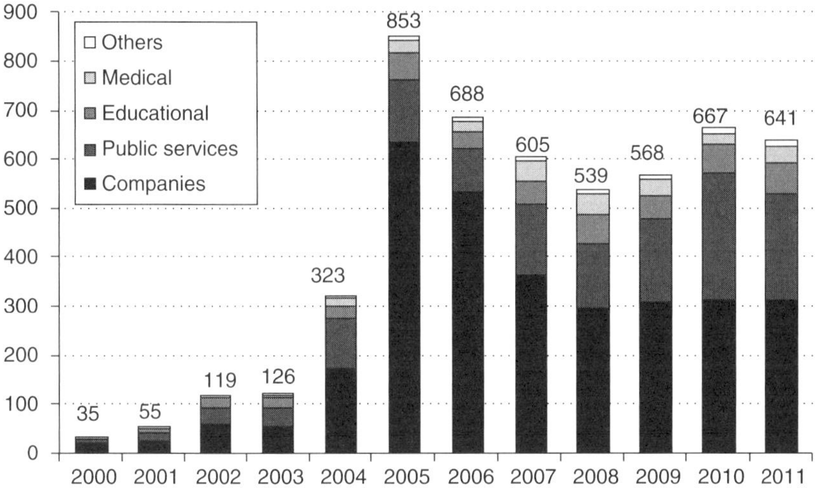

Figure 9.1 Yearly number of personal information leakage incidents by industry

4.2 Causes of Personal Information Leakage Incidents

Table 9.2 lists the number of incidents divided by causes for each year. Overall, during the 12-year sample period, 414 (7.9 per cent) malicious and intentional incidents occurred. An additional 4805 (92.1 per cent) were accidental and/or negligent incidents. Of the author's previous studies that relied on 1511 samples collected between 2000 and 2005, 16.4 per cent were malicious and intentional incidents. The remaining 83.6 per cent were accidental and/or negligent incidents. The differences between these results reveal that the percentage of accidental and/or negligent incidents has increased.

The most common cause was described as 'lost' (33.7 per cent of the total number of incidents). An additional 23.3 per cent were described as 'accidental and/or negligence leaks that occurred over the network'. The third most common cause was described as 'stolen' (19.1 per cent). Causes based on the previous 1511 samples were ranked as 1) lost, 2) stolen and 3) accidental and/or negligence leaks that occurred over the network, respectively. These results reveal an increase in the number of leaks that occurred over the network. It was notable that 'accidental and/or negligence paper-based leaks' increased because city halls and ward offices have frequently published incidents on their websites since 2009.

After 2005, incidents that occurred owing to USB flash drive use were

Table 9.2 Yearly number of personal information leakage incidents by cause

Cause \ Year	2000	2001	2002	2003	2004	2005	2006	2007	2008	2009	2010	2011	Total	
Malicious and Intentional														
Theft	1	0	0	2	2	9	1	0	1	0	0	0	16	0.3%
Leak that occurred over the network	1	1	6	6	16	24	10	11	12	15	25	19	146	2.8%
Paper-based leak	17	19	21	32	41	43	9	8	13	13	8	14	238	4.6%
Inappropriate process	0	0	0	0	7	0	0	0	2	2	1	2	14	0.3%
Total	19	20	27	40	66	76	20	19	28	30	34	35	414	
%	54.3	36.4	22.7	31.7	20.4	8.9	2.9	3.1	5.2	5.3	5.1	5.5		7.9%
Accidental and/or Negligent														
Lost	1	5	6	21	78	328	194	176	195	248	251	255	1758	33.7%
Stolen	1	7	16	22	117	234	147	103	92	93	87	76	995	19.1%
Leak that occurred over the network	12	20	64	27	38	114	206	204	164	113	134	121	1217	23.3%
Paper-based leak	0	2	2	13	16	86	63	23	46	78	158	146	633	12.1%
Inappropriate process	2	1	4	3	8	15	58	80	14	6	3	8	202	3.9%
Total	16	35	92	86	257	777	668	586	511	538	633	606	4805	
%	45.7	63.6	77.3	68.3	79.6	91.1	97.1	96.9	94.8	94.7	94.9	94.5		92.1%
Total	35	55	119	126	323	853	688	605	539	568	667	641	5219	

Figure 9.2 *Number of leakage incidents caused by the use of USB flash drives*

noticeable. The use of USB flash drives has expanded rapidly because these devices are convenient to carry; however, their convenience led to an increase in leakage incidents (see Figure 9.2). One half of all USB-related incidents were described as 'stolen'. The remaining incidents were described as 'lost'. The problem is that 78 cases were caused by the removal of data, even though removal was prohibited.

4.3 Industry Sectors

Figure 9.3 lists the percentage of personal information leakage incidents that occurred by industry sectors. The largest percentage (59.7 per cent) and the highest number of incidents (3117) were attributed to 'companies'. The second largest percentage (25.5 per cent) and the second highest number of incidents (1333) were attributed to 'public services', such as public offices, local governments and the police. Within companies, the largest percentage (15.3 per cent) and the highest number of incidents (797) were attributed to 'financial and insurance'. The second largest percentage (12.4 per cent) and the second highest number of incidents (647) were attributed to 'information and communications'. The third largest percentage (9.0 per cent) and the third highest number of incidents (468) were attributed to 'commercial'. Because 'financial and insurance' and 'public services' were monitored by administrative guidance focused on the protection of personal information, they appeared to have published all incidents that occurred, even small incidents. The total number of incidents has again increased since 2009, because the number of incidents attributed to 'public services' increased. The number of incidents attributed to other industries did not change significantly.

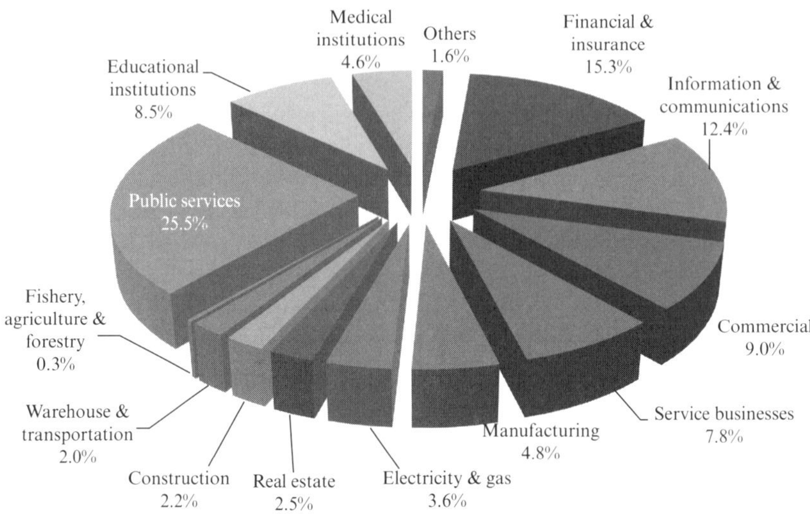

Figure 9.3 Percentage of personal information leakage incidents by industry sector

4.4 Types of Information

Classification of three types of information revealed that the highest number of incidents (3120) and the largest percentage (59.8 per cent) were described as 'basic information'. An additional 1742 (33.4 per cent) were described as 'credit information'. The remaining 357 (6.8 per cent) were described as 'sensitive information' (see Table 9.3).

An examination of the types of information provided by industry sectors revealed that significant characteristics could be described as 'sensitive information'. Although 59.7 per cent of all incidents were attributed to companies, 195 (54.6 per cent) of incidents described as 'sensitive information' were attributed to 'medical institutions' and 120 (33.6 per cent) were attributed to 'public services'. Incidents recorded for these two industries accounted for 88.2 per cent of all incidents described as 'sensitive information'. In particular, the fact that 80.6 per cent of 242 incidents attributed to medical institutions were described as 'sensitive information' indicates the weight of information handled by medical institutions.

Out of the 1742 incidents described as 'credit information', the highest number of incidents (1016, 58.3 per cent) was attributed to 'companies'. In companies, the highest number of incidents (574, 33.0 per cent) was attributed to 'financial and insurance'. The second highest number of incidents (419, 24.1 per cent) was attributed to 'public services'. The third largest

Table 9.3 Number of personal information leakage incidents by type of information

Industry sector	Basic information	Credit information	Sensitive information	Total
Public services	795	419	120	1334
Educational institutions	154	276	13	443
Medical institutions	32	15	195	242
Others	67	16	0	83
Companies				
Fishery, agriculture & forestry	8	10	0	18
Mining	0	0	0	0
Construction	99	14	0	113
Manufacturing	207	40	2	249
Electricity & gas	153	34	0	187
Warehouse & transportation	78	26	1	105
Information & communications	562	79	6	647
Commercial	357	110	1	468
Financial & insurance	215	574	8	797
Real estate	109	19	0	128
Service businesses	284	110	11	405
Subtotal	2072	1016	29	3117
Total	3120	1742	357	5219

number of incidents (276, 15.8 per cent) was attributed to 'educational institutions'. The total percentage of incidents attributed to these three industries equalled 72.9 per cent. 'Financial and insurance' institutions manage information related to finances (e.g. account information). 'Public services' institutions manage information related to household registrations, taxes, nursing care and so on. 'Educational institutions' manage information related to academic papers, grades and so on. Therefore serious personal information is at risk of being leaked by these three industries.

With respect to 'basic information', 2072 (66.4 per cent) out of 3120 cases were attributed to 'companies'. Their related numbers and percentages were ranked as follows: 1) 'information and communications' (562, 18.0 per cent), 2) 'commercial' (357, 11.4 per cent), 3) 'service businesses' (284, 9.1 per cent), 4) 'financial and insurance' (215, 6.9 per cent) and 5) 'manufacturing' (207, 6.6 per cent).

4.5 Number of Victims of Leaked Personal Information

In most cases, the number of victims of personal information leakage per incident was small. As shown in Table 9.4, fewer than 100 victims

Table 9.4 Number of victims by cause

Cause / Number of Victims	Malicious and Intentional				Accidental and/or Negligent					Total	%
	Theft	Leak that occurred over the network	Paper based leak	Inappropriate process	Lost	Stolen	Leak that occurred over the network	Paper based leak	Inappropriate process		
1–99	1	18	82	2	804	367	383	548	53	2258	43.3
100–499	3	18	36	4	377	250	332	34	28	1082	20.7
500–999	0	8	8	0	118	84	115	15	9	357	6.8
1000–4999	3	20	21	2	186	125	159	11	34	561	10.7
5000–9999	2	11	6	1	53	30	28	3	7	141	2.7
10,000–99,999	1	31	16	4	103	46	60	5	34	300	5.7
100,000–999,999	1	17	14	0	35	12	11	0	20	110	2.1
1,000,000–	0	6	3	0	2	0	1	0	2	14	0.3
Unclear	5	17	52	1	80	81	128	17	15	396	7.6
Total	16	146	238	14	1758	995	1217	633	202	5219	100
	414				4805						

were involved in 43.3 per cent of all incidents. Of these incidents, 513 (9.8 per cent) affected only one victim. On the other hand, 14 incidents involved more than 1 million victims, nine of which were caused by 'malicious and intentional' reasons. Six of those nine cases were 'malicious and intentional leaks that occurred over the network'. Until 2010, most of these incidents could be divided into two types: 'malicious and intentional leaks that occurred over the network' and 'accidental disposals'. In 2011, 'unauthorised access to information', which affected a significant number of victims, was experienced by Sony Corporation and other companies.

Of the 14 incidents that involved more than 1 million victims, 35.7 per cent were attributed to 'financial and insurance' companies, 21.4 per cent to 'information and communications' companies and 14.3 per cent to 'public services'. An additional 110 incidents that involved more than 100,000 but fewer than 1 million victims were reported by the same industry sectors (i.e. 10.9 per cent were attributed to 'financial and insurance', 15.5 per cent to 'information and communications' and 12.7 per cent to 'public services').

4.6 Recurrences of Incidents

Out of all business operators, 2217 published only one personal information leakage incident, while 37 operators published more than ten incidents. This total included 13 business operators in the 'public services' sector, five companies that operated in the 'information and communications' and 'financial and insurance' sectors, four companies that operated in the 'construction' and 'electric and gas' sectors, three companies that operated in the 'commercial' sector, two companies that operated in the 'warehouse and transportation' sector and one company that operated in the 'service businesses' sector.

In Osaka City the instances of published leakages have increased since 2009, reaching an all-time high of 248 incidents during the study period. Although the Osaka Municipal Policy Planning Office gave these matters serious consideration and prescribed measures to prevent the occurrence of similar incidents, it is possible to state that the measures adopted were inadequate. Out of all reported instances, 76 per cent occurred because of simple errors (e.g. provision of a resident certificate or document to the wrong person). Many of these incidents might have been prevented if business operators had double-checked the names listed on documents as well as on other identification at the times the documents were issued.

In addition, Yokohama City, the Tokyo Metropolitan Government, Shizuoka City and Saitama Prefecture publicised more than 30 incidents during the sample period. Many local governments tended to close their websites relatively quickly after they publicised incidents; however, Osaka City and Yokohama City have publicised incidents in detail on their websites for some time. This is one reason for the high numbers of incidents in these two cities that appeared in the database. Based on this fact, in all likelihood, other local governments may have experienced many more incidents.

Out of 1211 incidents recorded by 37 business operators that publicised more than ten incidents, 95.2 per cent were caused by 'accidental and/or negligent' reasons. 'Lost' was the reason for 37.2 per cent of incidents, 28.3 per cent were described as 'paper-based leaks', 13.4 per cent were described as 'leaks that occurred over the network' and 13.0 per cent were described as 'stolen'. Many incidents that occurred involved meter readers employed by 'power and gas' companies or individuals employed as fee collectors by the Japan Broadcasting Corporation (NHK). These incidents involved lost or stolen documents, receipts, mobile computers and so on. The highest number of causes for 95 incidents publicised by Nippon Telegraph and Telephone (NTT) East and West Japan were described as 'accidental and/or negligent leaks that occurred over the network'. These leaks may have occurred because of industry characteristics. In addition, 19 incidents were caused by file swapping and software viruses. Although those incidents were not publicised by other entities, as many as 14 incidents occurred because of 'malicious and intentional' causes, such as the inappropriate provision of information.

4.7 Summary of the Primary Analysis

Based on the results of the primary analysis described above, the following findings were obtained:

1. Causes: 'Accidental and/or negligent leaks that occurred over the network' increased in comparison to the results of the previous studies. 'Paper-based leaks' also increased owing to city halls and ward offices frequently publishing incidents on their websites since 2009.
2. Industry sectors: 'Companies' caused the largest percentage of incidents (59.7 per cent) and 'public services' caused the second largest percentage of incidents (25.5 per cent). Within 'companies': 'financial and insurance', 'information and communications' and 'commercial' were the business operators that caused the three highest numbers of incidents, respectively.
3. Number of victims: Fourteen incidents involved more than 1 million victims. Nine cases were caused by 'malicious and intentional' reasons,

six of which were described as 'malicious and intentional leaks that occurred over the network'.

4. Types of information: Among 'sensitive information' leakages, 88.2 per cent were attributed to 'medical institutions' and 'public services'. With respect to 'credit information', 72.9 per cent of all incidents were attributed to 'financial and insurance', 'public services' and 'educational institutions'.
5. Recurrences of incidents: Out of all business operations, 2217 published only one incident, while 37 business operators published more than ten incidents. Industry sectors with the highest rates of recurrence were as follows: 1) 'public services', 2) 'information and communications', 3) 'financial and insurance', 4) 'construction', 5) 'electric and gas', 6) 'commercial', 7) 'warehouse and transportation' and 8) 'service businesses'.

5. RESULTS OF THE ANALYSIS BASED ON THE USE OF CORRESPONDENCE ANALYSIS

Verification of the above-noted characteristics was achieved through the use of correspondence analysis. Correspondence analysis is an analytical method that places attributes with similar patterns in a cross-tabulation. It can reveal the relationships that exist between two categories by using a visually intelligible form of mapping.

5.1 Relationships that Exist between Industry Sectors and Causes

Based on an analysis of 1511 samples collected between 2000 and 2005, relationships were grouped into the following three major patterns:

1. Accidental and/or negligent causes: 'educational institutions', 'commercial', 'electricity and gas', 'construction' and 'fishery, agriculture and forestry'.
2. Malicious and intentional causes: 'public services', 'medical institutions' and 'warehouse and transportation'.
3. Leaks that occurred over the network: 'information and communications', 'service businesses', 'manufacturing', 'real estate' and 'others'.

A positioning map based on this analysis of 5219 samples is shown in Figure 9.4. The axes of Dimensions 1 and 2 describe relationships that occurred between two categories. Figure 9.4 reveals that some relationships occurred that differed from the 1511 data patterns.

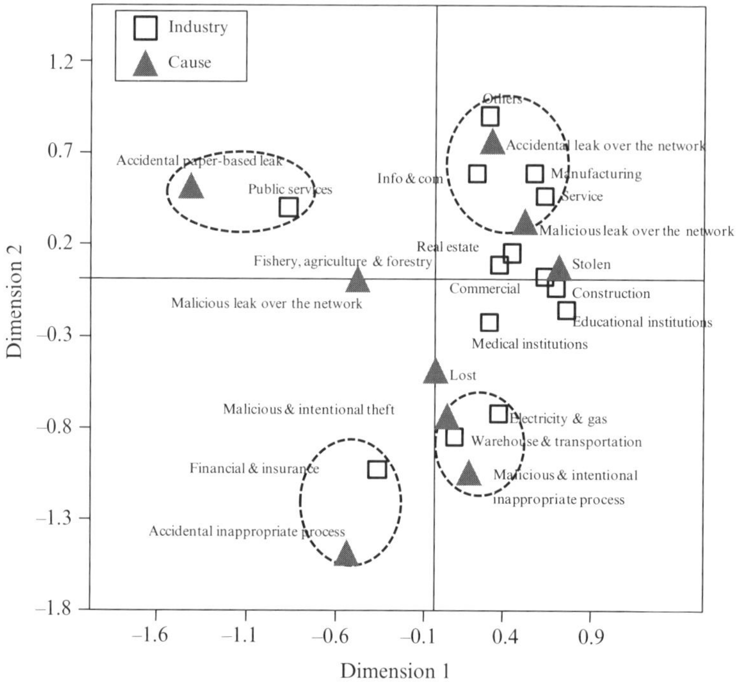

Figure 9.4 Positioning map based on industry sectors and causes

1. 'Accidental and/or negligent paper-based leaks' and 'public services' are shown in the upper left quadrant.
2. 'Accidental and/or negligent inappropriate processes' and 'financial and insurance' are shown in the lower left quadrant.
3. 'Leaks that occurred over the network', 'information and communications', 'service businesses', 'manufacturing' and 'others' are shown in the upper right quadrant.
4. 'Malicious and intentional theft', 'malicious and intentionally inappropriate processes', 'electricity and gas' and 'warehouse and transportation' are shown in the lower right quadrant.

5.2 Relationships that Exist between Industry Sectors and the Number of Victims

As shown in Figure 9.5, Dimension 1 describes the number of victims of leaked personal information per incident. The positive direction may reveal smaller numbers and the negative direction may reveal

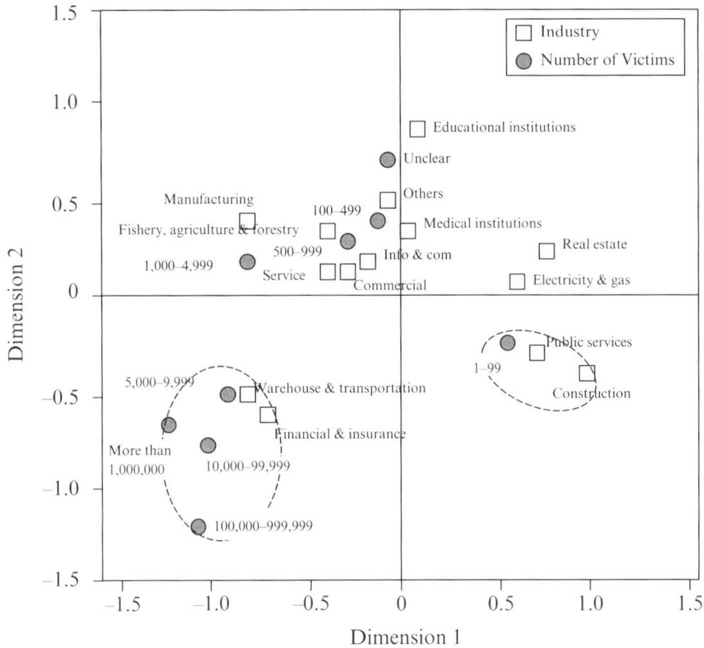

Figure 9.5 Positioning map based on industry sectors and the number of victims

larger numbers. It is possible that no significant findings appear in Dimension 2.

In the lower right quadrant is '1 to 99'. 'More than 5000' appears in the lower left quadrant. While '1 to 99' appears to maintain close relationships with 'public services' and 'construction', 'more than 5000' appears to maintain close relationships with 'financial and insurance' and 'warehouse and transportation'.

Of the 14 incidents that involved more than 1 million victims, 'financial and insurance' companies registered the highest number of incidents. 'Information and communications' companies registered the second highest number and 'public services' organisations registered the third highest number of incidents. Based on this analysis, 'warehouse and transportation' appears to maintain a close relationship with more than 5000 victims.

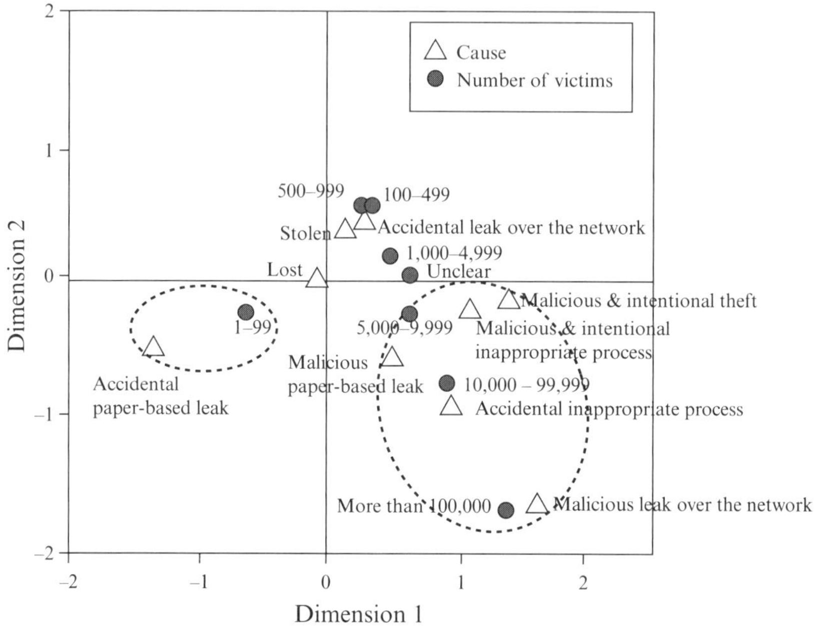

Figure 9.6 Positioning map based on causes and the number of victims

5.3 Relationships that Exist between Causes and the Number of Victims

In Figure 9.6, Dimension 1 describes the number of victims of leaked personal information per incident. The positive direction may show larger numbers and the negative direction may show smaller numbers. It is possible that no significant findings appear in Dimension 2.

'Accidental and/or negligent paper-based leaks' and '1 to 99' appear in the lower left quadrant. 'More than 5000' and most 'malicious and intentional causes' appear in the lower right quadrant. The results reveal that 'malicious and intentional causes' demonstrate a strong relationship with a large number of victims of leaked personal information.

5.4 Relationships that Exist between Industry Sectors and Recurrences of Incidents

In Figure 9.7, Dimension 1 describes the frequency of incidents. The positive direction may reveal larger numbers and the negative direction may reveal smaller numbers. It is possible that no significant findings appear in Dimension 2.

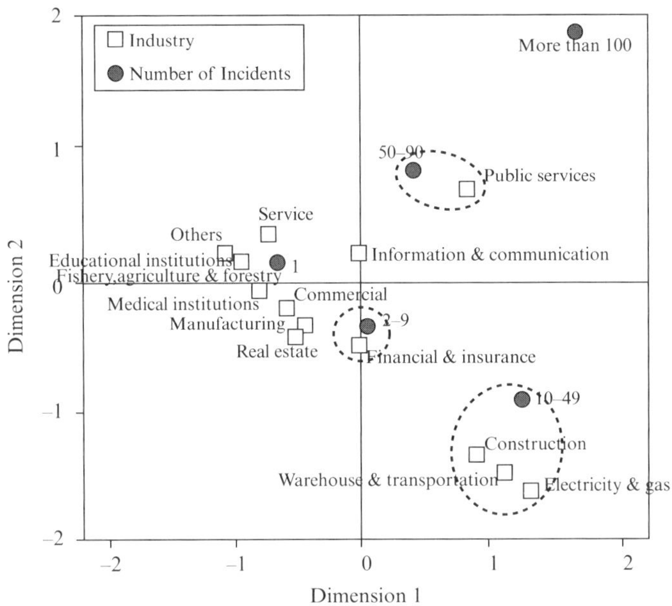

Figure 9.7 Positioning map based on industry sectors and recurrences of incidents

'More than 50' and 'public services' are located in the upper right quadrant. This means that 'public services' appears to maintain a close relationship with the recurrence of incidents. Then '10 to 49', 'construction', 'warehouse and transportation' and 'electricity and gas' appear in the lower right quadrant, and '2 to 9' and 'financial and insurance' are placed near the zero point in Dimension 1. Although no significant findings may appear in '1', '2 to 9' may maintain a relationship with 'financial and insurance' and '10 to 49' may maintain relationships with 'construction', 'warehouse and transportation' and 'electricity and gas'.

Out of all industry sectors that publicised more than ten incidents, 'information and communications', 'commercial' and 'service businesses' did not demonstrate any significant patterns.

6. SUMMARY AND OBSERVATIONS RELATED TO THE OCCURRENCE OF PERSONAL INFORMATION LEAKAGE INCIDENTS

6.1 Causes

In comparison with the results of the analysis of 1511 samples, 'leaks that occurred over the network' increased during this coverage period. Industry operators that demonstrated patterns of 'leaks that occurred over the network' included 'information and communications', 'service businesses', 'manufacturing' and 'others'. These companies demonstrated the same patterns revealed in the previous studies. 'Financial and insurance' demonstrated a clear pattern related to 'accidental and/or negligent inappropriate processes'. In fact, 316 out of 797 incidents that involved 'financial and insurance' companies appear to have been caused by improper disposal of microfilm, microfiche and so on. Although improper disposal decreased in comparison with the 66 incidents that occurred in 2005, 37 incidents occurred in 2011. Paper-based leakages increased significantly because city halls and ward offices have frequently publicised incidents since 2009. 'Malicious and intentional theft' and 'malicious and intentional inappropriate processes' appear to maintain relationships with 'electricity and gas' and 'warehouse and transportation'. 'Warehouse and transportation' demonstrated the same pattern revealed in the previous studies; however, 'electricity and gas' did not. Additional verification should be performed because the number of 'malicious and intentional' incidents caused by 'electricity and gas' was small.

6.2 Number of Victims of Leaked Personal Information

The results show that a large number of victims of leaked personal information per incident were caused by 'malicious and intentional' reasons. By industry, incidents that involved 'public services' organisations and 'construction' companies tended to affect only a limited number of victims per incident. Alternatively, victims of incidents caused by 'financial and insurance' and 'warehouse and transportation' companies appeared to be especially high. Based on the analysis, 14 business operators that caused incidents that involved more than 1 million victims operated in 'financial and insurance', 'information and communications' and 'public services' sectors. However, the results of the current study reveal that 'warehouse and transportation' appears to maintain a relationship with incidents that involve more than 5000 victims. 'Public services' appears to maintain a stronger relationship with '1 to 99', rather than with higher numbers of victims, because the percentage of the 16 incidents that involved more than 100,000 victims was only 1.2 per cent of the total number of incidents that involved 'public services'.

6.3 Recurrences of Incidents

The results reveal that relationships exist between '2 to 9 times' and 'financial and insurance'; between '10 to 49 times' and 'construction', 'warehouse and transportation' and 'electric and gas'; and between 'more than 50 times' and 'public services'. According to the analysis, 'warehouse and transportation' and 'public services' were new additions to the pattern of recurrence in comparison with previous analyses. The tendency towards incident recurrence by the 'information and communications' sector appeared to decline. 'Warehouse and transportation' may have developed a pattern of numerous incidents because three postal services (mail, postal savings and postal insurance) were privatised in Japan in April 2007. The fact that postal savings operations transferred from the 'public services' to the 'warehouse and transportation' sector may have caused results that differed from those obtained in the previous studies. It might also be assumed that this increase was caused by the growing importance of transportation in the 'warehouse and transportation' sector, as well as by changes that occurred in purchasing, such as online shopping. 'Public services' appeared to demonstrate a pattern of numerous recurring incidents, although the amount of information leaked in each incident was small.

7. CONCLUSION

As explained above, the analysis revealed industry specific patterns of personal information leakage incidents. The 'financial and insurance' and 'warehouse and transportation' sectors demonstrated patterns of a large number of victims of leaked personal information per incident. Patterns of recurrence of incidents were demonstrated by the 'public services', 'construction', 'warehouse and transportation', 'electric and gas' and 'financial and insurance' sectors. Thus, business operators that demonstrated patterns that included a large number of victims and recurrence of incidents were the 'financial and insurance' and 'warehouse and transportation' sectors. With respect to these two sectors, the 'financial and insurance' sector should pay particular attention to the management of personal information because it primarily handles 'credit information'. By comparison, the 'warehouse and transportation' sector handles 'basic information'.

The 'public services' sector should act in accordance with the Law for the Protection of Personal Data Held by Administrative Organs, the Law for the Protection of Personal Information Retained by Independent Administrative Institutions and/or with local government ordinances, depending on the organisation. Companies must be answerable to the Act on the Protection of Personal Information, within which enactments of individual methods for each field should focus on medical, financial and credit, and information and communications technologies, because personal information leakage incidents that occur in these industries might result in serious invasions of privacy.[5] The results of the analysis revealed differences that exist among industries. This suggests the need for a significant focus on careful treatment of personal information in the 'financial and insurance' sector. In addition, the fact that 80.6 per cent of incidents caused by 'medical institutions' consisted of leakages of 'sensitive information' indicates that incidents that occur in this sector must be addressed in a careful manner. Although medical institutions, such as hospitals, may represent all national, public and private cases, only private medical institutions are answerable to the Act for the Protection of Personal Information. Similar to the case of 'public services', public medical institutions must act in accordance with the Law for the Protection of Personal Data Held by Administrative Organs, the Law for the Protection of Personal Information Retained by Independent Administrative Institutions and/or with local government ordinances. If a patient transfers from a private hospital to a public hospital, personal information treatment often causes problems because of the law. Thus, the enactment of individual methods based on industry sectors, rather than methods appropriate for the overall establishment, would be desirable to avoid excessive complexity and to increase careful management of 'sensitive information'. The previous and current analyses failed to extract particular characteristics of the 'information and communications' sector. This may have occurred because informatisation in all industries other than 'information and communications' has currently advanced.

A detailed examination of personal information leakage incidents that occurred over the 12-year study period revealed that the majority of incidents occurred because of human error. In fact, the percentage that occurred because of willful wrongdoing was small. The results of this analysis revealed that intentionally caused incidents tend to include the leakage of personal information of a large number of victims. This implies that only one incident poses a risk of serious damage for a company. Moreover, information security threats continue to evolve rapidly. Currently, these attacks have changed from theft of information to large targeted attacks that involve international hacker groups.[6] For example, the large amount of personal information leakage caused by the Sony Corporation occurred because a cyber attack was made on the PlayStation Network Service in April 2011. This attack resulted in significant losses for the Sony Corporation.

New risks, such as large targeted attacks, have increased rapidly and are fuelling society's fears. Although business operators suffer cyber attacks, they do not become victims themselves; rather, they become perpetrators that leak personal information. Further actions are required to fight against evolving information security threats. Nine years have passed since the Act for the Protection of Personal Information was enacted. Actual conditions related to personal information leakage incidents that have occurred since its passage have been described previously. Rapid developments in information and communications technology (ICT) and changes in information security threats have occurred. The EU presented the General Data Protection Regulation proposal to tighten data protection systems. In addition, the United States, which maintains conceptions of privacy policies that contrast with the EU's policies, announced the passage of the Consumer Privacy Bill of Rights in February 2012. The EU and the United States attempted to coordinate their views on privacy management and issued a joint statement in March 2012. Japan, however, does not currently possess the same authority as the EU or the United States and has not yet been accredited at a sufficient protection level that would permit Japan to engage in the transfer of personal data with EU countries.

The principles espoused by the EU and the United States have become global standards; therefore, it may be time to establish specific principles to be imposed on business operators and independent monitoring agencies, because the absence of these entities explains why Japan has not yet been accredited at a sufficient protection level that would permit the transfer of personal data with EU countries. In addition, Japan must develop a personal information protection law to address problems resulting from the number of personal information leakage incidents that have occurred and because of society's strong reactions to the publication of these incidents. As explained previously, it would be desirable to change the law to include an individual method based on each industry's needs, rather than on the overall establishment's needs. This would enable the development of detailed individual treatments for each sector that would allow appropriate and careful management of information (e.g. 'medical institutions' and 'finance and insurance' sectors). Otherwise, it would be desirable to change to methods used by EU member nations, in which one law covers both public and private sectors.

The law should be developed to focus on the protection of personal information within the industry sectors, and pay particular attention to the management of personal information. If measures are taken to strengthen corporations against cyber attacks in cooperation with public and private sectors, then based on an increased sense of security, people in Japan might not become overly cautious about providing personal information.

New information security threats continue to occur annually, resulting in changes in the characteristics of personal information leakage incidents between 2005 and 2011. The author hopes to accumulate additional empirical data related to personal information leakage incidents and information security without delaying the speed of ICT.

NOTES

1. Ministry of Internal Affairs and Communications, Japan (2012).
2. KDDI Research Institute, Inc. (2012).
3. Japan Electronics and Information Technology Industries Association, pp. 22, 87 (2012).
4. Sakurai and Mitomo (2006, 2007).
5. Japan Federation of Bar Associations (2001).
6. Information-Technology Promotion Agency, Japan (2012).

REFERENCES

Information-Technology Promotion Agency, Japan (2012), 'The major 10 information security threats in 2012' (in Japanese), available at http://www.ipa.go.jp/security/vuln/documents/10threats2012.pdf.

Institute for Internet Privacy (n.d.) (in Japanese), available at http://jtrustc.co.jp/pnews/readnews.php?logno=201112.

Japan Electronics and Information Technology Industries Association (2012), 'Analysis Report on the EU General Data Protection Regulation Proposal', pp. 22, 87 (in Japanese), available at http://www.i-ise.com/jp/report/EUdata_protection.pdf.

Japan Federation of Bar Associations (2001), 'Written Opinion Concerning the "Act on the Protection of Personal Information"' (in Japanese), available at http://www.nichibenren.or.jp/ja/opinion/report/2001_15.html.

KDDI Research Institute, Inc., Japan (2012), 'Big data', *Nextcom*, **12**.

Ministry of Economy, Trade, and Industry (n.d.), 'Personal information protection management systems requirements' (in Japanese), available at http://privacymark.jp/reference/pdf/guideline_V2.0_120907.pdf.

Ministry of Internal Affairs and Communications, Japan (2012), 'White Paper 2012: Information and Communications Japan', p. 311 (in Japanese), available at http://www.soumu.go.jp/johotsusintokei/whitepaper/ja/h24/pdf/n4030000.pdf.

Sakurai, N. (2013), 'A study on industry-specific patterns of incidents of personal information leakage based on the incident database', *Journal of Asia-Pacific Studies*, **19**, 185–99.

Sakurai, N. and H. Mitomo (2007), 'Incidents of personal information leakage and their industry-specific patterns', *Japan Society of Public Utility Economics*, **58** (4), 53–62.
Sakurai, N. and H. Mitomo (2006), 'An empirical analysis of incidents of personal information leakage', Graduate School of Global Information and Telecommunication Studies, Waseda University, Japan.
Securities Identification Code Committee (n.d.) (in Japanese), available at http://www.tse.or.jp/sicc/category/ct_chart.html.

PART IV

Regulation required for the smart revolution

10. Smart devices, fixed/mobile convergence and the cloud: some medium-term regulatory challenges

Martin Cave and Keiko Hatta

1. INTRODUCTION

A single journey to work on public transport will acquaint the traveller with the sheer number and variety of smart devices use by households: smartphones being employed for the primitive tasks of voice calls and email; more modern uses such as social media and downloading web versions of newspapers and magazines or facsimiles of the paper versions; viewing on tablets of television programmes that may have been downloaded by satellite or cable the previous night; and so on.

In addition, largely unseen, millions of devices are engaged in machine to machine communication. In all, whereas today a family with two teenagers may have ten devices connected to the Internet, in 2022 this may well grow to 50 or more devices. Across the OECD area, for example, the number of connected devices in households may rise from an estimated 1.7 billion today to 14 billion by 2022 (OECD, 2013).

The account so far only really covers existing devices, some of them used ubiquitously, some of them in the hands of early adopters. More radical definitions of 'smartness', such as those discussed by the OECD, define it in terms of 'an application or service that is able to learn from previous situations and to communicate the results of these situations to other devices and users'. They may cause one to reflect anxiously on the famous remark of John Seely Brown and Paul Duiguid that while everyone thinks computers are predictable and people are unpredictable, it is actually the other way round (Seely Brown and Duiguid, 2000).

In this chapter we will keep our feet firmly on the ground and focus on smart devices as we know them, at least in outline today. The focus will be on identifying the problems that they are already creating for regulators. The underlying thread is that portable smart devices are a contributory or accelerating factor in two trends that have the capacity to vex regulators:

the tendency for fixed and mobile markets to converge and the tendency for data increasingly to be stored away from the device in the cloud.

Accordingly, Section 2 sets out some facts and projections about the spread of smart devices; Section 3 discusses how regulators will seek to cope with converging markets; and Section 4 examines regulatory issues involving the cloud.

2. THE SPREAD OF SMART DEVICES

The spread of smart devices has become synonymous with ubiquitous Internet connectivity. Undoubtedly, widespread wireless Internet connection laid the foundation for smart device expansion. The number of mobile phone subscriptions worldwide has more than doubled since 2005. Then wireless Internet connections overtook fixed broadband subscriptions in 2009 and in 2011 the number of wireless Internet connections became more than double the number of fixed broadband connections (OECD, 2012).

Smart devices have revolutionised the trend of mobile data traffic. As of 2012, laptops, among all mobile devices, generated the largest amount of traffic, but smartphones and tablets have been rapidly increasing their shares (Figure 10.1).[1] Cisco estimates that even though traditional (non-smart) mobile phones will represent around half of the mobile devices in 2017 (Figure 10.2), nearly 99 per cent of global mobile traffic will be generated by smart devices. By then, the Pacific and North America regions will account for almost two-thirds of global mobile traffic (Cisco, 2013).

Smartphones can be thought of as portable computers, capable of providing voice and messaging services. Tablets offer the same features but in larger screen sizes, with a little less portability. Each of these devices is mainly used for data applications, and individual consumers choose different screen sizes to fit their needs. Smartphones and tablets make it easy for consumers to access the Internet regardless of where they are. The use of cloud services from smart devices has made it possible for a whole host of new services and applications to develop, which can be tailored to the users' requirements. This is why the industry is finding a huge growth potential in the emerging markets. Among the applications, mobile video, the most data consuming application, is expected to account for two-thirds of the global mobile data traffic by 2017 (Cisco, 2013).

Smart devices are changing the balance of consumption as well as creating new categories of service that were not available before. There is clear demand substitution from traditional PCs, therefore fixed network

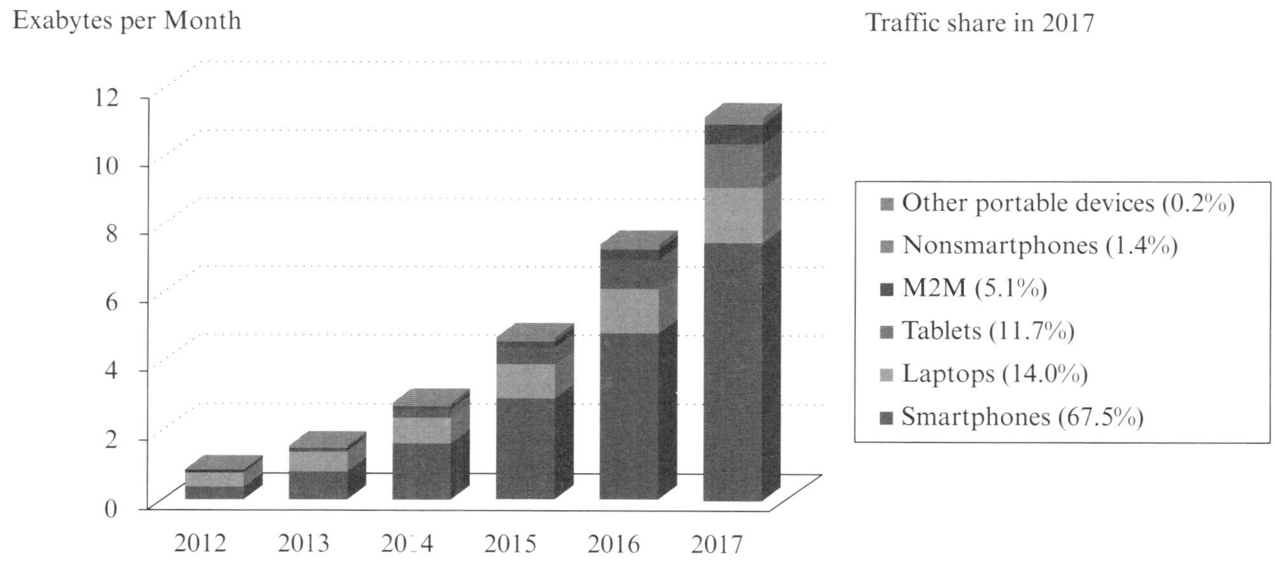

Figure 10.1 Global mobile traffic forecast 2012–17

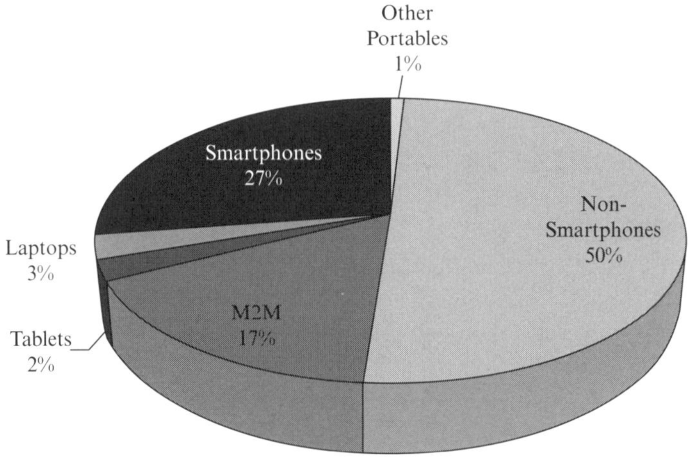

Figure 10.2 Global mobile devices and connections 2017

connections, to smart devices. The traditional PC market for desktop computers and notebooks, already showing a decline in 2012, is giving way to smartphones and tablets.

On the other hand, the smart device market is also attracting new types of consumer, e.g. very young and/or old age groups. This could further expand the mobile communications market without causing outright fixed-to-mobile substitution. In fact, in emerging markets 'consumer spending typically starts with mobile phones and, in many cases, moves to tablets before PCs' (IDC, 2013).

However, the shift of consumer demand for devices may not directly translate into a parallel decrease in demand for use of a fixed network. Indeed, the mobile and fixed networks may also remain complementary to each other for some time. It is expected that a significant proportion of traffic will still have to be offloaded from smartphones and tablets in 2017. This is because earlier generation devices were limited to Wi-Fi connectivity only and migration to new generation network does not happen overnight. The European Commission (2012a, p. 5) notes: 'More than half of all smartphone traffic appears to be routed over Wi-Fi networks and this nomadic traffic is growing 4–6 times faster than mobile traffic.'

3. THE REGULATORY CONSEQUENCES OF FIXED/MOBILE MARKET CONVERGENCE

3.1 Introduction

The premise of this section of the chapter is that the proliferation of fast devices combined with the availability of high speed mobile (and nomadic) broadband will change the relationship between fixed and mobile access and usage, and call into question the general assumption made by regulators to date that fixed and mobile retail and wholesale services do not fall in the same markets: or, if they do, it is a bundled market of complementary services.

We first examine the consequences of abandoning that assumption of separated markets. We do so without making any assumption about when or exactly how it will happen. One major change contributing to this event for the purposes of this chapter is the short- to medium-term one of further proliferation of smart devices.

As things now stand, the conventional wisdom is to regard fixed and mobile access as being to a large degree complementary, but fixed and mobile calls and (increasingly) broadband downloads are becoming substitutes (though not yet to the extent necessary for them to be found in the same market by the formal tests applied).

There are important nuances to this view. Clearly a significant number of people, bolstered by the growing sophistication of smart devices, are abandoning their fixed access service in favour of a wireless one. For them both access and usage are substitutes. On the other hand, there are some broadband applications for which current wireless speeds are inadequate and some calls with respect to which the vagaries of tariffing in a given jurisdiction make it clearly preferable to make using either a fixed or a mobile line.

These things are likely to change. In particular wireless speeds will rise to allow a wider range of video services to be received on smart devices on the move. However, taste differences across consumers may become more pronounced, with some seeing fixed and mobile access services as complements, some seeing fixed and mobile access and usage services as full substitutes, and others relying entirely on one technology or the other.

This section discusses the implications of these changes, after first outlining the now standard approach to deciding where and how to intervene in markets.

3.2 The Basis for Regulation

The philosophy underlying regulatory intervention in most jurisdictions today is remarkably similar: regulatory intervention is confined to markets that exhibit significant market power; these are usually wholesale markets, as unbundling network assets such as the local loop, making them available to competitors, is normally considered to make retail price regulation unnecessary.

However, this approach rests on a particular articulation between the analysis of retail and wholesale markets. The starting point is the protection of end users from abuse of market power. The more services are substitutes in any retail market, the wider the substitution between services in the corresponding wholesale market. Thus in very stark terms, if fixed and mobile access services are found to fall in the same retail market, the fixed and mobile wholesale access products that support those retail products are likely to lie in the same wholesale market. The convergence of markets tends to change market power at retail and wholesale levels in the same direction, if not to the same degree.

This creates the possibility that the whole underpinning of current regulation (wholesale regulation designed to protect retail competition) might collapse in response to a widening of the retail market. If previously there were four end-to-end mobile networks offering retail services in a country, the likelihood is that they were unregulated – except for termination. But the solitary fixed network would be subject to extensive regulation. When the retail markets come together, there are at least five retail and five wholesale networks, apparently reducing at a stroke the need for regulation of any of them – again with the exception of termination.

This outcome may seem unpalatable to regulatory agencies, which may have painstakingly built up a regime of fixed retail competition based on unbundling. In these circumstances, it is natural to ask if fixed regulation will in fact fall off a cliff in the way suggested.

For concreteness, we will frame the discussion that follows in terms of the European regulatory regime; but, as noted above, similar regimes operate throughout the world. The approach to regulation is set out in a 2006 document, prepared as an input to a subsequent Recommendation of the European Commission (European Commission, 2007).

> Thus, for the purposes of identifying any market the characteristics of which may be such as to justify ex ante regulation, this report has started from the proposition that it should be demonstrable [. . .] that the related end-to-end retail market would, absent regulation, be characterised by a substantial degree of market power, causing harm to consumers. This involves looking at competition on the retail market in the absence of upstream or wholesale regulation. The purpose of this is to ensure that regulation is only applied in those circumstances where there is a significant benefit to final consumers that cannot be achieved under competition law.
>
> If a competition problem is found, then a set of wholesale markets within the value chain is identified and each is examined to see if it is likely to exhibit

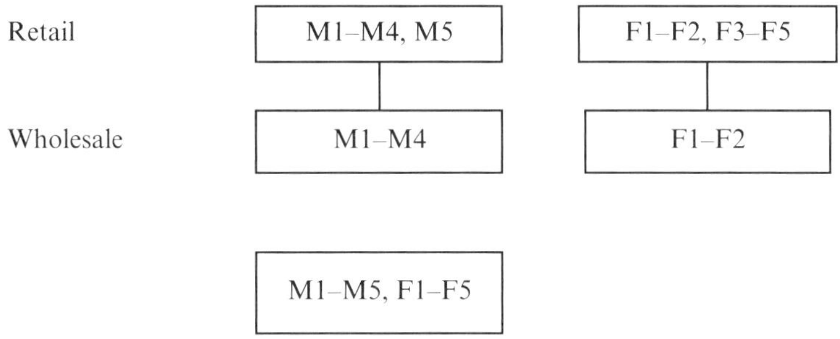

Figure 10.3 *Separate market for fixed and mobile*

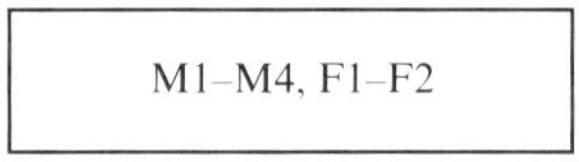

Figure 10.4 *A fully converged fixed/mobile market*

> [significant market power]. If it does, the analysis of the end-to-end market is repeated, now subject to appropriate regulation where SMP has been found. If regulation solves the problem in the end-to-end market, the process ceases. If not, the process continues based on the analysis of other inputs.
>
> If all wholesale inputs are regulated or found not to have SMP and dominance or SMP persists at the end-to-end level, then intervention in the retail market under Article 17 of the Universal Service Directive is necessary (Cave et al., 2006, p. 4).

The discussion that follows simplifies matters by considering a single wholesale product, which can be illustrated by wholesale broadband access.

The starting point pre-convergence is illustrated in Figure 10.3, where four mobile end-to-end companies (M1–M4) operate in both wholesale access and retail markets, together with an MVNO (M5) in the retail market. There is no significant market power (SMP). In the fixed sector, a telecommunications network and a cable company (F1–F2) provide network services and three other fixed suppliers (F3–F5) provide retail services.

Now suppose (Figure 10.4) that fixed and mobile retail services are found to be in the same market. We now have ten retailers (M1–M5 and F1–F5) and six suppliers (including self-suppliers) of wholesale services (M1–M4 and F1–F2). The chance that F1 or F2 will be found to have SMP in the wider wholesale market is much smaller. Under the European regulatory regime, if there is no SMP, there can be no remedy. Is this conclusion – no SMP and no remedy – inevitable?

3.3 Convergence Austrian Style

There is a worked example of such convergence, which we now describe. In 2009, the Austrian regulator RTR made a notification to the European Commission that it had defined a retail market for *residential* broadband customers that included both fixed and mobile services. As was noted at the time, Austria was an outlier (or trailblazer) of converging fixed and mobile residential markets, shown by the following facts:

- mobile operators had 35 per cent of the market, as opposed to 40 per cent for fixed DSL and 22 per cent for cable;
- the four mobile operators covered between 70 per cent and 94 per cent of the population;
- there was evidence of fixed and mobile broadband prices converging; and
- 75 per cent of mobile subscribers used their connection mainly on a standalone basis, rather than coupled with a fixed connection, and most mobile broadband connections were at a fixed location.

The basis for drawing a distinction between residential and business broadband was the presence of significant variations in product characteristics, prices and service levels between the two, including between residential and business DSL services.

After some delay, the European Commission was prepared to accept that Austria had demonstrated convergence of the residential market, by showing similar functionalities between fixed and mobile services, similar coverage and similar patterns of use and prices (European Commission, 2009).

Simultaneously, with respect to business services, the regulator found that the retail market contained only DSL-based services, with mobile, cable and fibre excluded. This separation of fixed and mobile retail markets was (and remains) the norm.

The situation with respect to the wholesale market was more complicated. The regulator identified a relevant market in bitstream connections for the use of business customers. This required argumentation since such services are identical with bitstream connections for the use of residential customers, which by implication from the wider residential retail market definition are in the same market as self-supplied mobile and cable wholesale products. RTR argued that there was in practice no supply-side substitutability between bitstream connections destined for the two separate retail markets.

The outcome is illustrated in Figure 10.5, which shows a distinction

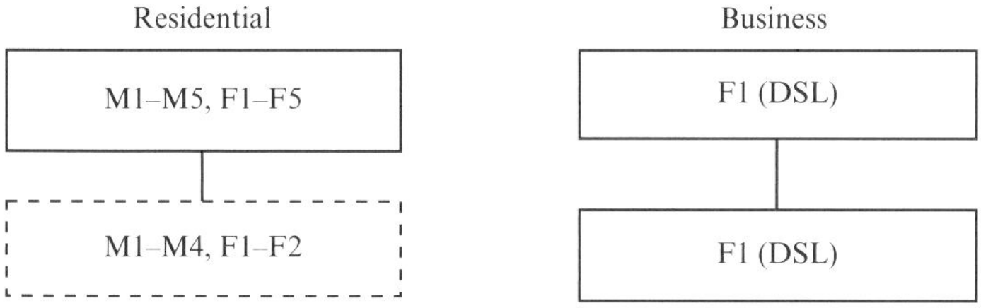

Figure 10.5 The Austrian special case

between business and residential retail markets. Residential retail markets are supported by a market (not explicitly defined, hence in a box with dotted lines) embracing self-supplied mobile and cable wholesale inputs and DSL wholesale products for residential use. Indirect constraints from the retail market are responsible for depriving the DSL provider from having SMP.

The retail market for business broadband, however, is supported by a wholesale market for DSL-based services that exhibits SMP and is thus subject to regulation. The consequences of fixed mobile market convergence are thus contained, as a result of the disconnection between wholesale product markets. If those markets were united, it is more likely that SMP would disappear.

On this basis it looks as if it was only a rather special set of circumstances that prevented fixed regulation in Austria from being threatened by the convergence of fixed and mobile retail residential markets.

3.4 Differentiation of Consumers: Complementarity and Substitution

It was noted above in the Austrian example that differentiation of retail markets is, under certain conditions, one means of limiting the effects of fixed/mobile convergence. The finding of mobile/fixed convergence is found to apply to residential but not to business demand. However, we note above that retail demand might split in more complex ways – in particular between customers buying the services, in a complementary way (often as part of a bundle), and those choosing between only fixed or only mobile. The second group will confusingly consist partly of people for whom fixed and mobile access and usage are genuine substitutes and who have chosen the better option for them and partly of people who require one service only.

It is recognised that if customers purchase a bundle of services, on a scale that makes it appropriate to define the retail services purchased as a bundle rather than a set of individual services, it is necessary to investigate the wholesale market from that perspective too. This may change the scope of firms with SMP.

How widespread is retail bundling today and how widespread will it have to be for the focus of regulation to fall on bundles? The general understanding is that it is appropriate to define a separate market when it satisfies the hypothetical monopolist test: a 5 per cent to 10 per cent increase in the bundle's price above the competitive price is profit-enhancing.[2] Or, to put it another way, this increase in price would not cause customers of such a bundle to 'unpick' it and switch to retail products bought separately in large enough numbers to cause profit to fall.

A separate retail bundled market might be created if there were an economy of scale or scope in selling a bundle or if customers valued savings in transaction costs associated with buying a bundle highly enough. Similar questions arise at the wholesale level. But the above-noted factors leading to identification of a retail bundle may not operate at wholesale.

Thus the most likely outcome is the coexistence of two types of retail customer. Some consumers purchase fixed and mobile services as a complementary bundle; some are disposed to consider only fixed services; some are disposed to consider only mobile services; a fourth group treats fixed and mobile services as full substitutes.

Only the last group of customers generate a wholesale market that contains both the fixed and the mobile components illustrated in Figure 10.4. The preference at retail of the other three groups generates instead the configuration of wholesale markets shown in Figure 10.3. Unless the fourth group swallows up everyone, by this argument the need for wholesale fixed regulation does not go away.

3.5 Anticipations and Lags in the Regulatory Response to Fixed Mobile Market Convergence

Ex ante wholesale or retail price controls are what they say they are: they are imposed in advance on the basis of expectations. One facet of this that is the subject of current controversy is that the costs underlying a price control should be based on the modern equivalent asset or MEA – which is defined as the asset embodying the technology that would now be chosen. This issue arises in relation to pricing copper loops, which, in the world of fibre, have been supplanted as the new network of choice. But this leaves room for selection among different fibre technologies.

In the presence of full fixed mobile substitution, a regulator would be able in principle to denominate one single wireline or wireless network as the MEA and might seek to base the access price of both technologies on that asset base. This might have the paradoxical effect of driving one or other type of network out of business, if its costs exceeded those of the other. For this to apply a very high degree of substitutability would have to exist, which is unlikely to arise as differences in technological features – for example, the higher speeds of wireline and the mobility associated with wireless – must limit substitutability. For this reason use of the MEA argument is likely to be confined to cases where one technology clearly dominates, for example as wireless may do in very sparsely populated areas.

The opposite approach is to delay the recognition of the market convergence. A well-established method of gradual withdrawal of price controls is to set what is sometimes known as a safeguard cap. This is essentially an insurance policy against the regulator having misjudged the degree of market power held by a firm being (partially) released from price control. An obligation to supply continues, but the price is set not on the basis of an estimate of cost, or another standard formula, but to allow some latitude to the access provider without placing the access seeker wholly at its mercy. This formula can be applied for a trial price control period and controls can then be removed completely or reinstated. Removing controls completely means removing the provider's obligation to supply. But if prices in the intervening period have remained at a cost-reflective level, rather than being taken right up to the (elevated) safeguard level, competitors' anxieties may be assuaged. Because this method involves a prior test of its effects, it is probably better than simply giving notice of the withdrawal of the cap.

A further alleviating factor is to grandfather (continue for a period) the access rights to fixed networks of unbundling operators in respect of customers to whom they are already supplying retail services. This would require fuller specification: would the right apply to the aggregate of customers at the time when deregulation occurred or only in respect of named customers continuing to rely upon the fixed access product? Would it apply to copper customers switching to fibre? And so on. Thus implementing this approach presents some difficulties.

3.6 Summary

We have seen that the implications for regulation of fixed mobile convergence are complex. On one hand, the move makes it harder to justify fixed regulation. On the other hand, the coexistence of growing substitutability (shown by the growing number of mobile only customers) and growing complementarity between fixed and mobile (shown by growth in fixed/mobile bundles) is likely to delay the abandonment of fixed regulation.

4. THE IMPACT OF CLOUD COMPUTING[3]

4.1 The Importance of the Cloud

Renda (2012) describes three generations of consumer experience of personal computing: the 'offline' decade starting in the mid 1980s; the advent of the World Wide Web from the mid 1990s; and finally, from about 2005, a transition towards high-speed Internet access by users who are almost always online, using multiple smart devices, exploiting various access forms.

As this transition occurs, it is accompanied and facilitated by replacement of hardware and software purchase and downloading of software and applications by customers 'buying everything as a service'. As Renda says (p. 24):

> This is essentially the cloud era: an era in which personal devices become increasingly light and users are able to lease software located in the cloud, as well as access their files that are stored somewhere in cyberspace and managed by a cloud provider – put more simply, a limitless 'office LAN' where the main server is not located downstairs but potentially on the other side of the globe.

For example, Apple's iCloud allows the 'syncing' of various devices with the cloud, such that the user always enters the same environment regardless of the device used to connect to the network.

4.2 The Cloud Structure[4]

While the above description resembles in structure those of other platform industries, an important feature has to be added. For all these services to be valuable, users have to be connected to the cloud. Thus, the cloud environment can be summarised by five layers of player: users, access providers, cloud infrastructure providers, cloud operating systems providers and cloud application providers.

As Figure 10.6 illustrates, users connect to cloud infrastructure and possibly software and services in the cloud through an access network. Various partially integrated services are possible. For instance, access network operators may offer cloud computing infrastructure. Also, providers of cloud operating systems may offer integrated software and offer their own cloud infrastructure. The role of policy is to provide an environment in which cloud services can flourish and in which various types of industry player who may use partially vertically integrated business models can operate in a level playing field. In 2012, Vice President Neelie Kroes (European Commission, 2012c) said: 'Cloud computing is a

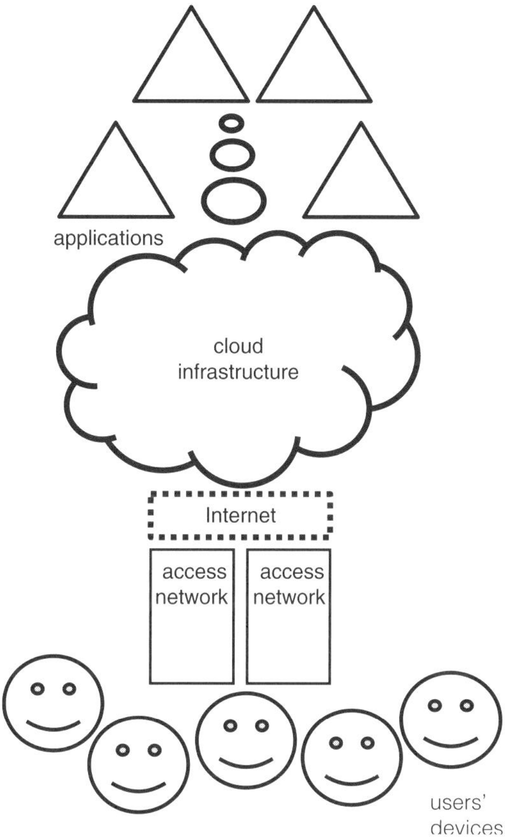

Figure 10.6 The cloud computing environment

game-changer for our economy. Without EU action, we will stay stuck in national fortresses and miss out on billions in economic gains. We must achieve critical mass and a single set of rules across Europe.'

4.3 Competition in the Cloud

A different concern is that global players may be able to dominate the most profitable segments of the cloud computing business, as other players may lack the possibility to offer vertically integrated solutions of cloud computing infrastructure and applications. By adding services to their cloud computing activities firms such as Apple, Google and Microsoft may be able to lock users into their platforms.

There is a dilemma here. Interoperability may be desirable features of a mature market, but imposing them at an early stage may chill incentives for innovative solutions. There are also varying 'depths' of interoperability, ranging from the adoption of harmonised cloud standards and architectures, to the incorporation of common interfaces.

A related issue that will shape the market and its success is the portability of data between cloud providers. Non-portability leads to large switching costs, which may make users hesitant to become locked in. However, it is *a priori* unclear whether this risk will actually materialise, since the interaction between competition and portability is a complex one. This suggests that monitoring the market may be appropriate, while enforcing portability or particular standards to ensure interoperability risks making misinformed decisions that may become the source of major inefficiencies in the market.

There is also the 'compared with what?' question. In the two earlier worlds noted above, lock-in based on past purchasing of on-premises equipment and applications was much more intense. The other players in the value chain (telecommunications and cable companies, media firms offering content) are also powerful. Perhaps more likely is a world of competing alliances, varying degrees of vertical integration and an innovative sector of application providers featuring low entry costs.

Concern has been expressed, however, about one facet of integration: that between providers of the mobile operating systems in smart devices and cloud computing. Companies such as Apple, Google and Microsoft – providers of mobile operating systems – are largely exempt from 'net neutrality' type restrictions imposed on ISPs: they can give priority (or exclusivity) to their own websites and applications (O'Brien, 2013). Given that the mobile world exhibits less dominance and more competition than the fixed, this is not unreasonable. But the lack of interoperability creates significant switching costs; as you change your supplier, the apps you have purchased become valueless. But if this stickiness in the market is overlaid with switching costs at the cloud level, the risk of customer detriment grows. At first sight, this seems an area where competition law rather than regulation is more appropriate.

It can also be argued that net neutrality has the desirable property that it enables cloud computing providers to make uniform service proposals. This applies to any rule that is harmonised across networks, but it might not be the case for differentiated quality of service (QoS). To the extent that enforcing a version of net neutrality does not allow for highly reliable services, certain business models are prone to fail. Thus, cloud-computing providers may come to rely on ISPs for service guarantees that cannot (profitably) be provided for all services but only for those that are particularly time-sensitive and require high priority. This holds, in particular, in the light of the foreseeable growth of the mobile use of cloud services and other Internet usage that is likely to lead to occasional capacity problems. In this context, it appears to be important to discuss the pros and cons of harmonised versus decentralised solutions of QoS.

5. SUMMARY AND CONCLUSIONS

This chapter assessed the prospects for the continuing diffusion of smart devices in the medium term and noted that most commentators are confident that they will continue to spread at a very significant rate. It then discussed two major short- to medium-term regulatory developments associated with the growing use of smart devices – their impact on fixed/mobile convergence and their impact on cloud computing.

In relation to fixed/mobile convergence, it is noted that full integration at the retail level of fixed and mobile markets would have a transformational effect on, in particular, the regulation of fixed networks, which would lose a substantial amount of the power that they would exercise, absent regulation, in retail markets; this is itself the justification of the present widespread intervention in wholesale markets, through such measures as unbundled local loops. If such a development did occur, there might be a basis for a warning period, or for interim measures to protect existing customers relying upon unbundlers, but the logic of such a development would seem to lead inexorably to deregulation.

However, it is not obvious that market developments are in the direction of this limitless increase in fixed/mobile substitution. Another key trend is towards complementary use of fixed and mobile infrastructure platforms via a number of devices, some of which operate both in mobile and in fixed or nomadic mode, often switching between them without the user's knowledge. Customers increasingly buy such service in bundles. The result is a more complex differentiation of consumption patterns, which generates a growing variety of wider or narrower retail markets that seems set to sustain significant market power for fixed network providers for some time to come.

The second topic is the effect of smart devices on cloud computing. Here the proliferation of increasingly light smart devices allows users to be 'always on', to lease software and apps, and to use a server in the cloud. The greatest regulatory challenge is to time interventions directed at promoting interoperability and permitting portability of data. Too soon and innovation will be stifled; too late and barriers to entry and switching will diminish competition. Because suppliers of cloud services are also providers of devices and their operating systems, vertical issues may also arise. This makes it necessary to monitor these developments carefully.

Taken together the two examples show how innovation via smart devices contributes to the need frequently to revisit the regulatory regime, which is simultaneously a means of promoting competition and (sometimes) stifling innovation. So far, little of the latter tendency has been observed in relation to smart devices. But regulators need to exercise constant diligence to prevent that happening.

NOTES

1. For a recent review of M2M connections, see GSMA Intelligence (2013).
2. Unfortunately, strategic pricing behaviour by a firm using bundling to sort customers and to price discriminate among them can cause observed prices to depart from competitive ones and lead to misleading results when standard market definition tests are applied. See Belleflamme and Peitz (2000, pp. 260–69).
3. The focus of this section is economic regulation of the cloud; other issues, including privacy, are not.
4. This and the following subsection are based in part on contributions by Martin Peitz to Cave and Peitz (2013).

REFERENCES

Belleflamme, P. and M. Peitz (2010), *Industrial Organization: Markets and Strategies*, Cambridge, UK: Cambridge University Press.

BEREC (2010), *BEREC Report on Impact of Bundled Offers in Retail and Wholesale Market Definition*, Riga, Latvia: BEREC.

Cave, M. and M. Peitz (2013), 'Network regulation and market regulation in European telecoms', available at http://www.cerre.eu/publications/network-regulation-and-market-development-european-telecoms (accessed 26 February 2015).

Cave, M., U. Stumpf and T. Valletti (2006), 'A review of certain markets included in the Commission's recommendation on relevant markets subject to ex ante regulation', available at http://broadband.cti.gr/download/review_regulation.pdf.

Cisco (2013), *Cisco Visual Networking Index: Global Mobile Data Traffic Forecast Update, 2012–2017*.

European Commission (2007), Explanatory Note Accompanying Commission Recommendation on Relevant Product and Service Markets, 2nd edn, SEC(2007)1483 Final.

European Commission (2009), Case AT/2009/0970: Wholesale Broadband Access (WBA) in Austria, 7 December.

European Commission (2012a), Communication: Promoting the Shared Use of Radio Spectrum Resources in the Internal Market, Brussels, 3.9.2012, COM(2012)478 Final.

European Commission (2012b), Communication from the Commission to the European Parliament, the Council, the European Economic and Social Committee and the Committee of the Regions, Unleashing the Potential of Cloud Computing in Europe (Text with EEA Relevance), SWD(2012)271 Final.

European Commission (2012c), Digital Agenda: New Strategy to Drive European Business and Government Productivity via Cloud Computing, Press Release, 27 September.

GSMA Intelligence (2013), 'Mobile operators' global M2M footprint: 391 mobile operators offer M2M services across 171 countries', available at https://gsmaintelligence.com/research/2013/06/mobile-operators-global-m2m-footprint/388/ (accessed 26 February 2015).

IDC (2013), IDC's Worldwide Smart Connected Device Tracker Forecast Data, 28 February.

O'Brien, K. (2013), '"Neutrality" is not so easy to define', *International Herald Tribune*, **17 June**, 14–15.

OECD (2012), 'Internet economy outlook 2012', available at http://www.oecdilibrary.org/science-and-technology/oecd-internet-economy-outlook-2012_9789264086463-en (accessed 26 February 2015).

OECD (2013), 'Building blocks for smart networks', available at http://dx.doi.org/10.1787/5k4dkhvnzv35-en (accessed 26 February 2015).

Seely Brown, J. and P. Duiguid (2000), *The Social Life of Information*, Cambridge, MA: Harvard University Press.

11. Problems with international mobile roaming: excessive deregulation works against users' interests

Hidenori Fuke

1. BACKGROUND

International mobile roaming services (IMRS) offer services enabling users travelling abroad to make and receive calls and to send and receive emails to and from their home countries by handsets bought and contracted in their home countries. With the diffusion of IMRS, the number of consumers who complain of unexpectedly high IMRS charges has been increasing in Japan. The National Consumer Affairs Centre of Japan (NCAC) has raised concerns about these problems (NCAC, 2006) and the Telecommunications Carriers Association (TCA) formed by Japanese telecommunications carriers has published solutions to reduce the problems (TCA, 2006, 2007). Japanese carriers advised customers on IMRS through their websites and leaflets and introduced a flat rate for data communications for overseas use.[1]

Despite these measures, new types of problem were reported with the diffusion of smartphones and tablet terminals with mobile functions. For example, the NCAC published this incident reported by a user in December 2011 (NCAC, 2011):

> I carried a smartphone for a few days of overseas travel. Although use during travel was very infrequent, I found the charge for it was about ¥13,000 ($123.97)[2] when I checked the bill after I returned to Japan. I believe that I didn't use my smartphone sufficiently to incur such a large bill. When I called my carrier's customer centre, I was told that if the automatic downloading of application software was switched on, downloading would start in both overseas countries and Japan without customer notification and the downloading charge would be added to their bills. I was not given that kind of explanation earlier and I would not agree to the payment.

This chapter analyses the IMRS system and proposes solutions to prevent these problems based on the finding that excessive deregulation works against users' interests.

2. BASIC PROBLEMS

The causes behind the aforementioned problems are not only the very high IMRS charges but also the complicated IMRS structure, which is not easily understandable for users. According to Organisation for Economic Co-operation and Development (OECD) data (Table 11.1), Japanese retail roaming charges are less expensive compared to the OECD average or the USA's. However, they are considerably higher than Korea's charges. Further, they are much higher than normal international call charges. According to NTT DOCOMO's World Wing,[3] its IMRS, the charge for a call to Japan by users visiting the USA is ¥140 ($1.34) per minute and the charge for receiving an incoming call in the USA is ¥175 ($1.67) per minute.[4] These charges are significantly more expensive than normal international charges. Users accustomed to flat rate domestic charges might unconsciously speak longer during calls. If a call lasts, for example, for 10 minutes, it will cost more than $15. Furthermore, Japan has adopted, as have many other countries, a caller-pay system so users are not familiar with the IMRS called party pay system. The Survey (see note 1) reported that 87.4 per cent of respondents stated that they understood the system. However, they tend to disregard the actual charges they incur for received calls and find unexpectedly big bills. In some countries, such as the USA, carriers charge for calling visiting users' numbers for more than 30 seconds even if called parties fail to answer.

In the case of data use, NTT DOCOMO's IMRS flat rate charge is ¥1980 ($18.88) per day up to 200,000 packets (24.4 MB) and users will not be charged more than ¥2980 ($28.42). If you travel abroad for about a week, the service should thus cost from ¥13,860 ($132.18) to ¥20,860 ($198.93). When I visited Thailand for a week, the bill was ¥13,860 ($132.18) for only

Table 11.1 Retail charges for 3 minute calls

	Country	$
Making a call from roaming countries to a home country	Japan	6.10
	Korea	5.46
	USA	5.16
	Average of 31 countries	7.79
Receiving a call from a home country to roaming countries	Japan	3.70
	Korea	1.36
	USA	5.16
	Average of 31 countries	4.49

Source: OECD (2009), pp. 56, 58.

Table 11.2 User evaluation of IMRS tariff

	n	High	Appropriate	Cheap	N/A
1. Voice call	2691	79.9%	11.6%	0.3%	8.3%
2. Data communication	1502	82.2%	11.4%	0.4%	6.1%
3. SMS	1180	68.9%	18.1%	1.4%	11.7%

Source: Author Survey.

checking email. If you do not subscribe to the flat rate, the ceiling is not applicable and you might incur very high costs. One loophole is that data volume count is based on Japan time, not local time. Another loophole is that smartphone OS or apps updates will start automatically unless users turn off automatic update settings and users will be charged separately from the flat rate for updating. According to the Survey, 50.2 per cent of respondents who used IMRS do not understand this charging system.

Thus, most Japanese IMRS users find that IMRS retail charges are high, as shown in Table 11.2.

The NCAC advises users that they should check smartphone settings and the charging system when they travel abroad (NCAC, 2011). However, problems regarding the use of smartphones will not decrease, despite this advice. One cause is that the established inter-carrier practices differ from an international settlement system of international calls or inter-ISPs interconnection of the Internet. The other reason is that individual users seldom use IMRS[5] and they normally do not investigate IMRS tariffs when they subscribe to mobile services. Our Survey found that only 29.1 per cent of users regard IMRS charges as important when they select a carrier.

Although the EU has regulated both retail and wholesale charges for IMRS since 2007, Japan's Ministry of Internal Affairs and Communications (MIC) has not introduced IMRS regulation and entrusted self-regulation to mobile carriers.

3. THE IMRS STRUCTURE

3.1 Voice Services

This section explains the basic IMRS structure of voice and data services. Figures 11.1 and 11.2[6] depict the voice IMRS structure. Each mobile carrier sets subscriber retail charges for IMRS. To offer IMRS, each

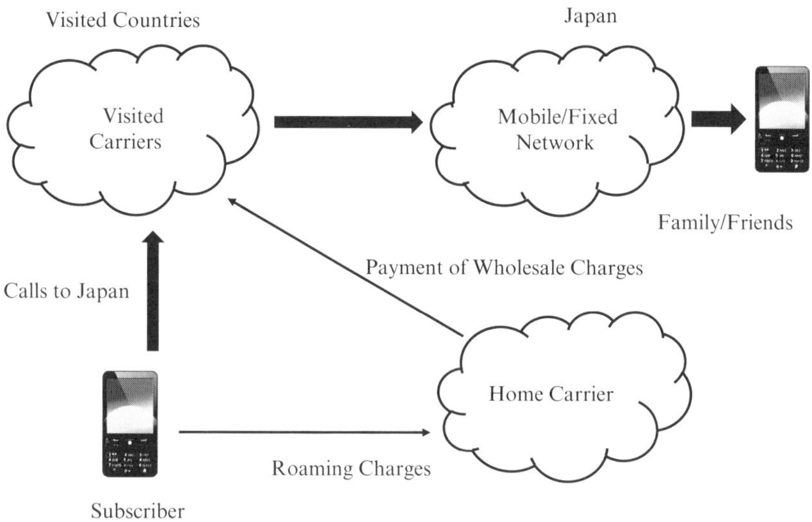

Figure 11.1 IMRS: calls from visited countries to a home country

carrier must conclude an agreement with overseas mobile carriers for access to their wholesale services.

Figure 11.1 depicts the case in which Japanese users call home cellular or landline phones from overseas. Calls are terminated at inland phones via overseas cellular networks. Users pay roaming charges (retail charges) to their home carriers and their home carriers pay wholesale charges based on inter-operator tariffs (IOTs: wholesale charges)[7] to overseas carriers. Wholesale charges are pre-negotiated between carriers and settled through clearing houses.

The system is similar to the settlement system of international calls by which inland callers pay international call charges to their home carriers and the home carriers pay settlement charges to carriers that terminate the call. The differences are that home carriers charge their subscribers for calls, including those made or received in overseas countries, and that carriers pay wholesale charges capped by IOTs to overseas carriers.

Figure 11.2 depicts the cases in which family or friends in the home country call users travelling overseas. Calls from family or friends once terminated at the home network of travelling users are then transferred to networks in visited countries. Family or friends in the home country pay normal inland charges and called users pay termination charges (retail charges) to their home carrier. The home carrier pays wholesale charges to visited carriers that terminate the calls. Both callers and called parties

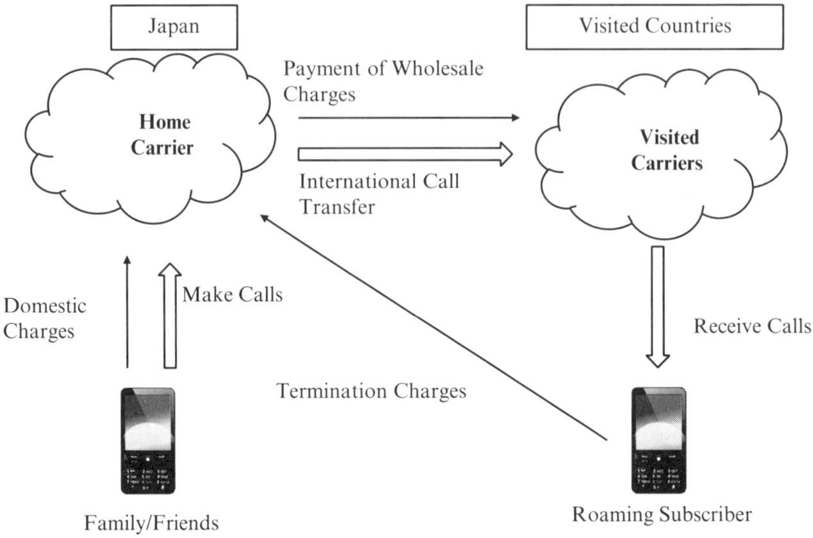

Figure 11.2 IMRS: receiving calls in visited countries

must pay charges, which is contrary to the normal principle in Japan, in that only the calling party pays. Furthermore, although international calls are normally charged per 30 seconds, IMRS calls are charged per minute. Even if a call last only 10 seconds, a charge of 1 minute is applied. Users and their family or friends who are accustomed to the caller-pay system often do not understand the international system and people tend to make calls unaware of the fees and thus incur unexpectedly high charges.

3.2 Data Services

Data services are offered via the rather complex system depicted in Figure 11.3. The data sent or received by users travelling abroad are first transferred from overseas carriers' IP networks to subscribed carriers' networks in the home country (route 'a' of Figure 11.3) and then sent to their upper ISPs to which the carriers connect and finally to the Internet. Wholesale charges are payable to overseas carriers from home carriers. If data are routed via route 'b' of Figure 11.3, routing can be the same as normal Internet service. One reason for IMRS to adopt a different system is that IMRS data transmission is offered on a volume basis and therefore requires measurement of data transmission volume.

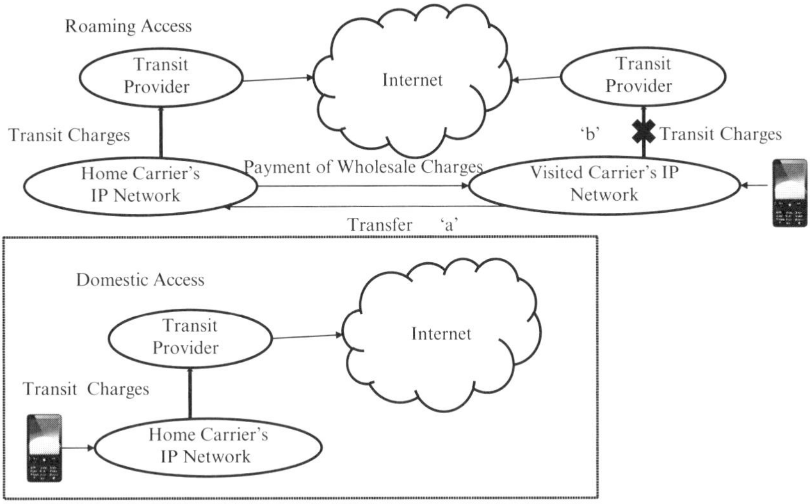

Figure 11.3 IMRS: data communication

3.3 Short Message Service (SMS)

Short message service (SMS) is a service whereby short messages can be sent to phone numbers. The charge from Japan is ¥100 ($1.02) per message and no extra charge is levied in visited countries. The charge for messages sent from visited countries to the home country is also ¥100 ($1.02). The home carrier settles accounts on the basis of wholesale charges.

4. CONTRIBUTING FACTORS FOR HIGH IMRS CHARGES

4.1 High IMRS Charges Compared to International Charges

IMRS charges are considerably higher than normal international charges as shown in Table 11.3, comparing NTT DOCOMO's international and IMRS charges for calls between neighbouring countries.

If family or friends of a user visiting the USA call the user for 1 minute, the IMRS charges should be totalled to between ¥204.4 ($1.94) and ¥217 ($2.07) per minute by adding from ¥29.4 ($0.28) to ¥42 ($0.40) of domestic charges to the IMRS charges. This method charges more than three times greater than a normal international call. Even if the call duration is less than 1 minute, IMRS calls are charged per minute. In this case,

Table 11.3 International and IMRS charges (NTT DOCOMO)

	Voice		International calls	Data communication		SMS	International SMS
	Send	* Receive		Fixed	Metered		
China	¥175/min	¥145/min	¥57/30sec (8–19 weekdays)				
Korea	¥125/min	¥70/min	¥49/30sec (9–8 weekdays/ weekends, holidays)	¥1980 (up to 200,000 packets)	¥0.2/packet	¥100/mail	¥50/mail
USA	¥140/min	¥175/min	¥34/30sec (8–19 weekdays) ¥31/30sec (9–8 weekdays/ weekends, holidays)	¥2980 (over 200,000 packets)			

Note: * Receiving charges payable by roaming subscribers. Callers pay normal inland call charges ¥14.7~¥21/30sec, as of 26 August 2013.

the IMRS charge is six times higher than an international call charge. The comparison for Korea or China also shows that IMRS charges are considerably higher than international charges.[8]

Data transmission is more controversial. If you subscribe to NTT DOCOMO's 'Pake-hodai', you have global access by paying ¥5400 ($51.50) per month when you remain in Japan. However, the 'Overseas Pake-hodai' applied to IMRS costs you ¥1980 ($18.88) or ¥2980 ($28.42) per day, depending on the upper ceiling of data transmitted. These charges total ¥56,700 ($540.72) or ¥86,700 ($826.82) per month, which is much higher than the upper limit of €50 (¥7143) set by the EC (EC, 2009, Article 6a). NTT DOCOMO advocates this flat rate service as 'simple, free from huge bills and convenient', regardless of charges being ten to 15 times higher than normal Internet service. I doubt the reasonableness of such high charges.

The factors behind higher IMRS charges are categorised into the IMRS service structure and the demand- and supply-side characteristics of IMRS.

4.2 Service Structure of IMRS Charges

First, although domestic cellular services are offered on a facilities basis except for small mobile virtual network operators (MVNOs), IMRS are primarily offered by interconnecting with overseas cellular networks.[9] Domestic cellular carriers offer IMRS by leasing overseas carriers' networks. They offer consumers retail IMRS on the wholesale IMRS offered by overseas carriers. IOTs negotiated between carriers form the basis of wholesale charges. As domestic carriers cannot offer IMRS without wholesale services and the authority of domestic regulators cannot extend to overseas carriers, these wholesale charges tend to remain at higher levels. Domestic carriers mark up their marketing expenses on wholesale charges, driving up the retail charges. The same structure applies to data transmission. As shown in Figure 11.3, data packets from visited countries are transmitted first to home carriers' networks and then via the Internet. This structure causes high-data transmission charges.

Second, IMRS is offered in a bundle usually including domestic mobile access, local calls, SMS and handset subsidies (OECD, 2009, p. 5). As the Japanese mobile market is vertically integrated, that is, mobile carriers sell handsets together with their networks and services, users accept bundled services as natural. Further, SIM cards lock is a common practice except for NTT DOCOMO[10] and users do not expect that they can select an IMRS provider other than their subscribed carrier.

4.3 Demand-side IMRS Characteristics

First, IMRS exhibits low price elasticity of demand.[11] IMRS is offered in a bundle with domestic services and users do not consider roaming charges important when they subscribe to a particular mobile carrier. Japanese users, in particular, ignore roaming charges because most do not often travel overseas (Table 11.4) and thus, only a small percentage of people actually use IMRS.[12] Few users care about IMRS charges at the time of subscribing to a particular carrier; they care more about available cellular handsets and domestic services and charges when choosing domestic carriers. Only when they plan to travel overseas do they care about IMRS. By that time, users are locked in by their subscribed carriers through tactics such as a discount for a long-term subscription and a handset subsidy and are thus obliged to accept their carrier's IMRS charges. During overseas travel, IMRS is a type of necessity good and thus, price elasticity of demand is very low.

Second, few alternative services to IMRS exist. Although users may use VoIP over Wi-Fi instead of IMRS, only business travellers carry their PCs and they must learn the necessary skills to set up their PCs for Wi-Fi services, whereas a smartphone has embedded Wi-Fi functionality. However, it is difficult to find available Wi-Fi access points and to set up for those networks. Our Survey found that 12.6 per cent of users chose Wi-Fi instead of IMRS. However, about 43.6 per cent of those that used Wi-Fi indicated difficulties, such as limited number of access points, finding access points and setting up for Wi-Fi. Further, the use of Wi-Fi networks may have limitations on the reception of incoming calls, as it is not an always-on service. Users cannot receive calls when they are beyond the reach of Wi-Fi access points.

Third, average users lack sufficient knowledge of IMRS charges. The explanation on carriers' websites is not easily understandable and thus, users tend to use mobile phones overseas in the same manner as in their home country.

4.4 Supply-side IMRS Characteristics

Mobile service providers encounter entry barriers. Since it is essential to obtain spectrum allocation to offer cellular services, and building cellular networks incurs high fixed costs, the cellular market is characterised by oligopoly. Japan's four cellular carriers dwindled to three in 2013 by Softbank's acquisition of E-MOBILE. The HHI index at the end of FY2012 was 3609, although that figure had decreased from 3902 in FY2006 (Table 11.5). The figure is nonetheless higher than 1800, which indicates a highly concentrated market per the USA's Merger Guideline.

Table 11.4 Numbers travelling overseas within the last three years

n	27,424
None	66.0%
Once	13.4%
Twice	7.1%
Three times	5.3%
Four times	2.0%
Five times	2.9%
Six times and over	3.2%

Source: Author Survey.

Table 11.5 HHI of the Japanese cellular market

	Fiscal year	2007	2012
Share (%)	NTT DOCOMO	52.00	46.70
	KDDI	29.50	28.60
	Softbank	18.10	24.70
	E-MOBILE	0.40	–
	HHI	3902	3609

Source: Calculated on the basis of Telecommunications Carriers Association (TCA) data.

Exacerbating this oligopolistic cellular market structure, the Japanese mobile market is vertically integrated as users accept a bundled offering of IMRS without the awareness of its fees or their options, enabling mobile carriers to maintain high IMRS charges without losing users. IOTs and wholesale charges are negotiated between oligopolistic carriers, necessitating little competition in the wholesale IMRS market.

The Japanese retail IMRS market is also not competitive because of the vertical integration of mobile carriers and the continuing practice of SIM cards lock. As IMRS is auxiliary to basic cellular services, cellular carriers compete in acquiring new subscribers but not in the IMRS market, rendering significant new entry by MVNOs virtually impossible.

As both wholesale and retail IMRS markets are not competitive, mobile carriers have little incentive to lower wholesale and retail charges. They can therefore set high retail IMRS charges by adding markup on high wholesale charges without losing their subscribers.

5. EU MOVES TOWARDS IMRS REGULATION

5.1 Regulatory Framework of 2003 and 2006

The OECD report stated that 'IMRS prices are unreasonably and ineffectively high' (OECD, 2009, p. 5) and that high wholesale rates charged by foreign operators are the major contributor to high retail charges (ibid., p. 5). However, many countries outside the EU have been reluctant to introduce IMRS regulation. Let us survey the outline of EU regulations and explore lessons for other countries including Japan.

The EU has promoted policies to integrate the EU economy, with one such policy being the creation of a single digital market as outlined in its Digital Agenda (EC, 2010a). One key performance target is to eliminate the difference between roaming and national tariffs to achieve a 'single market for telecom services' (EC, 2010b). The EC aggressively pursued IMRS regulation to achieve this objective.

The EU IMRS regulation traces back to the 2003 regulatory framework, which adopted the principle that ex ante regulatory obligations should be imposed only where effective competition is absent (EC, 2007b, Preamble, 5). Member states were to analyse whether a particular market was competitive per the EU Recommendation (EC, 2003). The Recommendation identified 'the wholesale national market for international roaming on public mobile networks' as a relevant market appropriate for ex ante regulation (EC, 2007b, Preamble, 6). National regulatory authorities (NRAs) demonstrated that 'it had not been possible for NRAs to effectively address the high level of wholesale community-wide roaming charges because of the difficulty in identifying undertakings with significant market power' because of the cross-border nature of IMRS (ibid., Preamble, 6). The retail IMRS market was not identified as a relevant market as the EC understood that IMRS did not constitute an independent market and that it is an element of a broader retail package (EC, 2006, p. 4). The second edition of the Recommendation (EC, 2007a) excluded the market from ex ante regulation as it introduced IMRS regulation as described in Section 5.2. One problem with IMRS is that NRAs cannot regulate operators in other member states.

5.2 IMRS Regulation 2007

The EC reached a conclusion that IMRS retail charges were too high and that the 2003 regulatory framework was insufficient in protecting consumers (EC, 2007b, Preamble, 4). The EC found that 'the excessive retail charges were resulting from high wholesale charges levied by the foreign host network operator' and 'also from high retail markups charged by the customers own network operator' (ibid., Preamble, 1). On the basis of this understanding, the EC introduced the following measures for IMRS calls:

1. Set price caps on average wholesale voice IMRS charges (ibid., Article 3).
2. Introduce Eurotariff (safeguard caps)[13] on retail voice IMRS charges (ibid., Article 4).
3. Require mobile operators to provide users with information on charges by a message service when users enter member states other than their home country (ibid., Article 6).

The regulation was to expire on 30 June 2010 and the EC would review the regulation by 30 December 2008.

5.3 Revision of IMRS Regulation 2007

The EU revised the 2007 regulation in June 2009 and extended the term of the regulation until 30 June 2012. Data and SMS were included in the regulation, summarised as follows:

1. Decrease price caps on average wholesale voice IMRS charges and change from a per-minute to a per-second basis while allowing a fixed charge up to 30 seconds (EC, 2009, Article 3 (amended)).
2. Decrease Eurotariff on retail IMRS charges and change to a per-second basis while allowing fixed charges up to 30 seconds for call origination (ibid., Article 4a (amended)).
3. Introduce price caps on wholesale SMS charges (ibid., Article 4a (amended)), introduce a Euro-SMS tariff for retail SMS charges and prohibit home carriers to levy any charge for the receipt of SMS by their users (ibid., Article 4b (inserted)).
4. Include maximum charges on retail voice and SMS IMRS charges in the basic information to be provided to users (ibid., Article 6 (amended)).
5. Ensure transparency and a safeguard mechanism for data IMRS services (ibid., Article 6 (replaced)).
6. Introduce a safeguard cap set per megabyte of data transmitted on average wholesale data IMRS charges (ibid., Article 6a paragraph 4 (inserted)).
7. Introduce measures to ensure the transparency of retail data IMRS charges (ibid., Article 6a paragraph 1–3 (inserted)):
   a. Inform customers about the risk of automatic and uncontrolled data connection and download.

b. Send an automatic message to a roaming user notifying that he is roaming and provide basic personalised information on charges.
c. Grant the selection of a facility that provides information on accumulated consumption expressed in volume or in currency.
d. Make available maximum financial limits for specified periods, one of which shall be €50 per monthly billing period.
e. Inform users when the amount of charges reaches 80 per cent of the decided financial or volume limit and immediately cease the service unless users wish to continue.

Here, it is important that the EU introduced the regulation on wholesale data IMRS charges and adopted measures to avoid 'bill-shock' for data IMRS charges.

Although these regulations contributed to the reduction in high IMRS charges (Table 11.6), the EU determined that these are insufficient and decided to introduce further regulations.

5.4 Structural Measures Added in 2012

With the understanding that IMRS charges have remained high and that these high charges result from high wholesale charges and high retail markups, the EU decided to add structural measures on 30 June 2012 (EC, 2012b). The EC also decided to lower further both retail and wholesale price caps. The measures included are as follows:

1. Require carriers to meet all reasonable requests for wholesale IMRS access to promote the entry of MVNOs (ibid., Article 3).
2. Require carriers to enable users to access IMRS by any alternative roaming provider (separate sale of retail IMRS) (ibid., Article 4).
3. Retain the safeguard price caps on retail charges until 30 June 2017 and on wholesale charges until 2022 (ibid., Articles 7 and 8).
4. Retain the safeguard caps on wholesale and retail charges until structural measures have become effective and competition has developed sufficiently (ibid., Article 19).

The EC will review the regulation's functioning by 30 June 2016. If the structural measures prove insufficient, the EC will make appropriate proposals.

Here structural measures, especially separate sale of retail IMRS, are very challenging as it is necessary to develop technical solutions to implement the regulation. The EC will implement the regulation beginning on 1 July 2014.

Table 11.6 EU regulation on IMRS

Unit: cent

Year of regulation			2007			2009*			2012		
	Regulation applied		30 Aug 2007	30 Aug 2008	30 Aug 2009	1 Jul 2009	1 Jul 2010	1 Jul 2011	1 Jul 2012	1 Jul 2013	1 Jul 2014
Voice	Retail charges	Origination (min)	49	46	43	–	39	35	29	24	19
		Termination (min)	24	22	19	–	15	11	8	7	5
	Wholesale charges		30	28	26		22	18	14	10	5
SMS	Retail charges	Mail	–	–	–	11	–	–	9	8	6
	Wholesale charges	Mail	–	–	–	4	–	–	3	2	2
Data	Retail charges	MB	–	–	–	–	–	–	70	45	20
communication	Wholesale charges	MB	–	–	–	100	80	50	25	15	5

Note: Voice calls should be charged per second while allowing fixed charges up to 30 seconds.

5.5 Limitations of the EU's Approach and Implications for Other Countries

The IMRS regulations that the EU introduced are based on its power to enforce its rules among its member states. In other countries such as Japan, it is more difficult to introduce IMRS regulation as they have no jurisdiction on overseas carriers. The EU itself admits that it cannot regulate IMRS between non-member countries. The only measures that it adopted that are relevant for non-member countries are the provision of transparency and safeguard measures.

As analysed in the preceding sections, the provision of transparency has been enforced and a form of safeguard measures, such as flat rate IMRS charges, has been introduced in Japan. However, customers continue to raise many claims against IMRS charges despite these measures. It is important to design measures that are workable in other countries. One option suggested by the EU regulation is the separate sale of IMRS from domestic cellular services. If the EC succeeds in finding reasonable technical means to implement this regulation, it can pave the way for other countries to introduce it. To rely on market forces as much as possible, it is desirable not to adopt direct price controls, and EU development is worth consideration.

Section 6 describes the options in Japan on the basis of the EU regulation experience and recent developments in the IMRS market.

6. CHANGES IN THE IMRS MARKET AND THE NEED TO INTRODUCE REGULATION

6.1 Changes in the IMRS Market

Changes in the IMRS market include global carriers such as Vodafone reducing charges between their subscribers. Several carriers have formed global alliances, such as the Conexus Mobile Alliance,[14] and reduced IOTs between member carriers. New services such as overseas plus number offered by NTT DOCOMO and KT enable users visiting each country to call as if using an inland service based on a dual SIM cards system. LINE offers free VoIP service over smartphones.[15] However, these changes are limited when considered within the entire IMRS market, with many users remaining subject to high IMRS charges and thus suffering anxiety and generating complaints.

Cellular services in Japan are totally deregulated as far as tariff is concerned although the market is oligopolistic as described. This excessive

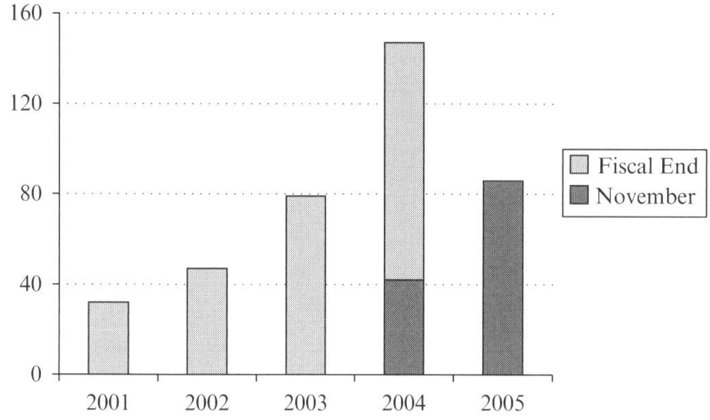

Figure 11.4 IMRS problems

deregulation is one reason behind the IMRS problems. The NCAC has reported many problems with IMRS since 2001. Although recent data are not published, the number of reported problems has increased since 2001. The number of problems as of November 2005 was more than double that in 2004 (Figure 11.4).

Table 11.7 reports NTT DOCOMO's revenue from IMRS for Q1–Q3 in FY2007.[16] The number of users was 780,000 and IMRS revenue was ¥14.6 billion ($139.2 million). The company reported that its ARPU from international services for FY2011 was ¥90 ($0.85) per month. Its international service revenue for FY2011 was estimated to be ¥63.8 billion ($608.4 million) (Table 11.8) as the IMRS share of international services for FY2007 was 56.4 per cent and IMRS revenue for Q1–Q3 was estimated to be at least ¥36.0 billion ($343.3 million).

As stated, the number of IMRS users as of the end of Q3 of FY2007 was 780,000. Since then, the number of users has not been published. When NTT DOCOMO's IMRS was suspended because of system errors on 13–14 August 2012, the number of users affected was reported as 80,000. Judging from this figure, the number of users has increased considerably.

Figure 11.4 presents the numbers reported to the NCAC. If the increase in the number of users and IMRS revenue is considered, it is reasonable to estimate that more latent problems exist.

6.2 Addressing IMRS Problems

Many factors contribute to the increasing problems with IMRS. The OECD lists several factors as summarised below (OECD, 2010, p. 8):

Table 11.7 NTT DOCOMO's IMRS data

	2006 Q1–Q3	2007 Q1–Q3	Growth rate (%)
International services revenue (hundred million ¥)	245	335	36.7
IMR (hundred million ¥)	125	189	51.2
International calls (hundred million ¥)	120	146	21.7
IMP share (%)	51	56.4	10.6
Number of users (thousand)	–	780	–

Source: Author's estimation based on NTT DOCOMO's financial data.

Table 11.8 NTT DOCOMO's IMRS revenue (fiscal 2011)

Number of users (thousand)	60,129
Number of users at the end of FY2010 (thousand)	58,010
International services ARPU (¥/month)	90
International service revenue (million ¥)	63,795
Estimated IMR revenue (million ¥)	35,992

Source: Author's estimation based on NTT DOCOMO's data.

1. High retail charges based on high wholesale charges and inelastic demand.
2. Lack of transparency in IMRS charges to users.
3. Bundled offering of IMRS with domestic services, including national calls, access and handset subsidy.
4. Lack of contestability in the IMRS market.
5. Lack of a perfect substitute for IMRS.

We should start from the understanding that high IMRS charges are the fundamental causes of problems. One factor behind high IMRS retail charges is high wholesale charges. I found that the EU regulation on retail and wholesale charges has contributed to a decrease in IMRS charges in EU member states (Table 11.6).

6.3 Regulation of Retail Charges

First, as both the EU and the OECD found, retail IMRS charges are considered to be higher than the underlying costs of providing the services and they are considerably higher than the normal international charges (Table 11.3) that constitute the basis for IMRS costs. IMRS are offered in a bundle with domestic services and demand for IMRS is inelastic to price. Further, the IMRS market is not contestable and perfect substitutes for IMRS do not exist. Presently, cellular services in Japan are totally deregulated on the basis of the belief that competition will protect consumer interests. However, current problems with IMRS have not been solved by market forces alone. Therefore regulation is necessary for retail IMRS charges by setting price caps as the EU is doing.

Second, Japan must change from a per-minute charge to a per-second charge as the EU implemented.

Retail regulation presents one problem. In its 2006 Proposal for a Regulation on IMRS (EC, 2006, p. 6), the EU acknowledged that regulating only retail charges may squeeze the carriers' profit margins. Therefore it is necessary to regulate wholesale charges together with retail charges.

6.4 Regulation of Wholesale Charges in Japan

The regulation of IMRS wholesale charges is more complex than that of retail charges. The regulation intends to protect the home country's consumers. The MIC can regulate only IMRS retail charges set by the carriers of its own country and wholesale charges offered by Japanese carriers. However, the wholesale services are used by overseas carriers for users visiting Japan and do not affect Japanese users. When home carriers offer retail services to their customers, they must buy wholesale services from carriers of foreign countries visited by Japanese travellers. The jurisdiction of the MIC cannot extend to the wholesale charges set by carriers of visited countries (Figure 11.5).

The EU introduced regulations on wholesale IMRS charges to avoid margin squeeze by retail charge regulation. This strategy was possible because the EU has jurisdictional authority over member states. The authority of Japan's MIC, however, cannot extend over IOTs and wholesale charges set by overseas carriers. Nevertheless, several institutional measures exist in Japan.

First, Section 40 of the Japanese Telecommunications Business Law (hereafter, 'TBL') stipulates that any telecommunications carrier must obtain approval from the MIC when they conclude an important agreement or a contract with overseas carriers. IOTs are considered as an agreement or contract requiring the Minister's approval. However, the MIC Ministerial Rule that provides detailed regulations based on the TBL exempts agreements or contracts that do not increase IOTs from

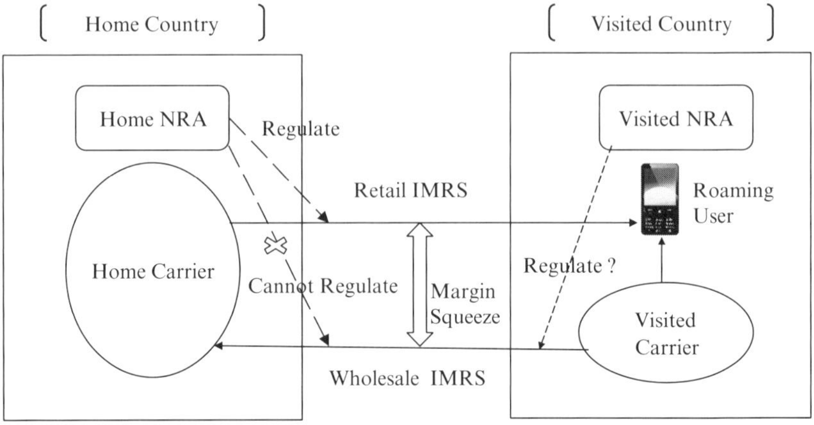

Figure 11.5 Difficulties in the wholesale charges regulation

such approvals. Here it is required to understand that the MIC is reluctant to impose regulation on IMRS. The MIC's basic approach towards IMRS is that the solutions to these problems should be left to the carriers' self-regulatory efforts. The TCA published reports on carriers' measures to cope with IMRS problems.[17] The measures included the detailed explanation of IMRS to users and the improvement of support desks, and they may be criticised as superficial. Thus, both the MIC and carriers are reluctant to introduce solutions to these problems. With the diffusion of smartphones, the MIC should introduce regulations to decrease high IMRS charges and solve other problems.

However, the aforementioned Section 40 of the Japanese TBL includes the IOTs and wholesale charges in the matters requiring the Minister's approval. Therefore the MIC can at least collect data on IOTs and publish them as the OECD recommends (OECD, 2010, p. 5). From the data collected, if the MIC finds that IOTs are excessively high, the MIC may ask Japanese carriers to decrease IOTs as a condition of approving agreements with overseas carriers.

Second, the aforementioned measure alone is difficult to implement as IOTs and wholesale charges must be negotiated with overseas carriers and the MIC cannot extend its jurisdiction to overseas carriers. As Japan has not formed a union with neighbouring countries such as the EU, Japan must obtain cooperation in controlling high IOTs. If the Japanese MIC succeeds in obtaining cooperation, it can ask Japanese carriers to lower IOTs with participating carriers, with the condition that other participating countries will also control IOTs with Japan. The OECD recommends having a mutual agreement between two or more countries to regulate wholesale and/or retail charges (OECD, 2010, p. 5).[18] Although this task is challenging, it is worth pursuing. In fact, the Singaporean and Malaysian governments have announced a mutual agreement to reduce the IMRS cost between their countries. The Arab regulators' network also agreed on a set of recommendations to reduce IMRS charges in April 2008 (OECD, 2009, p. 38). The MIC should study and learn from these experiences.

6.5 Structural Measures

One structural measure for consideration is the separate sale of retail IMRS from domestic cellular services. Although a nation cannot require overseas carriers to offer wholesale IMRS, a domestic regulation to promote competition could require the unbundling of retail IMRS from domestic cellular services. One difficulty with implementing this measure is the technical means to enable separate sales. Here it is important to observe and learn how the EU will implement their regulation. Even if reasonable technical implementation measures are found, Japan faces one further challenge: the treatment of SIM cards lock. Only NTT DOCOMO agrees to unlock SIM cards if so requested except for iPhones. Softbank is reluctant and charges for unlocking SIM cards. KDDI's 3G service is cdma2000, different from the W-CDMA of NTT DOCOMO and Softbank. Here, the regulation should intervene. If SIM cards lock is prohibited, separate sale of IMRS from domestic cellular services will work. The difference in 3G technical standard will be solved with the diffusion of LTE (Long Term Evolution).

6.6 Transparency

It is necessary to address the problem of the lack of transparency, especially on IMRS charges. The documents I examined from Japanese carriers and their websites provided only information that is too complicated for users to understand. As the OECD stresses, it is important to give users the information in a simple, comprehensible and transparent manner that enables comparisons across operators (OECD, 2009, p. 24). Here, the following measures adopted in the EU merit consideration:

1. Send information on IMRS charges when users cross a border.
2. Send warning notices when the amount of data transmitted approaches a certain upper limit.

7. CONCLUDING REMARKS

The IMRS system was used as an example to demonstrate that excessive deregulation works against users' interests. One factor behind the increasing problems with IMRS is the excessive deregulation of cellular services in Japan, although the market is characterised as an oligopoly. If the market is competitive, the matter can be left in the hands of market forces. However, the IMRS market is not currently competitive. If market forces are not working, some degree of regulation is necessary to protect users. I understand that the MIC's deregulatory measures by having implemented no regulations on tariff in the cellular market have gone too far and thus worked against users' interests. This simple worship of deregulation should be reconsidered. I hope that this chapter initiates discussions on IMRS and the pros and cons of excessive deregulation in Japan.

ACKNOWLEDGEMENT

This chapter is partly based on joint research with Associate Professor Shinya Kinukawa, Faculty of Global Media Studies, Komazawa University, supported by the Special Research Fund of Komazawa University.

NOTES

1. However, according to the September 2013 survey to identify the actual conditions of IMRS use (hereafter, 'the Survey'), 34.3% of IMRS users actually subscribed to the flat rate service. The Survey was designed by the author and Associate Professor Sinya Kinukawa, Faculty of Global Media Studies, and carried out by Rakuten Research, Inc. The number of respondents was 29,810.
2. ¥ is converted into $ at the rate of 0.010486 as of 6 January 2014.
3. As Japanese mobile carriers offer more or less similar IMRS, I use NTT DOCOMO as an example.
4. According to the Survey, 30.8% of actual IMRS users do not know that voice calls are charged per minute.
5. According to the Survey, 66.0% have not travelled abroad within the last three years.
6. Although there are more cases including calls made within visited countries and from visited countries to third countries, I discuss two typical cases.
7. IOTs do not represent the final outcome of wholesale negotiations, instead functioning as only a cap for wholesale roaming rates (OECD, 2010, p. 5).
8. As NTT DOCOMO forms the Conexus Mobile Alliance with overseas carriers such as KT of Korea and Smart of the Philippines, IOTs between allied carriers are cheaper and IMRS charges are relatively cheaper than the charges between these carriers and the USA.
9. Carriers such as Vodafone operating in many countries can offer the same types of services as domestic services.

10. NTT DOCOMO recently adopted a policy to allow the unlocking of SIM cards for users who so requested. KDDI and Softbank are overall keeping SIM cards lock. The MIC guided them to unlock SIM cards from May 2015 for newly sold terminals.
11. The WIK report conducted on behalf of the EC estimated demand elasticity for calls made to be between −0.2 and −0.3 (WIK, 2010, p.43).
12. According to the Survey, 21.3% of users have used IMRS.
13. The EC set an upper limit per minute for Eurotariff at 130% of the applicable maximum wholesale charge. Existing users were allowed to choose between Eurotariff and any other roaming tariff.
14. Refer to note 8.
15. Although the use of VoIP is free from call charges, it incurs data charges.
16. Somehow, NTT DOCOMO has stopped publishing data on IMRS since then.
17. TCA (2006) and TCA (2007).
18. Regrettably, the existence of two OECD reports is not widely known in Japan.

REFERENCES

EC (2003), Commission Recommendation of 11/02/2003 on relevant product and service markets within the electronic communications sector susceptible to ex ante regulation in accordance with Directive 2002/21/EC of the European Parliament and of the Council on a common regulatory framework for electronic communication networks and services (C (2003) 497).

EC (2006), Proposal for a Regulation of the European Parliament and of the Council on roaming on public mobile networks within the Community and amending Directive 2002/21/EC on a common regulatory framework for electronic communications networks and services, COM (2006) 382 final.

EC (2007a), Commission Recommendation on relevant product and service markets within the electronic communications sector susceptible to ex ante regulation in accordance with Directive 2002/21/EC of the European Parliament and of the Council on a common regulatory framework for electronic communications networks and services (2nd edn) (C (2007) 5406 rev 1).

EC (2007b), Regulation (EC) No 717/2007 of the European Parliament and the Council of June 2007 on roaming on public mobile telephone networks within the Community and amending Directive 2002/21/EC.

EC (2009), Regulation (EC) No 544/2009 of the European Parliament and of the Council of 18 June 2009 amending Regulation (EC) No 717/2007 on roaming on public mobile telephone networks within the Community and Directive 2002/21/EC on a common regulatory framework for electronic communications networks and services.

EC (2010a), Digital Agenda: Commission outlines action plan to boost Europe's prosperity and well-being (IP/10/581, 19 May 2010).

EC (2010b), Digital Agenda for Europe: key initiatives (MEMO/10/200, 19 May 2010).

EC (2011a), Proposal for a Regulation of the European Parliament and of the Council on roaming on public mobile communications networks within the Union (COM (2011) 402 final) (2011/0187 (COD)).

EC (2011b), Report from the Commission to the European Parliament, the Council, the European Economic and Social Committee and the Committee of the Regions on the outcome of the functioning of Regulation (EC) No 717/2007

of the European Parliament and of the Council 27 June 2007 on roaming on public mobile communications networks within the Community, as amended by Regulation (EC) No 544/2009, COM (2011) 407 final.
EC (2012a), Digital Agenda: EU reaches preliminary deal on future-proof roaming solution for mobile phone users.
EC (2012b), Regulation (EU) No 531/2012 of the European Parliament and of the Council of 13 June 2012 on roaming on public mobile communications networks within the Union.
NCAC (2006), Problems on mobile services that enable the use of handsets in overseas countries: mainly on international mobile roaming services (in Japanese).
NCAC (2011), Rapidly increasing problems on the use of smartphones (in Japanese).
OECD (2009), International mobile roaming charging in the OECD area, DSTI/ICCP/CISP (2009) 8/final.
OECD (2010), International mobile roaming services: analysis and policy recommendations, DSTI/ICCP/CISP (2009) 12/final.
TCA (2006), Problems on the use of international mobile roaming services and measures to cope with them (in Japanese).
TCA (2007), Measures to prevent problems on the use of international mobile roaming (in Japanese).
WIK (2010), Study on the options for addressing competition problems in the EU roaming market: study for the European Commission.

PART V

New features in the media industry

12. An estimation of marginal WTP for variety in the broadcasting platform

Manabu Shishikura and Norihiro Kasuga

1. INTRODUCTION

As a result of technological innovation in the information and communication industry, traditional media market categories such as 'communication' and 'broadcasting' have essentially become meaningless. The industry is rapidly being restructured and the so-called platform now plays a central role in the media market.

In this chapter, we examine whether information goods resources can be allocated via the platform. Our findings differ from those of previous studies based on utility functions. We then focus on the broadcasting industry as an example and empirically examine the subscription choice behaviour of consumers with respect to broadcasting platforms. Based on our results, we consider which types of utility function can adequately express consumers' preferences for information goods.

2. THE ECONOMIC CHARACTERISTICS OF THE MEDIA MARKET

The media is defined as a service that supplies information goods to consumers via both wireless and wired transmission technology. The economic characteristics of the media are: 1) an equal amount of consumption; 2) complementarity between software and hardware; and 3) two-sided markets.

2.1 Equal Amount of Consumption

Because we can consume an equal amount of information goods, the socially optimal quantity is the level at which the sum of users' marginal utility is equal to the marginal cost of the information goods. Therefore as the number of users increases, the social marginal benefit of information goods (i.e. the sum of users' marginal utility) increases as well. However, the marginal cost associated with an increase in the number of users is zero. Therefore, from a social welfare point of view, it is desirable that all users who have a positive marginal utility (i.e. a positive benefit) consume information goods.

However, the marginal utility for certain information goods differs among users. This means the willingness-to-pay (WTP) for information goods and the number of users are not necessarily linked. In some cases, users have a high marginal benefit, but the number of users is small. In other cases, users have a low marginal benefit, but the number of users is large. Therefore it is necessary that different types of consumer be charged different fees according to their marginal utility. However, to implement price discrimination in this way, suppliers must first have complete information on consumers' marginal benefits.

2.2 Complementarity between Software and Hardware

Media services are commonly provided as a combination of two components, namely hardware (i.e. access to the network) and software (i.e. programs). Although these components are often provided by different companies, both are crucial to supplying the media service to users. However, equal amounts of consumption (non-rival consumption) do not always hold for goods such as terminal equipment and transmission infrastructure. Therefore the social marginal cost of an increase in the number of users is positive and the service provided by the media industry is regarded as 'club goods'. When supplying club goods, it is necessary to control both the quantity of information goods and the number of users. If the software and hardware are provided by different companies (as is usually the case), there is no guarantee that the solution that maximises the profit of each supplier will be consistent with the solution that would maximise social welfare.

2.3 Two-sided Market

Suppliers of media services (i.e. information goods) typically have multiple sources of revenue, first, from the direct demand by consumers for the information goods, and second, from the indirect demand of advertising. The revenue of newspaper and magazine companies is usually split evenly between these two sources of revenue. However, pay-TV companies usually depend primarily on the user market for revenue, although they do receive revenue from the advertising market as well. In the case of terrestrial broadcasting systems, companies depend almost entirely on the advertising market.

Based on these different sources of revenue, we divide broadcasters in the Japanese market into three categories. First, we distinguish between public broadcasters, whose revenue depends on a reception fee from the audience and commercial broadcasters, whose revenue is derived from advertising and payments from viewers. Then, we further subdivide commercial broadcasters into ad-sponsored broadcasters, whose revenue depends primarily on advertising, and pay-TV carriers, whose revenue depends primarily on payments from users.

With respect to the characteristics described earlier, both the complementarity between software and hardware and the two-sided market bring ‘externalities’ to media demand. In addition, the equal amount of consumption of information goods means that price discrimination based on users’ marginal benefits is important. These characteristics complicate the demand for media services, which has resulted in increased attention to ‘platform business’, primarily responsible for managing such demand. The primary aims of platform business are to internalise the external effects of demand, to manage customer information and to create more sophisticated billing systems (Chou and Shy, 1990; Shishikura and Kasuga, 2008).

Traditionally, hardware suppliers have been engaged in demand management in both the user and advertising markets. However, the recent spread of the Internet and the progress of digitisation have seen the emergence of platform operators that are independent from hardware suppliers and information goods can now be supplied across many platforms. As an example of a platform business in the broadcasting market in Japan, we examine a multiple system operator (MSO) offered by a cable TV supplier and a platform operator offered by a CS satellite carrier.

The cable TV business in Japan began by retransmitting terrestrial broadcasting in areas that were not able to capture terrestrial TV signals because of mountains or high buildings. However, as these areas became less common and with the deregulation of cable TV after 1985, some cable TV companies began to add their own TV programmes to their services. As a result, a new type of company, operating ‘urban-type cable TV’, entered the cable TV market. This urban type usually has more than five self-produced channels and retransmits satellite TV programmes. The multichannel business continued to develop and MSOs, which integrate multiple cable carriers to benefit from economies of scale, played a role in improving cable TV business efficiency. MSOs provide the platform service for cable TV operators.

On the other hand, the CS satellite broadcasting business was started specifically to supply multi-channel services.[1] The carriage deal system was introduced to provide multiple channels and the policy framework for this business categorised market players into two groups, one being carriers for producing and programming TV programmes ('consignors') and the other being carriers for managing and operating broadcasting facilities ('trustees').[2] This framework reduced the initial costs associated with facility operation and promoted new entrants to the market.

As more carriers entered the platform business, traditional classifications in the media market began to disappear. Media fusion enabled progress and the number of supplied channels of information goods has increased.[3]

3. LITERATURE REVIEW

Since information goods are characterised by an equal amount of consumption, we calculate the social benefit of information goods by multiplying the number of users by each user's WTP. However, the current financing model used by broadcasters makes it difficult to consider both the number of users and their WTP simultaneously and, therefore, to maximise profit.

For example, ad-sponsored media try to maximise the number of users to benefit from the effect of advertising and often allow free access to their programming to attract users with a low WTP. As a result, all users who have positive marginal utility can consume the information goods (i.e. TV programmes) supplied by ad-sponsored media under the condition of equal amount of consumption. On the other hand, the user's WTP is not fully taken into account because there is no direct relationship between ad-revenue and the user's WTP. Even though there are multiple ad-sponsored operators in the media market, the difference in supplied information goods is minimal. This is because operators prefer to provide goods that are similar to the median customer preferences to maximise the number of users. As a result, the WTP of each customer is not always reflected in the supply of information goods.

On the other hand, pay-TV operators need to consider both the number of users and users' WTP to maximise revenue. Here, although the number of users is relatively small, the providers' market power is enhanced because there are few competitive providers in the market. Therefore when users have varied preferences, the incentive for product differentiation increases and so does the number of supplied channels. However, if it is difficult to implement price discrimination properly, the consumer is likely to be excluded from the market and an optimal solution becomes unlikely.

In the Japanese broadcasting market, carriers that depend on different revenue sources (i.e. ad-sponsored broadcasters and public broadcasters) have played an important role. Therefore previous studies tended to focus on the value of the social WTP for information goods supplied by these broadcasting carriers. Many previous studies use a cost–benefit analysis to measure the WTP for broadcasting services, including those of Arakawa (1995), Finn, Mcfadyen and Hoskins (2003) and Delaney and O'Toole (2004).

However, as we showed in the previous section, the platform has recently begun to play an important role in supplying information goods, for example, video distribution over the Internet. Accordingly, economic research has now shifted to the issues of excess/under entry over multiple platforms and under free-entry conditions.

The theoretical models determined by these studies depend on their assumptions on the type of utility function and the level of competition. If we assume a Salop-type utility function and consumer choice behaviour in a preference space model (i.e. the address model), free entry leads to excess variety. The reason for this is that individual firms do not consider the 'business-stealing effect', which describes how new entrants grab customers from incumbent carriers. Crampes, Haritchabalet and Jullean (2009) pointed out that excess entry occurs in a two-sided market, with both advertising revenue and a user fee, and that the volume of advertising reaches an optimal level if the consumer is assumed to have a Salop-type preference and if advertising revenue shows increasing or constant returns to scale.

On the other hand, if we assume a Dixit–Stiglitz type utility function, in which utility increases as channel availability (variety) increases, free entry can result in less than optimal variety. Hagiu (2008) extends the monopolistic competition model to the case of two-sided markets with hardware and software. He points out that the utility function determines the level of variety, even in the case of a two-sided market. He analyses two types of utility function. The first is a two-sided sponsored platform, in which revenue is generated from both markets. The second is a non-sponsored platform, in which free entry is allowed. The number of socially optimal applications is different. In short, a Salop-type utility function is closer to the optimal number of applications in the case of two-sided sponsored platforms, while a Dixit–Stiglitz type utility function better reflects the optimal number of applications in the case of non-sponsored platforms. Therefore from a social welfare point of view, the two-sided sponsored platform is preferable in the case of the Salop-type utility function, while a non-sponsored platform is preferable in the case of the Dixit–Stiglitz type utility function.

4. UTILITY FUNCTION, CONSTRAINT CONDITIONS AND BROADCASTING DEMAND

Based on the previous studies described in Section 3, we empirically analyse the consumers' choice behaviour in terms of broadcasting platforms. In this section, before reporting our empirical analysis, we describe the theoretical model for consumer choice behaviour using a utility function and constraint conditions.

Service users can only obtain positive utility by watching television, and the longer they do so, the higher their utility level becomes. Because broadcasting is classified as a time-consuming leisure activity, it has unique characteristics when compared to other goods or services with regards to its utility function and constraint conditions. However, if we assume a given amount of time spent watching television, the viewers' utility levels will be higher if they can watch programmes that better match their preferences. The probability of a match between the programme type and a user's preference is stochastically determined because broadcasting services are experience goods. The more channels a carrier makes available, the higher the probability of a match between the type of programme and users' preferences. Therefore we consider that an increase in the number of available channels (i.e. an increase in variety) increases users' utility per watching period. Therefore we assume that the utility obtained from a broadcasting service is a function of the number of channels and the time available to watch television.[4]

Based on these considerations, we regard the number of available channels as a lineup (i.e. the alternatives for an audience, or variety) and assume the following constant elasticity of substitution (CES)-type utility function with respect to the number of channels and TV watching time.

$$U = \left(\sum_{i=1}^{G} x_i^{\rho} \right)^{\frac{1}{\rho}} \text{ with } 0.5 < \rho < 1 \tag{12.1}$$

Here, x refers to the time spent watching each channel and G refers to the number of available channels. The user of a broadcasting service decides both the number of available channels (1,2,. . .,G) and the time spent watching each channel $(x_1, x_2, \cdots, x_G)$.

Now, suppose that

$$x_1 = x_2 = \cdots = x_G = x \tag{12.2}$$

Then, we obtain

$$U = G^{\frac{1}{\rho}}x \tag{12.3}$$

Because

$$\sum_{i=1}^{G} x_i = Gx = X \tag{12.4}$$

The utility function is expressed as a function of the number of channels, G and total viewing time, X.

$$U = G^{\frac{1-\rho}{\rho}}X \tag{12.5}$$

Under the utility function proposed here, even though the total watching time, X, is unchanged, users' utility levels can increase if the number of available channels, G, is increased.

Next, we examine the associated constraint conditions. We have to consider time as an extra constraint in addition to budget, the usual constraint, because TV watching behaviour is a time-consuming activity. Here we assume that consumers use all available leisure time (= 24 hours minus labour hours minus sleeping time) as TV watching time.

$$T = L + X \tag{12.6}$$

Here, T refers to the total time users have available during a certain period and L refers to labour hours. We further assume that users spend their total wage income on purchasing pay-TV channels. That is, we assume that

$$wL = pG \tag{12.7}$$

Here, w refers to a wage rate and p refers to the price per channel. We can rewrite the above two constraints in one equation, as follows:

$$T = \frac{p}{w}G + X \tag{12.8}$$

Each user decides both their TV watching time, X, and the number of channels to purchase, G, so as to maximise equation (12.5) under constraint equation (12.8).

From the first order condition (FOC), we obtain

$$\frac{G}{X} = \frac{1-\rho}{\rho}\frac{w}{p} \tag{12.9}$$

This means that the ratio of the number of available channels, G, to total TV watching time, X, depends on the real wage. Note that we can express the demand function for both total TV watching time and the purchasing of channels as follows if we substitute the constraint condition into the above equation.

$$G = (1 - \rho)\frac{w}{p}T, X = \rho T \tag{12.10}$$

This demand function shows that watching time is proportional to a user's available time, T, while wage and price do not affect a user's total TV watching time. On the other hand, the demand for channels depends on the real wage rate and a rise in wages increases a user's demand for the number of channels.[5]

However, since we assumed the viewing time of each channel to be divided equally by the number of channels, the time spent watching each channel decreases as the total available number of channels increases. That is,

$$x = \frac{X}{G} = \frac{\rho}{1 - \rho}\frac{p}{w} \tag{12.11}$$

5. THE METHOD OF EMPIRICAL ANALYSIS

Based on the discrete choice model, we implement an estimation of the marginal WTP for information goods supplied by the platforms with three different revenue systems, namely public broadcasting, ad-sponsored broadcasting and pay-TV broadcasting. We estimate the marginal WTP for an increase in the number of channels supplied by each type of broadcaster.

In the discrete choice model, it is assumed that consumers obtain their utility from various characteristics of goods (i.e. price, quality, content, etc.) and the utility function is expressed as a linear combination of the probability term, and the non-probability term, which is the attribute of the service.

We assume the user has a utility function for the channels supplied by public broadcasting, ad-sponsored broadcasting and pay-TV as follows:

Utility function obtained from public broadcasting:

$$V_{pub} = \alpha_{pub} G_{pub} - \beta\, p_{pub} \tag{12.12}$$

Utility function obtained from ad-sponsored broadcasting:

$$V_{com} = \alpha_{com} G_{com} - \beta\, p_{com} \tag{12.13}$$

Table 12.1 Definition of variables

Explanatory variables	Definition
PUB	Number of public channels
COM	Number of ad-sponsored channels
SPC	Number of pay-TV channels
METHOD	Signal transmission method ('Digital' = 1, 'Analogue' = 0)
FEE	Monthly fee
DISABLE	Dummy for record restriction ('Impossible' = 1, 'Non-restricted/Permitted sometimes' = 0)

Utility function obtained from pay-TV:

$$V_{spc} = \alpha_{spc} G_{spc} - \beta\, p_{spc} \tag{12.14}$$

Here, G_i shows the number of channels supplied by each platform and p_i represents the WTP for each platform.

Each consumer is assumed to select the situation that maximises the sum of the above utility functions, as follows:

$$V = V_{pub} + V_{com} + V_{spc} = \alpha_{pub} G_{pub} + \alpha_{com} G_{com} + \alpha_{spc} G_{spc} - \beta\, (p_{pub} + p_{com} + p_{spc})$$

$$V = \alpha_{pub} G_{pub} + \alpha_{com} G_{com} + \alpha_{spc} G_{spc} - \beta\, P \tag{12.15}$$

where P refers to total expenditure.

$$P = p_{pub} + p_{com} + p_{pay} \tag{12.16}$$

We assume the following linear non-probability term for the estimation.

$$V_i = \alpha_{pub} G_{pub\_i} + \alpha_{com} G_{com\_i} + \alpha_{spc} G_{spc\_i} + \beta_1\, METHOD_i + \beta_2 P_i + \beta_3\, DISABLE_i \tag{12.17}$$

The meaning of each variable is described in Table 12.1.

6. DATA

In our study, we use stated preference (SP) data (i.e. conjoint data) rather than real preference (RP) data. The reason is that there is an asymmetry between the channels offered by the cable TV and CS platforms in the Japanese pay-TV market. In particular, cable TV can provide terrestrial broadcasting channels, which are restricted within prefectural areas, as well as pay-TV channels, while the CS broadcasting carriers can only provide pay-TV channels.[6] If we were to use real data, it would be difficult to extract the effect of the offered channels on consumers' choices, because a consumer's choice of platform includes the combined effects of their preferences for a platform and for the offered channels.

A questionnaire was used as a part of the survey implemented by the Institute for Information and Communications Policy of the MIC (2005). The questionnaire was completed by a representative of each household (males and females between the ages of 15 and 79) in June 2005. Of the 4500 households that received the survey via email, 2035 households returned an answer sheet (giving a collection rate of 45.2 per cent). In our survey for the conjoint analysis, we asked six questions per person, yielding 12,210 responses in total. After eliminating non-respondents, the effective response rate was 89 per cent (10,857 responses). We adopted a choice-type answer format. Respondents were presented with three choices, created from combinations of the levels and attributes. They then chose one of the fictitious combinations (choices).[7] An example of the actual questionnaire is shown in Table 12.2.[8]

In the estimation, we use both a conditional logit model and a mixed logit model. Note that in the mixed logit model, because we could not anticipate which parameters should be treated as random, we first estimated under the assumption that all parameters are randomly and normally distributed, except for 'fee'. If the standard deviation of the estimated coefficient was not statistically significant, this parameter was then

Table 12.2 Example of questionnaire

	1	2	3
Public channel	6 ch	6 ch	2 ch
Ad-sponsored channel	3 ch	5 ch	3 ch
Pay-TV channel	0 ch	10 ch	20 ch
Signal transmission method	Analogue	Digital	Analogue
Monthly fee (yen)	6000	8000	10,000
Availability for picture record	Disable	No restriction	No restriction

treated as fixed. The estimation was then undertaken repeatedly until we identified the final model. We implemented 1000 repeated Halton draws to estimate the random parameters by simulation and to obtain their maximum likelihood values. Furthermore, because respondents answered six questions for the conjoint analysis, we also examined the estimation results under the assumption that the coefficient of certain respondents had the same value (i.e. each coefficient was not independent).

7. ESTIMATION AND RESULTS

Based on our estimation results, the parameters for variety preferences show quite different values among public, ad-sponsored and pay-TV channels. Therefore we can say that consumers' preferences for more channels differ according to the type of offered channels.

It is necessary that the following first order conditions be satisfied to maximise utility.

$$\frac{\partial V}{\partial G_i} = \alpha_i - \beta \frac{\partial P}{\partial G_i} = 0 \tag{12.18}$$

$$p_i = \frac{\partial P}{\partial G_i} = \frac{\alpha_i}{\beta} \tag{12.19}$$

Therefore we can calculate the marginal WTP for the increased number of channels of each platform from the estimated parameters. Table 12.4 summarises the marginal WTP for the number of channels supplied by each platform, as well as the other attributes.

The marginal WTP for additional channels in the case of pay-TV is lower than television ad-sponsored and public broadcasting. This is why the distribution of users' WTP for the pay-TV channel is large and also why the number of channels is relatively large compared to television ad-sponsored and public broadcasting.

The WTP of users for channels supplied by pay-TV shows a large variance. There exist users with both a high and a low WTP. Therefore, if there are fewer users with a high WTP, the marginal WTP for the average per person would be lower.

In addition, the number of channels offered by pay-TV is greater than is offered by ad-sponsored broadcasting or public broadcasting. However, the number of programmes per channel in television ad-sponsored and public broadcasting is greater than pay-TV because the same programme is re-broadcast repeatedly.[9] As described above, the difference in how programmes are organised in pay-TV and television

Table 12.3 Estimate result

	Conditional logit model		Mixed logit model		Mixed logit model (considering six times answer)	
Explanatory Variables	coefficients	t-value	coefficients	t-value	coefficients	t-value
PUB	0.081***	9.819	0.129***	9.690	0.095***	9.597
			(0)		(0.069)**	(1.975)
COM	0.087***	15.669	0.114***	13.969	0.129***	16.758
			(0.120)***	(3.265)	(0.086)***	(5.094)
SPC	0.012***	10.117	0.015***	9.853	0.012***	5.080
			(0)		(0.068)***	(28.503)
METHOD	0.416***	13.843	0.609***	11.507	0.598***	13.889
			(1.430)***	(9.549)	(0.847)***	(13.060)
FEE	−0.000194***	−37.128	−0.000248***	−20.665	−0.000251***	−34.153
			(0)		(0)	
DISABLE	−0.945***	−30.122	−1.248***	−17.681	−1.276***	−25.383
			(1.261)	(7.849)	(1.128***)	(17.957)
The Number of Samples	10533		10533		10533	
Log Likelihood	−9718.87		−9689.99		−9059.96	
Index of Likelihood Ratio	0.1601		0.1626		0.2171	

Notes: ***, ** and * represent p-values in the case of statistical significance at the 1%, 5% and 10% levels, respectively.
Parameters in brackets are the estimated standard deviations of the error terms from the mixed logit model, as well as the corresponding t-values.
An entry of 0 indicates that the parameters are not random, but fixed.

Table 12.4 Marginal WTP

	Conditional (¥)	Mixed (¥)	Mixed (with coeff. restriction) (¥)
Public channel (PUB)	414.9	519.6	379.3
Ad-sponsored channel (COM)	450.5	458.1	511.9
Pay-TV channel (SPC)	59.8	60.8	46.5
Signal transmission method (METHOD)	2143.6	2455.7	2382.8
Record restriction (DISABLE)	−4869.6	−5032.3	−5082.7

ad-sponsored and public broadcasting is thought to be the difference in the marginal WTP.

On the other hand, the marginal WTP for an increase in the number of channels for television ad-sponsored and public broadcasting is relatively high when compared to pay-TV.

For public broadcasting, given that there are only two channels available on the ground contract, the total WTP is between 1040 yen and 760 yen. There are four channels available on the satellite contract. Here, the total WTP is between about 2080 yen and 1520 yen. Since the viewing fee for public broadcasting is 1345 yen on the ground contract and 2290 yen on the satellite contract,[10] these values are slightly higher than the total estimated WTP.

The number of channels offered by the television ad-sponsored option is four to five terrestrial channels and eight to nine channels if satellite broadcasting is included. Despite the number of channels being greater than public broadcasting, the marginal WTP for more channels in television ad-sponsored is greater than that of public broadcasting.

In addition, the marginal WTPs for digital broadcasting and recording restrictions on TV programmes also show high values (with a positive and negative sign, respectively). This means that copy protection measures such as record restrictions may have a large effect on consumers' choice of platform.

8. CONCLUSION

The lack/excess of variety of information goods differ depending on the type of utility function assumed for each user. If utility is assumed to increase with diversity, there is then a possibility of a lack of variety in information goods supply.

The empirical results show that the assumption that the user has a positive effect on the increase in the number of channels is reasonable. In addition, the empirical results show that the marginal WTP for an increase in the number of channels is different between television ad-sponsored, public broadcasting and pay-TV. In addition, the marginal WTPs for additional channels of television ad-sponsored and public broadcasting were shown to be greater than that of pay-TV.

ACKNOWLEDGEMENT

We are very grateful for financial support from Grant-in-Aid for Scientific Research (C) (24530331) and Kinki University (SR01).

NOTES

1. CS analogue broadcasting began in FY1992, while CS digital broadcasting began in FY1996. Furthermore, digital broadcasting using CS at longitude 110 east, which is the same longitude used by BS broadcasting, began to provide broadcast services in 2002 to benefit from the receiving equipment already in place in the households.
2. SKY Perfect Communications Inc. is classified as a platform operator in Japan.
3. As regards the study on competition between cable TV and CS satellite platforms, refer to Karikari, Brown and Abramowitz (2003), Goolsbee and Petrin (2004), Wise and Duwadi (2005), Clements and Brown (2006) and Kasuga, Shishikura and Kondo (2008).
4. With programme sharing and re-broadcasting, an increase in the number of channels does not necessarily mean an increase in the number of information goods. However, programmes with the same content, time and location, are regarded as goods with different users, as per the supplying method. Therefore we believe it is reasonable to measure the diversity of goods supplied by a broadcaster using the number of channels they provide.
5. Note that this feature is derived from our setting the parameter with respect to total watching time of an assumed Cobb–Douglas-type utility function to 1. This feature, watching time, does not change even though wage or price changes and only the demand for the number of channels (= subscription demand for pay-TV) is affected by a restriction on the associated function. However, this assumption will be allowed, at least in the short run.
6. Shishikura, Kasuga and Torii (2006) analyse the factors for subscribing to pay-TV based on a discrete choice model and show that the basic channels offered by terrestrial broadcast carriers greatly affect the probability for subscribing to pay-TV. However, they could not show a significant effect of the special channel on subscribing behaviour. This is because their analysis is based on RP data and fails to extract only the effect of each channel on consumers.
7. The answer format, 'choice format', was developed by Louviere and Woodworth (1983). The format is used most frequently because it better reflects consumer choices/behaviour in the market than other formats such as the 'ordering format' or the 'revaluation format'. It also has the advantage that respondents find it easier to answer.
8. Because the diffusion rate of television is almost 100% and almost all households can receive signals for terrestrial broadcasts in Japan, we eliminated the alternative, 'do not

choose any alternatives' ('donnot choose any alte') at this time. Therefore our estimation results on choice probability may show higher values. However, this does not create a problem in confirming our hypothesis.

9. In this sense, there is a certain limit to evaluating the diversity of information goods in the number of channels.
10. These values are based on the amount of the subscription fee for public broadcasting until September 2012.

REFERENCES

Arakawa, K. (1995), *The Integration of TV Media: Focused on Cost Benefit Analysis*, Sohju sya (in Japanese).

Chou, C. and O. Shy (1990), 'Network effects without network externalities', *International Journal of Industrial Organization*, **8** (2), 259–70.

Clements, M. and S. Brown (2006), 'The Satellite Home Viewer Improvement Act: price and quality impact of direct satellite companies' provision of local broadcast station', *Telecommunications Policy*, **30** (2), 125–35.

Crampes, C., C. Haritchabalet and B. Jullean (2009), 'Advertising, competition and entry in media industries', *Journal of Industrial Economics*, **57** (1), 7–31.

Delaney, L.D. and F. O'Toole (2004), 'Irish public service broadcasting: a contingent valuation analysis', *Economic and Social Review*, **35** (2), 321–50.

Dixit, A. and J. Stiglitz (1977), 'Monopolistic competition and optimum product diversity', *American Economic Review*, **67** (3), 297–308.

Finn, A., S. Mcfadyen and C. Hoskins (2003), 'Valuing the Canadian Broadcasting Corporation', *Journal of Cultural Economics*, **27** (3–4), 177–92.

Goolsbee, A. and A. Petrin (2004), 'The consumer gains from direct broadcast satellites and the competition with cable TV', *Econometrica*, **72** (2), 351–81.

Hagiu, A. (2008), *Platforms, Pricing, Commitment and Variety in Two-sided Markets*, Saarbrücken, Germany: VDM Verlag.

Karikari, J., S. Brown and A. Abramowitz (2003), 'Subscriptions for direct broadcast satellite and cable television in the US: an empirical analysis', *Information Economics and Policy*, **15** (1), 1–15.

Kasuga, N., M. Shishikura and M. Kondo (2008), 'Platform competition in the pay-TV market', *Kobe University Economic Review*, **53**, 57–68.

Louviere, J. and G. Woodworth (1983), 'Design and analysis of simulated consumer choice or allocation experiments: an approach based on aggregate date', *Journal of Marketing Research*, **20** (4), 350–67.

Shishikura, M. and N. Kasuga (2008), 'The number of TV channels and platform competition in the pay-TV market: the effect of indirect network externality and channel compatibility', *Journal of Economics and Business Administration (Kokumin-Keizai Zasshi)*, **198** (5), 29–45 (in Japanese).

Shishikura, M., N. Kasuga and A. Torii (2006), 'Multimedia, multichannel and the future of the broadcasting market: an analysis of subscription demand for pay-TV', *Journal of Economic Policy Studies*, 60–63 (in Japanese).

Wise, A. and K. Duwadi (2005), 'Competition between cable television and direct broadcast satellite: the importance of switching costs and regional sports networks', *Journal of Competition Law and Economics*, **1** (4), 679–705.

13. The impact of newly emerging media on traditional media platforms in Taiwan: a co-opetition perspective

Yu-Li Liu

1. INTRODUCTION

Whenever a new medium appears, there is much discussion regarding its impact on the old media. The new technology makes it difficult to distinguish the differences between the new media and the old media. A new medium will become an old medium after a number of years. Furthermore, an old medium can be a new medium if a new function, new service, or new value is added to it. For instance, when cable TV first appeared, it was considered to be a kind of new medium. After digital TV appeared, analogue cable TV became an outdated medium. However, once cable TV adopted digital technology and transformed itself into digital cable TV, it once again became a new medium with its new functions.

Under digital convergence, there are other new media platforms such as Internet Protocol TV (IPTV), mobile TV and Over-the-Top (OTT) video services, which are joining the ranks of the competition along with traditional media. In recent years, users have witnessed the emergence of these new services. As technology advances, the impact of these new services is also starting to surface. Newell, Pilotta and Thomas (2008) found that new media such as the Internet grew at twice the rate of incumbent media, namely, television, radio, newspapers and magazines, and created an overall increase in the consumption of ad-supported media. A UK firm, Digital TV Research, has predicted that global IPTV penetration will reach 10.8 per cent of TV households by the end of 2017, up from 3.7 per cent at the end of 2011. The number of subscribers to IPTV services worldwide is predicted to rise from 51 million at the end of 2011 to 165 million by the end of 2017 (Beach, 2012). Informa Telecoms & Media (2012) predicted that over 380 million people will watch online videos

54 billion times via connected-TV devices globally in 2015. In recent years, users have witnessed the emergence of these new services. Not only does the new medium have an impact on the traditional media owing to displacement or complementary effects between media platforms, but it also has an impact on the users as a result of the added choices and flexibility provided by the new media (Liu and Hsu, 2011). In addition to the consumer and traditional media perspectives, there are also concerns regarding whether the current laws and policies are providing a level playing field for both newly emerging and traditional media.

Whenever there is a new medium, the traditional media will deliberate and react to the impact in different ways. Likewise, the traditional media platforms in Taiwan also have to pay attention to the development of the newly emerging media, because their ratings and revenues might decrease owing to the newly emerging media platforms. This chapter discusses the competition among the players in the related TV market, while also analysing the impact of the newly emerging media on the traditional media platforms.

2. MEDIA ECOSYSTEM AND VALUE CHAIN IN TRANSFORMATION

Digital convergence has had much of an impact on both the ecosystem and the players. While newly emerging media services such as IPTV and OTT video services have begun to be popularised among users, traditional media have also found ways to match the new media platforms. The players in the new ecosystem not only compete, but also cooperate. The value chain of the new media platforms also involves more players, such as content aggregators and content distribution networks. In Taiwan, the analogue terrestrial signal was switched off in June 2012. Since cable TV penetration is over 80 per cent, most people watch terrestrial TV via cable TV. As of June 2014, 60.02 per cent of cable subscribers had adopted digital cable (National Communications Commission, 2013). The government has urged cable operators to complete digitisation by the end of 2014 (Executive Yuan, 2012).

With the development of digital convergence, new media platforms have rapidly emerged in Taiwan. The first was IPTV, followed by mobile video platforms. The most recent one was the OTT video service, which appeared in 2010 in Taiwan (Liu, 2012). IPTV is defined by ITU's IPTV Focus Group as 'multimedia services such as television/video/audio/text/graphics/data delivered over IP-based networks managed to provide the required level of quality of service and experience, security, interactivity and reliability' (ITU-T Newslog, 2006). Mobile video platforms only refer to mobile multimedia services via mobile smart devices instead of mobile TV via broadcasting (because the government has not granted mobile TV licences). OTT video service is defined as the distribution of video content to a wide range of IP-enabled devices such as TVs, PCs, mobile phones and tablets over the unmanaged Internet (Venturini, 2011). At the present time, OTT video service may operate with a set-top box on a television. It can also be accessed through a connected TV, where the connection is already built into the television. With a built-in connection, OTT video service providers may choose to cooperate with the TV manufacturers via splitting revenue. Services available on OTT video may include games, children's programmes, e-learning, financial news, banking services, karaoke and even shopping. Service operators may also develop apps to attract more users.

In Taiwan, Chunghwa Telecom's (CHT) video service referred to as MOD is the major IPTV service provider. It commenced its service in 2004. As of August 2013, IPTV penetration had reached about 14.7 per cent (He, 2013). Although Japan, South Korea and other developed countries started to provide mobile TV services in 2004, the NCC hesitated to grant mobile TV licences because it did not find successful examples of mobile TV in the world. Therefore Taiwanese people can only watch mobile video via their smart devices. As in the United States, users of Taiwan's OTT video services do not need to apply for licences. Next VOD, a subsidiary of Next Media, started to provide the OTT service via broadband in December 2010. It gave away the set-top box free. It also allowed its new subscribers to watch video services free for the first three months. Owing to the unstable broadband service, Next VOD did not fly as it wished. It closed down its business in November 2012, because its parent company Next Media wanted to sell its Taiwan print and television businesses to other media conglomerates (UDN, 2012). Other OTT video services are not as big as Next VOD and have not been seen as constituting a threat to the traditional media platforms.

Nowadays, users are demanding multichannel delivery, time-shifting capabilities of services, content enhancement and on-demand personalisation (Küng, 2008). Therefore traditional television service providers are also turning to a multi-platform or time-shifting approach that provides Internet content in order to adapt to the ecosystem that is undergoing a process of transformation (Winslow, 2012). Players like the traditional media platforms and content providers are now seeking ways to respond to the changes in the ecosystem.

IPTV and OTT both require broadband to distribute content. The former has its own facility, while the latter usually uses other ISPs'

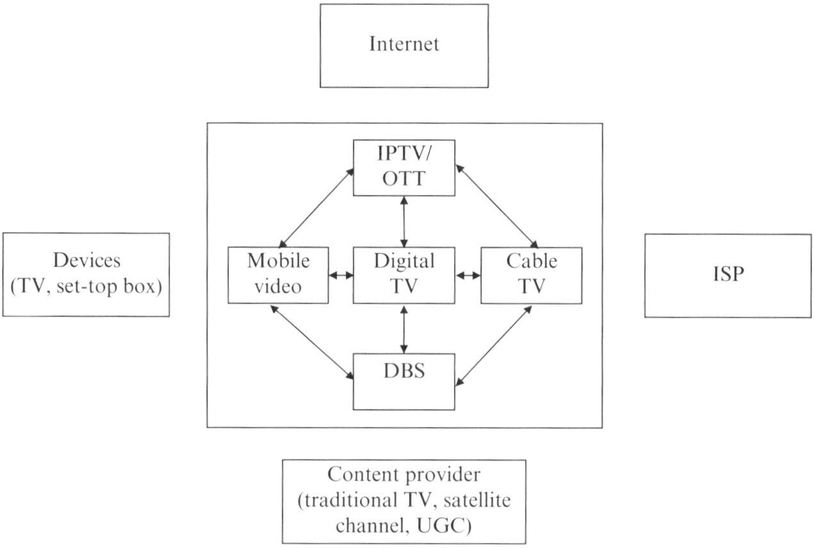

Figure 13.1 *Media platform ecology in Taiwan*

broadband. In Taiwan, the major TV platforms are terrestrial TV and cable TV. IPTV is the third biggest platform. Direct broadcasting service (DBS) is quite small. Mobile video platforms are mostly managed by the mobile operators. OTT TV players are very small as well, because they are still very new to the users. Aside from the various platforms available on the market, television and set-top box manufacturers, content providers, Internet and Internet service providers are also part of the media ecosystem and are interacting with the newly emerging service platforms (see Figure 13.1).

As illustrated in Figure 13.1, the interactions between platforms and other players are presented since they all have influences on each other to different extents. While traditional broadcasters serve as platforms, they may also produce their own content. Meanwhile, the open Internet also hosts a lot of user generated content (UGC) and content from other platforms. In addition, while IPTV and OTT video services disrupt the way in which the traditional broadcasters usually distribute content, IP delivery for content is in fact less costly than the traditional media platforms. However, since IP delivery often requires much bandwidth, it may cause these new media platforms to be discriminated against by ISPs. In addition, IPTV also competes for good content and viewership with cable TV. As for OTT service, it is not easy for it to grow at the infant stage because it also needs capital and a certain economic scale to acquire good content to compete with the traditional platforms. Content providers include independent producers, broadcasters, satellite channel operators and UGC providers. These players are definitely trying to find new business opportunities from the emergence of new media.

Examples of cooperation and competition between different players such as new media, traditional platforms and content providers can often be observed in the media value chain. With the transformation of the media ecosystem comes the change in the media value chain. Gershon (2000) has indicated that traditional barriers between these different media industries have collapsed and a more complex value chain has emerged. For instance, a traditional sequence of the value chain for media often includes *creation of content*, *production*, *publishing* and *broadcasting distribution* (Daidj and Jung, 2011). By getting involved in each component, media platforms are more able to adapt to the changing ecosystem and to refine its value chain. However, with new technology and changes in the ecosystem, the value chain has been modified to include a few more components. For instance, *creation of content* is now followed by *aggregation*. The next process is *distribution*, which is no longer limited to traditional broadcasters. Instead, content is also distributed to new *terminals* such as set-top boxes, personal computers and end consumers (Daidj and Jung, 2011). Therefore, as Venturini (2011) concluded, the new media platform has come to be characterised by the sequence of *creation*, *aggregation*, *distribution* and *consumption* in the value chain. With the digitalisation of various sectors, convergence in the new media industry makes the players in the value chain no longer independent. However, what remains a key feature is the component of content in the media value chain (Evens, 2010; Daidj and Jung, 2011). As Kompare (2010) described it, the current developments have changed how 'television', regardless of cable or broadcasting, is made, distributed and consumed.

3. IMPACT OF THE SMART REVOLUTION

The development of new media platforms in the smart revolution has had some impact on the consumers and the traditional media platforms. To the consumers, the new media provide more programme choices (linear and non-linear) and user flexibility (time and place). They could be regarded as supplementary or complementary to traditional media (Robinson et al., 2000; Stempel et al., 2000). Some OTT services can even transcend geographical boundaries and be received globally if the copyright issue is solved. To the traditional media platforms, the new media will spur them to consider adapting to time-shifting, on-demand, app delivery and innovative business models.

Since the amount of time people spend using media is both scarce and limited, the phenomenon of time displacement may appear. It states that the more time people spend on one medium, the less time they will spend on others (Dutta-Bergman, 2004). Since some new media platforms may provide different functions to the incumbent media such as time-shifting and app applications, the concept of functional displacement depicts that a new medium will likely displace an old medium if they share a similar function in a more effective way (Lee and Leung, 2006). Generally speaking, it is also suggested that Internet users tend to spend significantly less time on traditional media (such as TV) than non-Internet users (Tai-Quan and Zhu, 2011). According to Winslow (2012), there has been a heated debate over the long-term impact of OTT on traditional TV viewing and pay-TV subscription. In essence, OTT video services and traditional television provide similar services. While some argue that OTT TV marks the doomsday for traditional television, some disagree. Tai-Quan and Zhu (2011) concluded in their study that although there are clear signs of a displacement effect taking place, traditional media will still survive.

The impacts of IPTV on the traditional media platforms such as terrestrial TV and cable TV are different. Terrestrial TV stations consider IPTV to be a new media distribution platform. Although currently there is no must-carry obligation for IPTV, the terrestrial TV stations are willing to be carried by IPTV via third party. Cable operators, by contrast, regard CHT's MOD service as a threat to their business. They have thus urged the government to lessen the restrictions inherent in the Cable Radio and TV Act in order to compete with IPTV on a level playing field. The following is an analysis of IPTV's and OTT's impact on the incumbent media platforms and content providers.

3.1 IPTV's Impact on the Traditional Media Platforms and the Content Providers

In Taiwan, there is only one big IPTV player and one small IPTV player. The major player is Chunghwa Telecom (CHT), which launched its IPTV service, called Multimedia on Demand (MOD), in 2004. As of June 2014, it had 1.286 million subscribers. Vee TV, a regional cable TV operator, is the minor player. Taiwan was ranked as one of the top ten countries in the world for IPTV penetration in 2012 and the penetration rate of IPTV in Taiwan is expected to increase and reach 34 per cent by 2017 (Murray, 2012).

The impact IPTV has on terrestrial TV and cable TV is very different.

Before November 2007, CHT could carry the major channels of five terrestrial TV stations free of charge because the must-carry rule of the Cable Radio and TV Act was applicable. Since IPTV is a newly emerging media platform, the terrestrial televisions deem IPTV to be another means for content distribution in order to reach a larger audience. For this reason at the early stage there were no objections to CHT carrying these major channels without paying any programme fees. For the second and third digital channels of these stations, they were carried by CHT through retransmission consent via a third party. However, since CHT's MOD was ruled as an 'open platform' by the NCC, it was regulated by the Telecommunications Act and thus the must-carry rule could no longer be applied. A consequence of this is that all the terrestrial TV channels have been carried via a third party since then. Consequently, CHT has to pay for terrestrial TV content, regardless of the channels. This means that a rivalry between terrestrial television and IPTV does not exist. The terrestrial television stations can at least collect some royalty fees from CHT.

However, the relationship between IPTV and cable TV is very hostile. Cable operators used to argue that cable TV and IPTV look much the same, but cable TV can only run at a regional level, while CHT's IPTV can run nationwide. Cable TV's subscription rates are highly regulated, while CHT's IPTV's subscription rate is not regulated. Therefore they have urged the government to further relax the restrictions of the Cable Radio and TV Act in order for them to compete with IPTV on a level playing field. CHT has also argued that its management is severely regulated by the NCC because it has government shares. CHT is not allowed to produce programmes; neither can it act like a content aggregator. CHT can only act as an open platform. Therefore CHT's MOD service is highly restricted by the NCC rule. CHT has also appealed to the NCC to relax the controls.

According to Liu and Hsu (2011), players such as IPTV and cable TV operators are sometimes regarded as direct competitors of each other owing to the displacement effect and sometimes engage in head-to-head competition since IPTV can be similar to cable TV. In the case of viewers in Taiwan who frequently watch television, it is revealed in the study that cable TV and IPTV are complementary to each other. However, in the case of viewers in Taiwan who watch television infrequently, cable TV and IPTV may give rise to a platform displacement effect on each other. Therefore cable TV faces the threat of being replaced by IPTV given their similar functions.

NCC statistics indicate that, as of June 2014, Taiwan's cable TV subscribers had reached 4.989 million, which accounts for about 59.89 per cent of households.[1] Compared with NCC statistics for 2006, this shows that cable TV penetration has decreased by about 5 per cent. Therefore the displacement effect has occurred, although not to a great degree.

Since direct satellite broadcasting (DBS) is not a prevalent media platform in Taiwan, it has rarely been influenced by IPTV. As for IPTV's impact on the content providers, the situation differs widely among them. The independent producers are happy to see the extra distribution platform. As for the cable TV channels, they are glad to see the opportunities brought by the new platform, but they hesitate to be carried on CHT's MOD because they are afraid of being boycotted by the cable operators, although there has been no formal instance of this happening. Therefore CHT's MOD has encountered problems providing cable channels that appeal to its subscribers. It is also the reason why cable TV has only declined by 5 per cent since 2006.

3.2 OTT Video Services' Impact on the Traditional Media Platforms and Content Providers

In addition to IPTV, OTT video service is another newly emerging media platform. It is defined as the distribution of video content to a wide range of IP-enabled devices such as TVs, PCs, mobile phones and tablets over the unmanaged Internet (Venturini, 2011). An increasing number of users are accessing video services on their computers, tablets, mobile phones and other devices via the Internet. Users are now able to access the content in various locations and at various times (Strover and Moner, 2012). Online video platforms have similar functions to television. Essentially, both types of platform deliver video content to consumers and therefore the displacement effect might occur. A study conducted by Jiyoung (2013) has revealed that the time spent on the Internet watching user-generated videos or other video-sharing content does reduce the time spent watching TV. The study also suggests that understanding 'the functional uniqueness of an online video platform' when compared to a traditional platform is important, given the fact that both of these platforms distribute video content.

According to Berman, Battino and Feldman (2011), some of the key factors that drive consumers to enhance and replace traditional media with new media include value shifts and substitution. Value shifts refer to how consumers are now choosing new devices and services via new media platforms for the sake of simplicity of use, integration and convenience (Berman et al., 2011, p.45). For instance, aside from viewing content for free on Hulu or PPStream, viewers may also purchase films and TV programmes via sites like iTunes, or pay to stream them via Netflix and Hulu Plus. The rise of online video use has created a shift from a passive viewing audience to an active audience (Bondad-Brown et al., 2012). As Kompare (2010) puts it, traditional media platforms are still viable, 'but online distribution continues to grow and develop as a more varied and participatory means through which people use, rather than only view, rerun television'.

According to a widely publicised report by Nielsen Media Research in May 2009, the amount of time spent viewing TV continues to grow to over 153 hours a month, which is more than 50 times the amount of time spent watching online video (Kompare, 2010). However, in a recent survey, Belkin and Harris Interactive found that 30 per cent of US Internet users would consider abandoning their cable TV and watching television exclusively online. The fact that online video streaming has become popular over the past few years is a sign that people will quickly turn against traditional subscription TV (Kleinman, 2013).

In the case of Taiwan, OTT video service providers include network operators, ISP service providers, TV manufacturers and content aggregators. While CHT is an incumbent telco, it is banned from producing content as an IPTV service provider as the government currently holds company shares. Since OTT is not regulated by the NCC, CHT and HTC have set up a 50:50 joint venture to establish Huada that provides the OTT service. As a result, Huada has great advantages as it can distribute content through CHT's network, but has a major disadvantage in that it still lacks quality content to attract users.

Besides CHT, Kbro is another network operator that plans to provide OTT video services in the future. As a cable MSO operator, Kbro attempted to include Next VOD into its OTT services. But when Next VOD went out of business in October 2012, Kbro repositioned itself to provide OTT services in two stages. At the current stage, Kbro offers Super MOD to its customers. This is a VOD service via the set-top box. While Kbro has quality content, the next stage is to provide OTT services via the open Internet.

Another OTT video service provider, 5 TV, is in fact an ISP. Owing to its background as a service provider it has significant strengths in terms of distributing content. On the other hand, its operations are only regional.

Catchplay, a content aggregator, is also an OTT service provider in Taiwan. Catchplay has obvious strengths in terms of content variety.

Samsung Taiwan, as a TV manufacturer, has also joined the field to provide an OTT video service. The services are provided via the connected TV device known as smart TV, in which the connection to OTT services is already pre-configured.

Aside from the OTT players mentioned above, YouTube is also a very popular OTT video service provider in Taiwan, but is not owned by Taiwanese interests.

Elta TV, on the other hand, is a local OTT player that started out initially as a technology company. It is known for its sports programming. However, it still needs time to grow.

The other OTT player, hichannel is also supported by CHT. It has better network distribution owing to CHT's high broadband penetration rate throughout Taiwan. However, it should be noted that although consumers have access to the hichannel service they may not necessarily use it. In addition, hichannel has neither branded content nor secured exclusive content that can appeal to the audience (see Table 13.1).

Currently, the OTT video services are not regarded as a big threat to cable TV and IPTV, because their economic scale is small and adoption rate is low in Taiwan. However, they might influence some of the on-demand service revenues of the traditional platforms. Having observed the successful OTT examples such as Netflix and Hulu from the United States, CHT and Kbro have joined the market and tried to develop their own OTT services. As for the terrestrial TV stations, since they are not against airing their content on another platform for more exposure and viewership, they do not have a rival relationship with OTT video services. However, some of the content providers may wish to establish a profit-sharing relationship with OTT video platforms, although some channels with broad appeal may only be interested in collecting content fees at the early stage. Therefore they will hesitate to distribute their content through OTT platforms unless they can receive satisfactory payment for the content. Since the OTT video services rely on broadband for content delivery, if they do not own any ISPs, their quality of service often cannot be guaranteed. Meanwhile, their services might be discriminated against by ISPs if their content causes traffic congestion in the bandwidth. The OTT operators often complain that their businesses are also influenced by the OTT players in China, because Taiwanese people can watch Chinese OTT programmes, which may be pirated movies or programmes, free of charge through the Internet. It thus makes it hard for Taiwanese OTT operators to collect money from viewers for the same movies or programmes shown on Chinese OTT platforms.

4. COOPERATION, COMPETITION AND CO-OPETITION AMONG THE MEDIA PLATFORMS

Since the competition brought about by the newly emerging media is inevitable, the traditional media platforms can cooperate and compete

Table 13.1 OTT players in Taiwan

	Huada	Kbro (Super MOD)	5 TV	Catchplay	Samsung Taiwan	YouTube	Elta TV	hichannel
Background	CHT, HTC	Cable MSO	ISP	Content aggregator	TV	Google (not local player)	Technology company	CHT
Strength	Network, distribution	Network, content	Via connected TV	Content	Via connected TV	Network, content	Has niche programmes	Network, distribution
Weakness	Not enough content	OTT TV is not its own service	Regional	Management issue	Regional	More UGC content	Small scale	No exclusive content

with new platforms at the same time. In the real business world, today's rival can be tomorrow's partner. Therefore co-opetition exists in the media ecosystem and market. The concept of 'co-opetition' has many origins that can be found in game theory and applied to political science and economics (Walley, 2007; Daidj and Jung, 2011). It has been expanded by Brandenburger and Nalebuff (1996) who illustrate a basic structure for co-opetition as a single company and its connections to customers, competitors, suppliers and complementors. More precisely, co-opetition can be found between the company and its competitors, complementors and suppliers. The concept of 'co-opetition' is described as the situation where competition and cooperation co-exist simultaneously between rival firms or companies (Gnyawali and Park, 2011). Similarly, Daidj and Jung (2011) define this mode of cooperation as the strategic alliances between rival companies, which merely prolong the competition in a different form. As stated before, the interactions and dynamics in the media ecosystem have changed with the emergence of new media. Both competitive and complementary relationships can be observed among incumbent and newly emerging media platforms. While some of the players work as competitors against each other, some may cooperate by forming alliances. More importantly, some may even compete and cooperate at the same time.

According to Brandenburger and Nalebuff's (1996) Value Net, a player is your complementor 'if customers value your product more when they have the other player's product than when they have your product alone'. However, a player is your competitor 'if customers value your product less when they have the other player's product than when they have your product alone'. Take cable TV as an example. Before IPTV appeared in the market, cable TV had been competing with terrestrial TV for ratings and advertising revenues. In 2001, in order to have more budget for producing idol drama (trendy drama) and increase the audience share, a cable TV channel, SET TV, cooperated with a terrestrial TV station, Taiwan Television Enterprise (TTV). They formed a strategic partnership to co-produce a series of the idol drama *Lavender* by sharing the production cost, advertising revenue and copyright. SET TV allowed TTV to air the programme first. It aired the programme a few days later. Originally, SET TV was supposed to compete with TTV, but they cooperated in order to increase audience share and have a better economic outcome. The result was a great success (Chen, 2010). Therefore SET TV repeated this model by cooperating with another terrestrial TV channel, Chinese Television System (CTS), to co-produce another idol drama, *My MVP Valentine*, in 2002. This business strategy has been imitated by other cable TV channels such as GTV.

The next example is that of IPTV, whose direct competitor is cable TV. Its complementor can be other content aggregators or DVD rentals. For instance, Taiwan's biggest DVD rental is Asia 1, which can choose to go to CHT's IPTV platform or to remain as a DVD rental. If it goes to CHT, it will have extra revenue, but its rental revenue will be affected. Therefore if Asia 1 is willing to be carried by CHT's IPTV, it means that the accumulated revenue should be greater. In this way, Asia 1 can be considered to be IPTV's complementor. CHT's IPTV subscribers can purchase Asia 1's VOD programme on CHT's platform, while CHT also provides its own VOD service. If customers value CHT's IPTV more when they have the other player's product (such as Asia 1) than when they have CHT's IPTV alone, the relationship between CHT's IPTV and Asia 1 can be an example of co-opetition.

Taga, Schwaiger and Little (2012) suggest that traditional platforms should try to diversify into premium services and follow the on-demand platforms. In essence, the introduction of such services will increase the value of the platforms. For instance, the UK cable network Virgin Media welcomes both Netflix and Amazon's LoveFilm as part of its TiVo-based TV platform. US cable companies are also considering adding Netflix to their platforms through set-top boxes (STBs) (Patrizio, 2012). In Taiwan, an example can show how cable MSO Kbro incorporated the OTT video service to create a more competitive edge and increase more value for its platform. In September 2012, Kbro launched a so-called Super MOD service, which integrates multichannel, VOD and connected TV services. Its major purpose is to compete with CHT's IPTV.

In the real world, it is not easy to compete and cooperate at the same time. For instance, IPTV and digital cable operators both need to provide STBs to their customers. They can provide either simple STBs or hybrid STBs, while only the hybrid STBs can receive services from other platforms. For instance, in the local market, Kbro provides hybrid STBs referred to as Super MOD, but CHT only provides the proprietary STBs. The IPTVs in Hong Kong, South Korea and the UK each provide hybrid STBs to their customers, but their purposes in so doing are different. PCCW's Now TV in Hong Kong wants to provide convenience to its customers. KT's Qook TV wants to help its customers have easy access to Skylife, a DBS service it acquired (Dziadul, 2013). BT's BT Vision in the UK wants to reach more customers. BT even joined the BBC, ITV, Channel 4, Channel Five, TalkTalk and Arqiva to provide Internet connected TV called YouView, which is offered through a set-top box that can be purchased from a retailer for a one-off payment, or as part of a package from BT or TalkTalk. BT joined the consortium of YouView in order to attract more customers and become more competitive. BT also provides exclusive sports programmes and channels to its customers for free in order to maintain its viewership (McCabe, 2010). Even though CHT observes the business models of the IPTV services provided abroad, it still insists on providing its own STB.

The new media platforms not only compete or cooperate with the traditional platforms, but as we saw in Figure 13.1 they also need to interact with other players, such as manufacturers, IPSs and content providers. For instance, CHT formed an OTT company called Huada with HTC, a phone manufacturer in order to bypass the government regulation. There is a regulation for CHT's IPTV, but there is no regulation for OTT business. Therefore Huada has started to act as a content aggregator on CHT's media platform. In the international IPTV market, Verizon has a strategic relationship with DIRECTV. KT acquired Skylife, a direct broadcasting satellite (Dziadul, 2013). Orange, formerly known as France Telecom, has acquired shares from a video-sharing website named Dailymotion to sustain its growth in terms of IPTV subscriptions (Plesser, 2011). CHT, restricted by the regulations, cannot acquire content business directly, so it can only welcome content providers, especially terrestrial TV or cable TV channels, to come to its IPTV platform in order to provide branded content to its customers. Formosa TV (FTV), a terrestrial TV channel, is very active in the new media market. It has aggregated much content and has used CHT's platform as a base for reaching a larger audience and acquiring more users.

5. CONCLUSION

With the advent of new media platforms, the media ecosystem has to a great extent been changed. Although there have been many studies that have focused on the new media's impact on the old media, the displacement effect and complementary effect still co-exist, depending on the consumer demand and the media's ability to adapt. While some of the players work as competitors against each other, some may cooperate as alliances. Some may even compete and cooperate at the same time. New media platforms will give consumers more choices, flexibility and convenience. Some may even go global and transcend the geographical restraints. Such media platforms may even spur the traditional media platforms to consider adopting innovative business models characterised by time-shifting, on-demand and app delivery with different pricing options. Conversely, traditional media at the same time are also learning how to cooperate and compete with the new media platforms. They are also learning to join or to incorporate the new media platforms in order to maintain their competitive edge and withstand the rampaging effect of the new media. With the impact of convergence and the structural change taking place in the value chain, it is suggested that 'co-opetition' might be a viable strategy for the new and traditional media platforms.

NOTE

1. In fact, according to a report compiled by the Directorate-General of Budget, Accounting & Statistics of the Executive Yuan, the real cable TV penetration was about 82.9% in 2011, because some households with illegal reception were not counted.

REFERENCES

Beach, J. (2012), 'Global IPTV subs to triple by end-2017', *IP&TV News*, available at http://www.iptv-news.com/2012/08/global-iptv-subs-to-triple-by-end-2017/ (accessed 27 February 2015).

Berman, S.J., B. Battino and K. Feldman (2011), 'New business models for emerging media and entertainment revenue opportunities', *Strategy and Leadership*, **39** (3), 44–53.

Bondad-Brown, B.A., R.E. Rice and K.E. Pearce (2012), 'Influences on TV viewing and online user-shared video use: demographics, generations, contextual age, media use, motivations and audience activity', *Journal of Broadcasting and Electronic Media*, **56** (4), 471–93.

Brandenburger, A.M. and B.J. Nalebuff (1996), *Co-opetition*, New York: Doubleday.

Chen, Y.L. (2010), 'The application of game theory: a case study of coproduction and broadcasting between TTV and SETV', *2010 Chinese Communication Society Conference*, National Chung Cheng University, Chiayi.

Daidj, N. and J. Jung (2011), 'Strategies in the media industry: towards the development of co-opetition practice?', *Journal of Media Business Studies*, **8** (4), 37–57.

Dutta-Bergman, M.J. (2004), 'Complementarity in consumption of news types across traditional and new media', *Journal of Broadcasting and Electronic Media*, **48** (1), 41–60.

Dziadul, C. (2013), 'IPTV surges in Korea', *Broadband TV News*, available at http://www.broadbandtvnews.com/2013/03/20/iptv-surges-in-korea/ (accessed 27 February 2015).

Evens, T. (2010), 'Value networks and changing business model for the digital television industry', *Journal of Media Business Studies*, **7** (4), 41–58.

Executive Yuan (2012), 'Digital convergence development policy initiative (2010–2015)', 2nd edn, Taiwan, available at http://www.arcnet.org.tw/upload/files/dc7bc3d9.pdf (accessed 27 February 2015).

Gershon, R. (2000), 'The transnational media corporation: environmental scanning and strategy formulation', *Journal of Media Economics*, **13** (2), 81–101.

Gnyawali, D.R. and B.J.R. Park (2011), 'Co-opetition between giants: collaboration with competitors for technological innovation', *Research Policy*, **40** (5), 650–63.

He, P.J. (2013), 'CHT's MOD enters 4KTV market with the help of CHIME', *Economic Daily News*, 16 August.
Informa Telecoms & Media (2012), 'New horizon for cable: ten strategies in the connected era', available at http://www.informatandm.com/wp-content/uploads/2012/03/New-Horizons-for-cable-white-paper-6th-March.pdf (accessed 19 August 2013).
ITU-T Newslog (2006), 'IPTV standardization on track say industry experts', available at http://www.itu.int/ITU-T/newslog/IPTV+Standardization+On+Track+Say+Industry+Experts.aspx (accessed 21 August 2013).
Jiyoung, C. (2013), 'Do online video platforms cannibalize television? How viewers are moving from old screens to new ones', *Journal of Advertising Research*, **53** (1), 71–82.
Kleinman, A. (2013), '30% of internet users in the U.S. would consider becoming cable cutters: survey', *The Huffington Post*, 2 April, available at http://www.huffingtonpost.com/2013/04/02/cable-cutters-internet_n_3000576.html (accessed 27 February 2015).
Kompare, D. (2010), 'Reruns 2.0: revising repetition for multiplatform television distribution', *Journal of Popular Film and Television*, **38** (2), 79–83.
Küng, L. (2008), *Strategic Management in the Media*, London: Sage.
Lee, P.S.N. and L. Leung (2006), 'Assessing the displacement effects of the Internet', *Telematics and Informatics*, **25** (3), 145–55.
Liu, Y.L. (2012), *Law and Policy Research of Digital Convergence: Communications Platforms and Content-Related Issues*, unpublished report, National Science Council, Taiwan.
Liu, Y.L. and W.Y. Hsu (2011), 'Media platform competition: the displacement effect of IPTV on cable TV in Taiwan', *Adoption, Usage and Global Impact of Broadband Technology: Diffusion, Practice and Policy*, Hershey, PA: IGI Global.
McCabe, M. (2010), 'YouView will encourage competition, says chief', *Media Week*, 16 September, available at http://mediaweek.co.uk/news/1028931/YouView-will-encourage-competition-says-chief/ (accessed 27 February 2015).
Murray, S. (2012), 'Global IPTV subscribers to triple', *Global IPTV Forecasts*, Digital TV Research Ltd, available at https://www.digitaltvresearch.com/ugc/iptv12%20product%20copy%20PDF_sample_54.pdf (accessed 27 February 2015).
National Communications Commission (2013), '2013 Q4 cable subscribers and digital set-top-box penetration', available at www.ncc.gov.tw (accessed 27 February 2015).
Newell, J., J.J. Pilotta and J.C. Thomas (2008), 'Mass media displacement and saturation', *International Journal of Media Management*, **10** (4), 131–8.
Patrizio, B. (2012), 'How will OTT affect traditional pay-TV providers?', *The Guardian*, 16 March, available at http://www.theguardian.com/media-network/media-network-blog/2012/mar/16/ott-pay-tv-providers-subscribers (accessed 27 February 2015).
Plesser, A. (2011), 'France's Orange has global plans for IPTV expansion with Dailymotion', *Business Insider*, available at http://www.businessinsider.com/frances-orange-has-global-plans-for-iptv-expansion-with-dailymotion-2011-11 (accessed 27 February 2015).
Strover, S. and W. Moner (2012), 'Immersive television and the on-demand

audience', paper presented at the annual meeting of the International Communication Association, Sheraton Phoenix Downtown, Phoenix, AZ.

Taga, K., C. Schwaiger and A.D. Little (2012), 'OTT video: First to scale wins', available at http://www.telecomasia.net/content/ott-video-first-scale-wins (accessed 24 May 2013).

Tai-Quan, P. and J.H. Zhu (2011), 'A game of win-win or win-lose? Revisiting the internet's influence on sociability and use of traditional media', *New Media and Society*, **13** (4), 568–86.

The Digital TV Consultancy (2013), 'Chunghwa Telecom MOD/IPTV subscribers stable at 1.2 million in 1Q 2013', *Digital TV News*, available at http://www.digitaltvnews.net/content/?p=22601 (accessed 2 May 2013).

Venturini, F. (2011), 'The race to dominate the future of TV', New York: Accenture, available at http://www.accenture.com/SiteCollectionDocuments/PDF/Accenture_Communications_Media-Entertainment_OTTV_Future_of_TV.pdf (accessed 27 February 2015).

Walley, K. (2007), 'Co-opetition: an introduction to the subject and an agenda for research', *International Studies of Management and Organization*, **37** (2), 11–31.

Winslow, G. (2012), 'Over-the-Top platforms shake up ethnic market', *Broadcasting and Cable*, **142** (31), 22.

PART VI

The smart revolution in countries

14. The Australian digital market: opportunities and challenges

Gary Madden and Hasnat Ahmad

1. INTRODUCTION

The digital revolution of the last decade has led to the widespread expansion of broadband in Australia. High-quality communication infrastructures are crucial to seize the windows of opportunity offered by digital services. However, Australians have been slower in adopting digital services than other leading economies. As a result, the National Digital Economy Strategy (NDES) commits Australia to be among the world's leading digital economies by 2020.

Importantly, the government aims to assist business and the community to engage in the national growth process via the digital economy. The government further realises that seizing the window of opportunity created by the digital revolution requires broadband to be guaranteed for all Australians, regardless of where they live. Currently, a superfast broadband network is being developed through investment by the Australian government through the National Broadband Network (NBN).

At 2012, the World Economic Forum reported that digital technology accounted for an estimated 21 per cent of GDP growth in the world's most advanced economies during the past five years (World Economic Forum, 2012). Furthermore, the Boston Consulting Group (BCG) estimates the digital economy will contribute a total of $4.2 trillion to the G20's GDP by 2016 (BCG, 2012). According to the World Bank (Bowles, 2011) each 10 per cent increase in the population using broadband communication in Australia will probably add 1.2 per cent to the annual GDP.

This chapter provides an overview of the Australian digital market, including the history of Australian broadband, current market trends, and broadband opportunities and challenges. The chapter also provides selected information on ICT market players and institutions. The chapter is structured as follows: Section 2 provides an overview of government organisations sharing the responsibility of monitoring, supervising and regulating the Australian Internet market. Section 3 explores the evolution of Australian broadband and highlights key developments. Section 4 examines the nature and status of services delivered via the NBN. Section 5 and Section 6, respectively, provide an overview of Australian online consumers and business by focusing on the current market landscape. Section 7 discusses the opportunities available to consumers and business. Section 8 provides an overview of the e-government landscape. Finally, Section 9 discusses the challenges posed by an evolving digital market to the Internet market.

2. INSTITUTIONS

Several government organisations share the responsibility for monitoring, supervising and regulating the Internet market. An objective of the Department of Broadband Communications and Digital Economy (DBCDE) is to develop a vibrant, sustainable and internationally competitive broadband and communications sector that promotes the digital economy. The DBCDE develops policy advice and programme delivery to support the efficient and strategic management of public resources, facilitate the competitive provision of services, enable access to reliable and reasonably priced basic and essential services, and provide effective consumer protection.

Through these activities DBCDE supports and encourages the development of world class communications infrastructure that is competitively priced, widely accessible, highly reliable and innovative and that facilitates choice. DBCDE's Infrastructure Group is responsible for implementing the NBN and associated regulatory reform. It also provides advice on telecommunications infrastructure, spectrum management and related matters.

The Australian Communications and Media Authority (ACMA) is an agency responsible for regulating and monitoring Internet content, enforcing Australia's anti-spam law and making rules about accessing the Internet via premium mobile phone services. The ACMA is a statutory authority within the federal government portfolio of the DBCDE. ACMA's responsibilities include: promoting self-regulation and competition in the communications industry; fostering an environment in which electronic media respect community standards and respond to audience and user needs; managing access to the radio frequency spectrum; and representing Australia's communications interests internationally.

The ACMA's focus includes consumer experience in transacting with the NBN, specifically, with regard to transitioning and the ongoing service environment. The ACMA is focused on the current consumer protection framework and assists consumers during this transition period (ACMA, March 2012).

The Australian Competition and Consumer Commission (ACCC) is an independent statutory authority formed in 1995 to administer the Trade Practices Act 1974 and other relevant acts. The ACCC provides advice on access and technical convergence on communications markets. In preparing plans for the NBN, the government sought the ACCC's assistance. In 2008–09, ACCC provided competition-related advice to the government's NBN panel of experts in relation to tenders for the network. The ACCC acknowledged the NBN's potential to start a new wave of infrastructure investment, technological change and product innovation in the sector.

The ACCC's access regulation will continue in an NBN environment. The ACCC believes, in the NBN world, that defining access services as close as possible to basic physical infrastructure should maximise the ability of access seekers to control their own costs and supply chain, differentiate service offerings, innovate and improve service quality. The ACCC is committed to working closely with all parties to maximise the prospects for competition under the current regulatory regime. However, in the longer term, the new regulatory regime would be matched with the structure and design of the NBN to develop effective competition.

NBN Co Limited is the wholly owned Commonwealth Company established, on 7 April 2009, to design, build and operate the infrastructure to enable advanced digital services to be provided to all Australians. NBN Co aims to connect 93 per cent of homes, schools and workplaces with high-speed FTTP networks. The remaining 7 per cent will be connected using next generation fixed wireless and satellite, subject to final design.

3. AUSTRALIAN INTERNET EVOLUTION

Very low population density in most of regional, rural and remote areas inhibits access provision at affordable prices. According to the Australian Information Industry Association (AIIA, 2012), half of the Australian population resides in five cities and 75 per cent in 25 cities. Large distances and small populations mean the service provision requires substantial infrastructure investment with low return.

While Australia was late in introducing ADSL, Internet subscribers quickly migrated from dial-up service (96.4 per cent broadband connections at June 2012). Delayed ADSL introduction was because of limited interest in ADSL service until the ACCC allowed regulated access to the Telstra's copper loop in 1999. The ACCC also forced Telstra to allow competitors into the market by issuing a competition notice in 1999 to prohibit the launching of own services prior to competitors being provided with access to Telstra's exchanges. Around 2000, Telstra introduced ADSL broadband, with an ability to transmit data 26 times faster than by modem. ADSL required access to the telephone exchange and the Telstra-owned copper line network – this allowed Telstra to be dominant in the Internet market. During this period Telstra established a strong customer base. Other ISPs resold Telstra service.

Gradually, larger ISPs began taking advantage of regulated access to the unbundled local loop (ULL). ULL reduced costs and gave service providers control of their service networks, except the copper pair. In September 2000, competition to Telstra's DSLAMs was first provided by the then Optus subsidiary XYZed. In 2003, the installation of DSLAM by Adelaide-based ISP Internodes in Meningie initiated competition in the residential infrastructure market. Several other ISPs followed Internodes.

At 2005, second-generation ADSL, ADSL2+, increased Internet speed to as fast as 24 million bits per second – 400 times faster than modems. At November 2005, Telstra began a plan to substantially alter their copper network and set up fibre to the node (FTTN) networks. The proposal had intended to build an FTTN network to serve 4 million addresses in Sydney, Melbourne, Brisbane, Perth and Adelaide. However, Telstra indicated that the deployment was conditional on receiving concessions from the 'access rules' that apply under Part XIC of the Trade Practices Act 1974. With no regulatory holiday granted, Telstra's proposal was withdrawn in 2006.

At November 2007 the first Naked DSL product was announced by iiNet. Other providers followed, offering Naked DSL to customers, reducing line rental fees. Wireless broadband is growing, with many point-to-point fixed wireless broadband providers serving broadband-deficient regional and rural areas, predominantly via Motorola Canopy and WiMAX.

At December 2007, December 2008, December 2009, December 2010, December 2011 and June 2012 there were, respectively, 5.22 million, 6.64 million, 8.06 million, 9.74 million, 11.12 million and 12.03 million broadband subscribers. Table 14.1 shows many of the broadband subscribers migrated from dial-up services, with 96.4 per cent of connections being broadband at June 2012. This rapid increase in broadband subscription highlights the strong demand for broadband.

Mobile wireless Internet continued to be the fastest growing technology by subscriber numbers (5.86 million) at end-2012. Although Digital Subscriber Line (DSL) connections grew by 1.73 per cent to 4.63 million in the six months ending June 2012, the share of DSL continued to decline, accounting for 38 per cent of access connections. The number of satellite

Table 14.1 Internet subscribers by access connection

	03/2005	09/2006	12/2007	12/2008	12/2009	12/2010	12/2011	06/2012
Dial-up	4180	2750	1890	1286	891	707	473	439
Broadband	1800	3910	5220	6639	8059	9739	11,123	11,597
Total ('000)	5980	6660	7110	7925	8950	10,446	11,596	12,036
Dial-up	69.9	41.3	26.6	16.2	10.0	6.8	4.1	3.6
Broadband	30.1	58.7	73.4	83.8	90.0	93.2	95.9	96.4
Total (%)	100.0	100.0	100.0	100.0	100.0	100.0	100.0	100.0

Sources: ABS: Internet Activity, Australia, Mar. 2005, Cat. No. 8153.0; Internet Activity, Australia, Sep. 2006, Cat. No. 8153.0; Internet Activity, Australia, Dec. 2007, Cat. No. 8153.0; Internet Activity, Australia, Dec. 2008, Cat. No. 8153.0; Internet Activity, Australia, Dec. 2009, Cat. No. 8153.0; Internet Activity, Australia, Dec. 2010, Cat. No. 8153.0; Internet Activity, Australia, Dec. 2011, Cat. No. 8153.0; Internet Activity, Australia, Jun. 2012, Cat. No. 8153.0.

connections was around 94,000 at June 2012. While fibre subscription remained small at 52,000, this number grew by 40 per cent from December 2011 to June 2012. Figure 14.1 shows high-speed broadband is increasingly popular, with 96.4 per cent of Internet connections being broadband at June 2012. It is anticipated that broadband connections will soon phase out the dial-up connections.

4. NATIONAL BROADBAND NETWORK

High-speed broadband is an important agenda item in the government's development goals. During 2007–08 the DBCDE attempted to increase the availability of faster broadband. This is to be achieved by deploying the NBN. This commitment to the universal provision of high-speed broadband is a key objective for the government.

NBN Co aims to provide high-speed broadband services to Australians via fibre, fixed wireless and satellite, depending on location. The government proposes the NBN to be a national, wholesale-only, open access broadband network using fibre-to-the-home (90 per cent of the population, speeds up to 100 Mbps) and wireless and satellite technologies to more remote parts of rural Australia (average speeds of 12 Mbps). The government aims to implement legislative changes to govern NBN Co and to commence consultation on changes to the existing telecommunications regulatory regime.

On 6 May 2010, an NBN implementation study provided details of the interplay among potential commercial NBN stakeholders, mainly

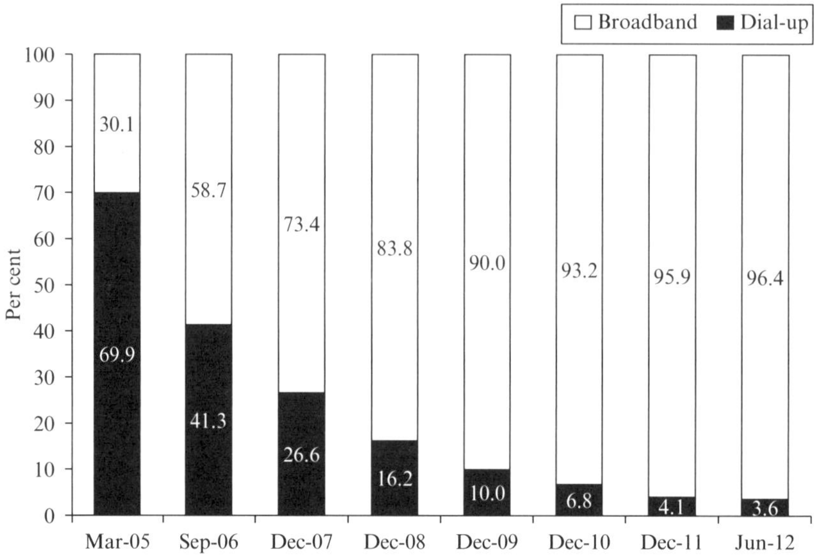

Sources: ABS: Internet Activity, Australia, Mar. 2005, Cat. No. 8153.0; Internet Activity, Australia, Sep. 2006, Cat. No. 8153.0; Internet Activity, Australia, Dec. 2007, Cat. No. 8153.0; Internet Activity, Australia, Dec. 2008, Cat. No. 8153.0; Internet Activity, Australia, Dec. 2009, Cat. No. 8153.0; Internet Activity, Australia, Dec. 2010, Cat. No. 8153.0; Internet Activity, Australia, Dec. 2011, Cat. No. 8153.0; Internet Activity, Australia, Jun. 2012, Cat. No. 8153.0.

Figure 14.1 Proportion of subscribers by connection

communications services, energy and utilities, property and construction, financial and information technology. The study confirmed the architecture of the NBN as optical fibre-dominant and suggested that 93 per cent of households should be connected to fibre-to-the-premises (FTTP) network. Another 4 per cent should access a fixed wireless network with the remaining 3 per cent offered access to satellite service.

The study recommended that the government requires NBN Co to anticipate Layer 2 competition (ULL) after the completion of network deployment. NBN Co is asked to develop a plan in consultation with the ACCC to identify an appropriate topology to enable physical unbundling and determine the extent of deployment. The ACCC works closely with policy makers to ensure that regulatory arrangements during the transition enhance competition in the telecommunications industry in the short to medium term.

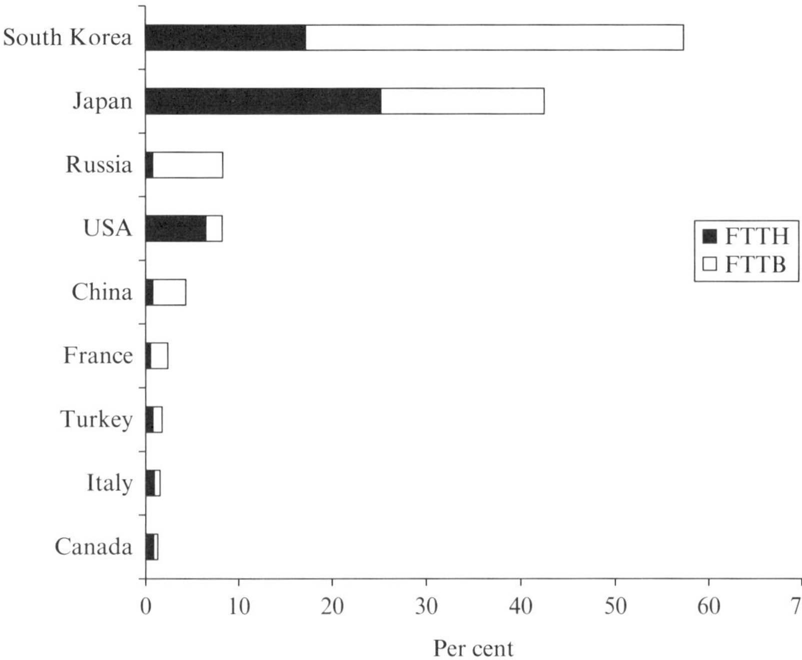

Source: Fibre to the Home Council, February 2012.

Figure 14.2 G20 FTTH/B and LAN penetrations

4.1 NBN Deployment

Senator Conroy announced the start of the deployment in Tasmania on 16 July 2009. Fibre-to-the-home (FTTH) and fibre-to-the-building (FTTB) are nascent. Indeed, only 30 countries have more than 1 per cent adoption of FTTH/B (Fibre to the Home Council, February 2012). Only nine (South Korea, Japan, Russia, the USA, China, France, Turkey, Italy and Canada) G20 economies have adopted more FTTH/B (Figure 14.2). Australia has total fibre customers of around 0.4 per cent of the total population at June 2012.

At August 2011, the first NBN services were being delivered in the Tasmanian (Stage 1) communities of Scottsdale, Smithton and Midway Point. Construction in the Stage 2 Sorell and Triabunna commenced. In Stage 3, another 90,000 premises will be connected in Hobart, Launceston, Devonport and Burnie. According to Senator Stephen Conroy, 40,000 premises in Hobart, 30,000 in Launceston and 10,000 in Burnie and Devonport are to be connected.

On 18 October 2011, NBN Co released a 12-month national deployment plan. The plan listed communities by state and territory where work was to begin (485,000 premises) between October 2011 and September 2012, as well as where work on the deployment had commenced (63,500 premises). At January 2012, more than 4000 homes and businesses were connected to NBN (NBN Co, 3 January 2012). They comprise 2315 premises using fibre-optic service and 1700 in rural and remote areas receiving broadband via interim satellite service. In March 2012, NBN Co released another three-year plan with work to commence in over 1500 communities (3.5 million premises) by 30 June 2015. The plan is expected to cover 2.7 million premises in addition to the January 2012 updated figures of 0.8 million.

On average deployment is expected to take 12 months from the start of fibre network deployment until individuals can receive high-speed broadband. On 1 July 2011, NBN Co launched commercial service on the Interim Satellite Service (ISS), providing rural and regional areas with immediate access to enhanced broadband services. At June 2012, NBN Co had 85 customers trialing the fixed-wireless service, with a further 1800 homes and businesses in the area able to order service as part of the trial.

At February 2013, the NBN was switched on in metropolitan Brisbane and Canberra. On 1 February 2013, more than 4000 homes and businesses in Canberra's northern suburbs (Gungahlin and Harrison) were connected to NBN. Construction also commenced to connect around 20,000 premises in the ACT, in areas including Civic, Palmerston and Amaroo. The Canberra rollout should be completed by 2015.

At March 2012, a $300 million contract was awarded to Visionstream Australia to connect all Tasmanian homes to broadband Internet by 2015. Completion in Tasmania is expected to create 800 jobs (Lara Giddings, March 2012). At June 2012, NBN Co launched NBN fibre services in Deloraine, Triabunna, Sorell and Kingston Beach, with George Town and St Helens. On 4 April 2013, NBN Co announced the deployment of fixed wireless in eight Northern Territory areas. On 21 March 2013, NBN Co announced a delay of three months in fibre optic cable deployment. The revised target for June 2013 is to connect between 190,000 and 220,000 premises.

The actual number of NBN connections was released by NBN Co on 10 April 2013. The company announced an average 34.4 per cent adoption in locations where the network has been operating for more than 12 months. Adoption across sites operating more than six months is above 28 per cent (NBN Co, 2013). Further, Kiama (NSW) and Willunga (South Australia) have the highest fibre adoption, with more than 50 per cent of premises connected. On 19 April 2013, NBN Co announced that Australia would be a Gigabit Nation by year end.

On 5 May 2013 the three-year deployment plan was updated by adding 1.3 million premises to the NBN construction timetable. This extends broadband deployment to 5 million premises by June 2016. Furthermore, Blacktown is the first Sydney suburb to switch on to the NBN on 5 May 2013. About 1300 homes and businesses are connected, with 9300 premises soon to be connected.

5. ONLINE CONSUMERS

Technological advance facilitated the development and delivery of consumer data services and applications. Consumers are purchasing goods and services online. About half of total Internet subscribers use mobile wireless for access (Table 14.2). The share of mobile wireless Internet continues to rise, with a 5 per cent increase during June 2011 to June 2012. The proportion of the subscribers using DSL connections recorded a 3 per cent decline over the same period. Cable is the third most popular mode of connecting, with 8 per cent of Internet connections. At June 2012, with around 1 per cent adoption of satellite connections, the mode is the fourth most popular mode of Internet access. While fibre subscribership recorded an increase of 40 per cent over the six months ending June 2012, the proportion of fibre-based connections remained small at around 0.05 per cent of connections.

In November 2012, EXA reported that 93 per cent of the Australian online community are active Internet users, with an average 22 hours

Table 14.2 Internet subscribers by access connection (per cent)

	June 2011	December 2011	June 2012
Dial-up	5.0	4.1	3.6
DSL	41.0	39.0	38.0
Cable	8.0	8.0	8.0
Satellite	1.0	1.0	1.0
Mobile wireless	44.0	47.0	49.0
Total broadband	95.0	95.9	96.4
Total	100.0	100.0	100.0

Source: ABS: Internet Activity, Australia, Jun. 2012, Cat. No. 8153.0.

spent online weekly. The report further states that 41 per cent of Internet users access when not at home or work. Further, 15.6 million online users visited 42 billion pages during July 2012, with total time spent estimated at 29 billion minutes. During July 2012, the most popular ten websites were: Google, Facebook, MSN, YouTube, Microsoft, Yahoo!7, Wikipedia, eBay, Apple and Blogger. Google is the most popular site and Blogger the least popular. At July 2012, the population aged over 50 years comprised 32 per cent of the online users. The proportions of the population aged 2–17 years, 18–24 years, 25–34 years and 35–49 years was, respectively, 10 per cent, 13 per cent, 18 per cent and 27 per cent of the online population. Jessica Irvine (News Limited Network, 24 March 2013) highlighted that Australians spent $12.8 billion shopping online during 2012.

The rapid adoption of mobile wireless broadband is argued to be owing to tablet and Internet-enabled mobile phone adoption. Currently, there are 108 mobile cellular subscribers per 100 persons at 2011 (ITU, 2012). At February 2013, 50 per cent of adults owned smartphones (ACMA, 2013). The ACMA reported 8.67 million smartphone and 4.4 million tablet owners at May 2012, with smartphone adoption more than doubling during May 2011 to May 2012 (smartphone ownership increased from 4.25 to 8.67 million units).

Relative to non-smartphone users, smartphone owners are: nine times more likely to conduct online activity via handsets; four times more likely to purchase goods online; three times more likely to stream or download audio or video content; three times more likely to pay bills online; and twice as likely to access social networking sites (ACMA, 2011–12). Table 14.3 shows online activity statistics for persons aged 18 years. The proliferation of smartphone applications has substantially improved the access to online services. During June 2012, 4.45 million adult smartphone users downloaded mobile applications.

According to Nielsen the main motivation for online search (71 per cent) is pricing, while searching for product information (54 per cent) is the second reason. PayPal and eBay research conducted by Nielsen notes that an estimated $5.6 billion was spent by online mobile device users in 2012 (PayPal Australia, 2012). Furthermore, Facebook is covering more than half of the population, with 11.8 million users at January 2013. Table 14.4 reports the ten most popular social media sites (Thang Ngo, 2013). Table 14.4 also reports statistics on social media traffic.

According to ACMA social media sites experienced significant growth in 2011–12. Interestingly, social media is not only a domain for young persons, with an increasing number of baby boomers using the services. Baby boomers represent almost 20 per cent of all page views and spend nearly 100 hours per month online (Nielsen Online Ratings, June 2012).

Table 14.3 Smartphone Internet activity, June 2012

Activity	Level (%)
Browsing websites or searching for information	90
Accessing news, sports, weather updates	84
Social networking	71
Streaming audio and/or video content	55
Paying bills	38
Downloading audio and/or video content	33
Purchasing goods and/or services	33

Source: ACMA, Communications Report 2011–12 Series, Report 3: Smartphones and tablets take-up and use in Australia.

Table 14.4 Social media statistics, January 2013

	Users (number)	Users (%)
Facebook	11,784,460	51.50
YouTube	10,790,715	47.11
LinkedIn	3,924,421	17.13
Tumblr	2,992,048	12.77
Twitter	2,925,383	12.76
WordPress	2,735,586	11.94
Pinterest	1,409,197	6.15
Flickr	1,202,510	5.25
Instagram	915,745	4.00
MySpace	615,334	2.69

Source: Adcorp, 'Social Media Statistics – Australia and New Zealand' (2013).

Facebook and LinkedIn were accessed, respectively, by 72 per cent and 27 per cent of the online baby boomers during June 2012.

The ACMA (Report 2) recorded an increase in online activity by mobile phone handsets compared to computers during June 2011 to June 2012 (2012). The highest proportional increases of mobile users were recorded in banking and finance (10 per cent), information, entertainment and amusement search (7 per cent rise) and communications (5 per cent rise). Table 14.5 shows the June 2011 and June 2012 statistics on the Internet activities by access device.

Table 14.5 highlights reasons for accessing the Internet via personal computers or mobile phones. For instance, 75 per cent of online users search or access information online via their personal computer, while

Table 14.5 Internet activity by access device (per cent)

	June 2011		June 2012	
	Computer	Mobile phone	Computer	Mobile phone
Communication (email/IM/VoIP)	75	57	78	62
Research and information	75	53	74	60
Banking and finance	68	28	67	38
Entertainment and amusement	53	42	58	49
General (browsing/surfing/ downloading)	61	46	59	50
Shopping (buying/selling)	46	17	55	22
Blogs and online communities	39	40	38	39
Advertising	33	12	32	13
Competitions, website registrations	26	6	24	8

Source: ACMA, Communications Report 2011–12 Series, Report 2: Australia's progress in the digital economy: participation, trust and confidence.

more than half of mobile online users access the Internet for online communications.

6. ONLINE BUSINESS

E-commerce provides the business with an opportunity to reduce costs and enhance efficiency. High-speed broadband offers benefits including: saves time, reduces the opportunity costs of providing goods, facilitates collaboration among business partners, aids audio-visual content development and enables high-definition video conferencing and cloud computing.

From 2011–12, the government aimed to provide $12.4 million over three years for a Digital Enterprise Initiative to assist SMEs and not-for-profit organisations in the 40 communities passed by the NBN. Several existing initiatives of the Department of Innovation, Industry, Science, Research and Tertiary Education provide assistance to SMEs, including Enterprise Connect; the Small Business Advisory Services; and the Small Business Support Line. The digitalbusiness.gov.au website assists SMEs to use online tools for promoting sales and business improvements. SMEs and their customers can also benefit from the ease of access through their online presence. Online business presence facilitates efficient business–customer interaction through faster online communication channels. Digital presence also enables businesses to place and receive orders online and to serve distantly located customers. Furthermore, businesses engaged with graphics, videos and other forms of digital media could use this channel to deliver products online.

The ABS reports there were 2,141,280 actively trading businesses in Australia at June 2012 (ABS 2013, 8165.0). Business establishments grew by 0.4 per cent from 2011. The percentage of businesses with Internet access rose from 90.1 per cent in 2009–10 to 91.2 per cent in 2010–11 (ABS 2013, 8166.0). Broadband Internet is the most popular mode of access with 99.1 per cent of business connection via broadband at June 2012.

Despite high levels of Internet access, less than half of Australian businesses (43 per cent) had an online presence, such as a website, eBay store or social media page (DBCDE, 2012). Australian business received online orders for $189 billion during 2010–11. E-commerce accounted for $57 billion, $81 billion, $123 billion and $143 billion, respectively, at June 2007, June 2008, June 2009 and June 2010 (ACMA Report 2, p. 29). The proportion of businesses that placed and received online orders is, respectively, 50.8 per cent and 28 per cent. It is reported that 75 per cent of Australian establishments believe in the positive impact of NBN infrastructure (IBIS, 2012). Similarly, McKinsey Global Institute's 2011 research study found that SMEs with an online presence grew more than twice as fast as those without Internet.

The NAB online retail trade index monthly update March 2013 indicates that online retail purchases in the past 12 months amounted to $13.3 billion. This level is equivalent to around 6.0 per cent of the traditional 'bricks and mortar' retail sector (excluding cafés, restaurants and takeaway food) for the same period. According to NAB, online sales were boosted in both March 2011 and March 2012 by flagship tablet releases. However, the lack of e-products launched during March 2013 slowed online sales growth. Table 14.6 presents online activity by SMEs in 2010, 2011 and 2012.

Inline to online product markets are becoming popular among customers. According to the IBIS World report, service exports, in the form of tourism, education, health, business services and IP, are growing. The report estimates that service exports will grow from 20 per cent to 25 per cent GDP by 2050. The report predicts that tourism can potentially contribute $175 billion to GDP by 2030.

The analysis also argues that the Internet use patterns of such businesses remained largely unchanged over the last two years. At June 2012, 95 per cent of SMEs reported owning a personal computer. Internet connectivity among SMEs remained high, with 92 per cent of SMEs connected to Internet. Of the 8 per cent of SMEs not connected, 3 per cent owned a computer but had no Internet connection, while the remaining 5 per cent

Table 14.6 Internet activity by SMEs (per cent), 2010–12

Activity	June 2010	June 2011	June 2012
Communicating via email	98	96	97
Banking	88	91	91
Products and services search	91	91	89
Reference information or research data	88	86	87
Looking for suppliers, products or services	88	88	86
Directories access	87	82	81
Payment for products and services	81	80	80
Placing product and service orders	79	79	77
Accessing online catalogues	79	76	76
Receiving payments	74	75	68
Advertising/business promotion	63	67	63
Streamlining communications	68	61	63
Receiving orders	62	62	55
Monitoring markets and competition	42	44	42
Email marketing	31	31	30
Online auction sites to sell	22	21	20
Business advertisement on other websites	21	20	15

Source: Sensis E-Business Report: The online experience of small and medium enterprises, August 2012.

did not own a computer. Interestingly, only 15 per cent of online SMEs reported having a strategy to promote digital business activity. Around 62 per cent of SMEs owned their websites at June 2012.

The proportion of businesses using social media sites rose from 18 per cent at June 2011 to 27 per cent at June 2012. And 86 per cent of SME online social network activity is via Facebook. Other important social networking sites include: Twitter, LinkedIn, Blog, YouTube, Google+, MySpace, Pinterest and Foursquare. Furthermore, 63 per cent of Internet-connected SMEs use smartphones at June 2012, while 29 per cent SMEs online report using tablets, compared to 16 per cent at June 2011.

Digitally engaged small businesses are experiencing enhanced growth (Deloitte Access Economics, 2013). The survey sorted businesses into categories based on digital engagement: very low (35 per cent), low (24 per cent), medium (25 per cent) and high (16 per cent). Only 28 per cent of SMEs with very low digital engagement recorded growth, compared to two-thirds of highly digitally engaged SMEs. The survey further indicates that highly digitally engaged businesses were very likely to hire new staff.

Table 14.7 SME growth (per cent) and digital engagement

Digital engagement	Growth	No growth	Steady
High	62	0	38
Medium	45	11	44
Low	43	9	48
Very low	28	11	61

Source: Deloitte Access Economics, Connected Small Businesses (2013).

7. ONLINE OPPORTUNITIES FOR BUSINESS AND CONSUMERS

7.1 Entertainment and Amusement

The availability of high-speed broadband coupled with higher download limits has enabled home entertainment and other home-based services. Subscribers can access high-definition video, multiple video and video-on-demand. The tripe-play bundle of voice, data and video is already offered. The NBN will create further changes to the entertainment, video and multimedia markets. The NBN will increase the range of home entertainment services including online TV shows, movies and gaming from the computer or Internet-connected television. Further, customers will be able to use many activities simultaneously through a single device. At June 2011 and June 2012, the share of users accessing the Internet for entertainment and amusement purposes via personal computer was, respectively, 53 per cent and 58 per cent of the online populations, while the proportion of mobile online users increased from 42 per cent to 49 per cent during the period.

7.2 Diversification of the Communication Channels

Society can now communicate using alternative modes according to technology choice and personal circumstances. Channels include: email, VoIP (Voice over Internet Protocol), instant messaging, audio and video conferencing, and social media. Social media is fast becoming a popular platform. As noted, 71 per cent of online users access social media sites. Facebook is being accessed by more than half of the Australian population, with 11.8 million users at January 2013. Further, Live Wire, an online community run by the Starlight Children's Foundation, connects young persons with serious illnesses, chronic health conditions or disability, and in a safe, supportive online environment. It is estimated there are

450,000 persons with access to three online communities featuring social networking tools and informative content, to reduce isolation and facilitate communication.

7.3 E-health

Tele-health attempts to reduce distance, time and cost, especially when patients are located in rural or remote area barriers to service access. The estimated ongoing benefits from tele-health are between $2 billion and $4 billion (Access Economics, 2010). On 8 May 2013, a joint media release by the Minister for Mental Health and Ageing and the Minister for DBCDE, Senator Stephen Conroy, announced that the government awarded $20.3 million to nine tele-health projects that will use the NBN infrastructure to pilot methods for health care delivery (Conroy, 2013). The projects are aimed to serve around 2500 patients in over 50 locations by June 2015. NBN-enabled services will include high-definition video conferencing, tele-monitoring of health indicators, medication management and rehabilitation.

The National E-Health Strategy (NEHS)[1] was developed by Deloitte with key stakeholders in 2008. NEHS provides a basis to guide e-health policy development. By end-2020, the NEHS aims for:

1. Completion deployment and broadband availability to 98 per cent of Australians;
2. 90 per cent of consumers or carers having access to electronic health records;
3. 50 per cent active access and use of e-health records;
4. 80 per cent of prescriptions, test orders/results and referrals performed electronically;
5. 20 per cent of consultations via electronic or tele-health means; and
6. 95 per cent of care providers having appropriate computing infrastructure.

These objectives are mainly to be achieved through NBN. On 22 January 2012, Minister for Health Tanya Plibersek and Minister for DBCDE Senator Stephen Conroy announced the NBN-enabled Telehealth Pilot Program. The government had initiated several other projects to improve e-health facilities, including Personally Controlled Electronic Health Record Systems, Medicare Benefits Schedule Expansion, NBN-enabled Telehealth Pilots Program, NSW Telehealth trials in Kiama and Armidale and the Townsville Telehealth trial.

7.4 E-education

E-education can potentially provide benefits that include the provision of remote education, online real-time lectures and online materials, and reduce time and location constraints. The NBN aims to connect Australian schools, TAFEs and universities to aid collaboration on innovative and flexible education services and resources by 2020. The NBN will allow schools in regional areas to connect with metropolitan schools to enable the delivery of courses otherwise not available. Educational institutions are increasingly moving content online, e.g. Open Universities Australia, iTunes U, YouTube University and Coursera are examples. Government initiatives include the NBN-enabled Education and Skills Program, EduONE, Virtual English Tuition for Migrants, ABC Education Portal, CSIRO's Mobile Robot.

International experience shows that increased use of online learning enabled by high-speed broadband can increase the reach of education services, availability of open-education resources and an improvement in education outcomes. Examples include:

- Carnegie Mellon University's Open Learning model has shown that online learning can substantially reduce the time required to learn a subject (50 per cent of the time required using traditional methods) while greatly increasing course completion rates (99 per cent compared to 41 per cent for traditional methods).
- High school students attending Florida Virtual Schools outscored the states' standardised assessment average by more than 15 percentage points in Grades 6 through 10 (FCC, 2010).
- A European Union study found broadband access in classrooms resulted in improvements to pupils' performance in national tests taken at age 16 (EC, 2006).

High-speed broadband services offered via the NBN aim to enrich the classroom experience by allowing greater use of online educational resources, including video, and increased collaboration beyond the classroom.

7.5 Transport

CCTV cameras provide updated traffic flows in major cities. Such information is helpful for trip planning and avoiding congestion. Faster NBN would improve these services by providing more detailed and timely information and by providing a centralised traffic monitoring and management system to roads authorities. Furthermore, the government requires that new major urban road projects have a minimum intelligent transport system technology incorporated into their design. For example, a $90 million investment through the Australian Rail Track Corporation for a trial of the Advanced Train Management System is intended to enable the digital monitoring of the interstate rail network. This aims to provide more reliable services, cost savings and improved safety.

Additionally, National ICT Australia (NICTA) is developing advanced sensing and surveillance technologies capable of providing rich information about traffic to smarter traffic control applications. The technology will enable:

- Accurate traffic flow measurement and vehicle classification together with stopped vehicle detection;
- Reliable, low cost vehicle tracking as well as queue length estimation; and
- Truly intelligent vision systems for traffic monitoring.

7.6 Teleworking

High-speed broadband offers more flexible work practices through teleworking. The NBN aims to equip subscribers with the ability to access video conferencing and virtual offices, and to send and receive large files. By 2020, the enhancement of telework will allow 12 per cent of the population to have flexible work arrangements. In particular, Access Economics revealed that NBN-enabled telework will add $8.3 billion to GDP by 2020, creating the equivalent of an additional 25,000 full-time jobs. Around 10,000 of these jobs will be in regional Australia.

With regard to the environment, it is estimated that should 10 per cent of the workforce telework 50 per cent of the time, 120 million litres of fuel and 320,000 tonnes of carbon emission would be saved. Teleworking would also deliver a productivity benefit of $1.4 billion to $1.9 billion annually. The survey of Household, Income and Labour Dynamics in Australia (HILDA) indicates that around 18 per cent of surveyed employees were undertaking some work from home in 2009. At 2011, Deloitte Access Economics reported that Australians only teleworked at half the rate of Canada and the US. The Digital Economy Initiative proposes to double the level of teleworking by 2020 (DBCDE, 2013).

7.7 Cloud Computing

Initial business cloud computing use was primarily as a data backup tool and for archiving files. Cloud computing was later used for web hosting, testing and development. On 27 February 2012, Business Software Alliance (BSA) announced that Australia now ranks as the second most prepared nation to adopt cloud computing. BSA President and CEO Robert Holleyman said the success of cloud computing 'requires laws and regulations that let data flow easily across borders' (Wolski, 2012).

Further, the Forrester Research (2012) report entitled, 'Sizing the Cloud Markets in Asia Pacific', states that the cloud market will grow from $732 million in 2011 to $3.2 billion by 2020 and that these segments of the virtual private cloud market will grow from $157 million in 2011 to $2.4 billion in 2020 (Barnes, 2012). According to ACMA (Communications Report 2), the broad deployment of services by Apple, IBM, Microsoft, Google and Amazon has enhanced the availability of cloud related applications, with 75 per cent of the adult population using a cloud computing service during the six months to May 2012. The email (70 per cent of the adult online population), social networking (62 per cent of adult online users) and photo-storage applications (47 per cent of adult online users) stood, respectively, at the first, second and third ranks in the list of most widely used cloud-related applications. The cloud computing industry generated $1.1 billion in revenue during 2011–12.

According to Citi Research, the NBN could enhance the demand for cloud services by promoting business and consumer confidence in solutions by means of the fast onshore data transfer facility. The NBN provides new ways of delivering business services, including smart infrastructure and high-tech services such as IPTV and online gaming services. On 14 November 2012, Senator Conroy said, 'Australian companies increasingly need to consider and adopt cloud services in an effort to move at least part of their IT into a data centre. On an international scale, companies wanting to do business in Australia can experience latency issues so we are seeing many of those making the decision to deploy at onshore data centres in Australia.' Further, Senator Conroy revealed the government's National Strategy for Cloud Computing on 29 May 2013 at the CeBIT conference in Sydney. The strategy aims to equip government agencies, businesses and consumers with tools and information to assist cloud service adoption.

7.8 E-commerce

E-commerce and m-commerce are becoming increasingly important to mobile operators, financial institutions and payment processing firms. At 2012, credit and debit cards are the most common online payment methods. Outside the banking sector, PayPal is an established e-payment facilitator. The sale of goods and services online is growing steadily, with $189 billion in goods and services traded online during 2010–11. E-commerce accounted for $57 billion, $81 billion, $123 billion and $143 billion, respectively, at June 2007, 2008, 2009 and 2010 (ACMA, Report 2). Australians use the Internet for direct (transactions) and indirect (checking product or service availability and price comparisons) purchase activities. An estimated 7.81 million persons aged over 14 years undertook buying, selling or other shopping activity online during June 2012, compared to 6.6 million transactions during June 2011. Further, online advertisements by firms are becoming popular. By end-2011 online advertising revenues were 17 per cent higher than the previous year.

The Econsultancy (2013) report entitled, 'Australia: Ecommerce Trends Briefing', states that online sales reached a high in the year to November 2012, with 71 per cent of consumers purchasing online during 2012. The report states that 26 per cent of online shoppers spend $200–$499 on goods. However, only 25 per cent of retailers currently integrate a database with their website. By 2015–17, nearly 75 per cent of retailers are expected to offer online purchasing and ordering. About 59 per cent of retailers are expected to provide online self-service applications during this period.

8. E-GOVERNMENT

The government provides e-government services to allow interaction with citizens. Services include health, education, transport, and online bill and fine payments. The Department of Finance and Deregulation, through the Australian Government Information Management Office (AGIMO), works across governments to maintain Australia's position as a leader in the productive application of ICTS to government administration, information and services. The AGIMO facilitates the efficient and effective use of ICT by government departments and agencies. The AGIMO also works with government and other bodies at the local, state, national and international levels.

Digital engagement can improve government productivity by facilitating efficient service delivery, for example, enabling medical practitioners to offer remote health services (i.e. monitoring, diagnosis, treatment and answering on-call queries of patients). To best seize the opportunities presented by the digital world, the government is encouraging the maximum use of NBN for government services. The Australian government has committed to Service Delivery Reform initiatives within the Human Services portfolio that will transform the way people receive services and interact with government.

At 19 January 2011, Special Minister of State Gary Gray released the report, 'Interacting with government: Australians' use and satisfaction with e-government services 2011' (AGIMO, 2011). The report states that 82 per cent of Australians prefer to access government services online. Further, overall satisfaction with e-government services is high (86 per cent). When choice is available between using the Internet or other channels, 46 per cent prefer the Internet. Since 2008 use of e-government has been stable, with 67 per cent of persons using e-government services in 2011. According to the AGIMO report (2011) 54 per cent of the population use e-government as a main way to communicate with government, 37 per cent communicate in person and the rest communicate via post. The most common reasons for contacting the government are: online information seeking (24 per cent), online payment receiving (23 per cent), information exchange (19 per cent), information provision (16 per cent) and online payment making (15 per cent).

9. CHALLENGES

How effective will Australia be in the global digital economy? Transforming narrowband into broadband platforms is challenging at the infrastructure and higher layers. Targeted strategy is required. An overview of challenges and the status of policy are provided.

9.1 Urban–Regional Digital Divide

Adequate service provision requires substantial infrastructure investment with low returns. The incidence of households with an Internet connection is increasing, but a gap exists between the regional and metropolitan areas. Government is actively working through the NBN to enhance digital economy capability by reducing the urban–regional digital divide. The NBN Regional Legal Assistance Program grants funds for the delivery of legal assistance services via the NBN and to attract and retain staff in selected regions. The NDES aims to bridge the gap by 2020.

9.2 Inadequate E-skills

The Innovation and Business Skills Australia (2013) study highlights skill deficiencies among Aboriginals, persons aged over 55 years, those on low incomes and the less educated. On 23 May 2013, the DBCDE Minister announced the government's plan to partner with ten national industry peak bodies in ten sectors to develop Digital Business Kits.

The kits are intended to address industry-specific gaps. The government is also offering $500,000 grants to national industry peak bodies to develop and promote tailored 'how-to' guides to assist businesses build skills and knowledge to take advantage of opportunities offered by the NBN.

9.3 Online Security

At June 2012, almost half of Australian SMEs are concerned about hacking problems. There were 1250 cyber-attacks against government systems in the year to September 2012. NBN Co's internal networks were penetrated by 'Trojans' that avoided detection by anti-virus software twice in 2012. An NBN Co spokesperson did, however, say that the attacks were ultimately detected and that no confidential information was stolen. Despite assurances, the government instigated National Cyber Security Awareness Week to protect personal and financial information. The site staysmartonline.gov.au advises on online security. Furthermore, the Australian Internet Security Initiative (AISI), developed by ACMA, intends to help address problems with 'compromised' computers. The AISI collects data from several sources on suspected IP addresses and provides these data to ISPs. ISPs can then inform customers about suspicious IP addresses along with the advice.

9.4 Spectrum Management

The delivery of a range of many services (including TV, radio broadcasts, mobile telephones, multimedia and data, air traffic control, emergency services, defence and management of power and water) is made via spectrum. Free-to-air TV is planned to progressively switch from analogue to digital-only transmission by end-2013 (DBCDE, 2012). The TV switchover means digital broadcasting will require far less spectrum. As analogue signals are switched off, the ACMA will be able to reallocate 'free' spectrum. In a converged environment, where there are multiple platforms capable of delivering broadcasting content, the current regulatory distinction between spectrum allocated for broadcasting services and spectrum allocated for other purposes is no longer relevant. A regulatory framework must be developed to manage and allocate spectrums to achieve increased efficiency. Such a framework will be more responsive to evolving technology and shifting spectrum demand.

9.5 Online Content Regulation

The government introduced the Online Content Co-Regulatory Scheme on 1 January 2000 to address concerns about the nature and accessibility of illegal and highly offensive Internet content. Key components of the scheme include: codes of practice for ISPs and online content service providers, a complaint mechanism, community education about online safety issues, and supporting activities including research and international liaison. However, the current content regulation scheme does not deal adequately with the challenges of media convergence. In February 2012, the Australian Law Reform Commission (ALRC) made recommendations for effective online content regulation.

9.6 Copyright Policy

The adoption of digital facilities has complicated the assurance of compliance with Australian copyright laws. Despite the changes introduced to the Australian copyright law, further strengthening of digital copyright laws is needed. On 3 March 2012, the government signalled a review of copyright laws to ensure relevance to an evolving digital market. On 21 August 2012, News Limited's Chief Executive (Kim Williams) asked the government to protect digital media from copyright violations.

9.7 Competition and Regulation

A revised regulatory framework is necessary to ensure effective competition and to provide market participants with favourable conditions and a level playing field. Currently, the Competition and Consumer ACT 2010 (CCA) prohibits market participants from any anti-competitive behaviour. At February 2012, the ACCC's submission to the Convergence Review highlighted the inability of current regulatory frameworks to address competition issues generated by emerging digital markets. The Convergence Review agreed to the ACCC's findings on the need for enhanced powers to deal with aspects of emerging online competition.

9.8 Network Neutrality

Winterford and Hill (2008) argue that local net neutrality is considered a commercial issue of USA-based ISPs, aiming to generate additional revenue. Johnson (2009) reports that no Australian ISPs are blocking access to competitor content or charging premium fees to content providers. Aaron Harwood (5 September 2011) argues that Australia is not net neutral and provides Telstra's 'Bigpond Unmetered' and iiNet's 'Freezone' as examples of non-neutrality. Harwood asserts that other countries' policies and directions (regarding the Internet) may have local effects and that it would be unwise to dismiss the debate as another country's problem. The Convergence Review in a section on 'content-related competition issues' mentions the need for new powers to deal with network neutrality and other competition-related issues. On the network neutrality issue, the Convergence Review states:

> Internet service providers (ISPs) may sometimes employ network management practices, such as slowing down some users' traffic to avoid or reduce network congestion. However, there are concerns that Internet traffic can be subject to management practices that are designed to limit competition and reduce innovation (Convergence Review, p. 30).

The Review proposes the establishment of a regulator with proactive powers to make rules and issue directions in the content market to administer a well-targeted, more responsive and effective regulation.

9.9 Interoperability

Interoperability is the ability of diverse systems and organisations to work together. Interoperability calls for the development of independent, 'siloed' systems to make businesses and government more interconnected. The government's Technical Interoperability Framework provides a step towards establishing technical compatibility for the exchange of data and harmonisation of business transactions within trusted environments. The most recent version of the Australian government's Technical Interoperability Framework was developed by the Interoperability Framework Working Group (IFWG), a reference group of senior technical architects nominated by the Chief Information Officers' Committee (CIOC). The Australian Government Information Management Office (AGIMO) supported the review. This new version responds to developments in the ICT industry which are supporting businesses and government to be more interconnected. The Distributed Systems Technology Centre (DSTC) provided independent, expert advice during the course of the review.

This ICT-focused framework defines a conceptual model and agreed technical standards that support collaboration between agencies. It asserts that adoption of common technical protocols and standards under the framework will ensure that government ICT systems interoperate in trusted ways with industry and government partners. The Framework's scope relates only to Australian government agency interoperability and it does not affect technologies deployed within an agency or constrain interactions outside government. The Framework is an initial step towards developing an online environment where government services are integrated to serve the needs of the community. The Framework recognises that interoperability will develop out of independent, 'siloed' systems, but with a common business need – to exchange data.

NOTE

1. More information is available at http://www.ahmac.gov.au/cms_documents/National per cent20E-Health per cent20Strategy.pdf.

REFERENCES

Adcorp (2013), 'Social Media Statistics – Australia and New Zealand', available at http://www.adcorp.com.au/Social-Media-Statistics-January-2013-Aust-NZ (accessed 15 May 2013).

Australian Bureau of Statistics (2005–12), Internet Activity, Australia, various (No. 8153.0), Canberra, ACT: ABS.

Australian Bureau of Statistics (2013), 'Counts of Australian Businesses, Including Entries and Exits', May 2013 (No. 8165.0), Canberra, ACT: ABS.

Australian Bureau of Statistics (2013), 'Summary of IT Use and Innovation in Australian Business', June 2012 (No. 8166.0), Canberra, ACT: ABS.

Australian Communications and Media Authority (2012), Communications Report 2011–12 Series, Report 2: Australia's progress in digital economy participation, trust and confidence, Canberra, ACT: ACMA.

Australian Communications and Media Authority (2013), 'Smartphone use doubles in 2012', media release, MR 5/2013, 1 February.

Australian Communications and Media Authority (2013), Communications Report 2011–12 Series, Report 3: Smartphone and tablets take-up and use in Australia, Canberra, ACT: ACMA.

Australian Government Information Management Office (2011), 'Australians' Use and Satisfaction with e-Government Services', available at agimo.gov.au/2012/01/19/2011-australians-use-and-satisfaction-with-e-government-services/ (accessed 19 February 2013).

Australian Information Industry Association (2012), 'AIIA Digital Economy Policy Statement', available at http://www.aiia.com.au/?page=de_policy (accessed 10 May 2013).

Barnes, M. (2012, February 3), 'Sizing the cloud markets in Asia Pacific', available at http://www.forrester.com/Sizing+The+Cloud+Markets+In+Asia+Pacific/fulltext/-/E-RES61125 (accessed 27 February 2015).

Boston Consulting Group (2012), 'Clicks grow like BRICS: G-20 Internet economy to expand at 10 per cent a year through 2016', available at http://www.bcg.com/media/PressReleaseDetails.aspx?id=tcm:12-100468 (accessed 19 March 2013).

Bowles, M. (2011, January), 'Realising the full socio-economic promise of the

National Broadband Network in preparing all regions of Australia for participation in the digital economy', available at http://www.derf.com.au/files/Bowles_NBN_Digital_Economy_paper_Pre-Press_Version_with_images_pre-press.pdf (accessed 7 February 2013).
Senator the Hon. Conroy, S. (2012), 'Book launch: transforming Australia: the broadband revolution and the digital economy', Parliament House, Canberra, available at http://www.minister.dbcde.gov.au/media/speeches/2012_-_minister_speeches/011.
Senator the Hon. Conroy, S. (2013), 'NBN to pilot new models of health care for 2500 patients', available at http://www.minister.dbcde.gov.au/media/media_releases/2013/070 (accessed 7 June 2013).
Deloitte Access Economics (2013), 'Connected small businesses: how Australian small businesses are growing in the digital economy', available at https://www.deloitteaccesseconomics.com.au/uploads/File/Connected per cent20Small per cent20Business.pdf (accessed 19 May 2013).
Department of Broadband, Communication and Digital Economy (2013), 'More Australian businesses getting online: latest ABS statistics', available at http://www.digitalbusiness.gov.au/2012/08/02/more-australian-businesses-getting-online-latest-abs-statistics/ (accessed 8 May 2013).
Department of Broadband, Communication and Digital Economy (2012, August), 'Creating jobs through NBN-enabled telework', available at http://www.deloitte.com/assets/DcomAustralia/Local per cent20Assets/Documents/Services/Corporate per cent20Finance/Deloitte_Creating_jobs_through_NBN_enabled_telework.pdf (accessed 12 February 2013).
Department of Broadband, Communication and Digital Economy (2012), Convergence Review: Final Report, Canberra: ACT.
Department of Finance and Deregulation (2006), 'Government Strategy, Responsive Government: A New Service Agenda', available at http://agimo.gov.au/policy-guides-procurement/previous-programs-reviews/e-government-2006/ (accessed 9 January 2013).
Econsultancy Digital Market Excellence (2013), 'Australian ecommerce: nine stats you might have missed', available at http://econsultancy.com/au/blog/62132-australian-ecommerce-nine-stats-you-might-have-missed (accessed 23 May 2013).
European Commission (2006, December), 'The ICT Impact Report: a review of studies of ICT impacts on schools in Europe', available at http://ec.europa.eu/education/pdf/doc254_en.pdf (accessed 3 January 2013).
EXA (2012, November), 'The Australian Online Consumer', [News Update], available at http://www.exa.com.au/articles/november_12/australian_online_consumer/ (accessed 27 February 2015).
Federal Communications Commission (2010), 'National Broadband Plan: Connecting America, Broadband.gov', available at http://www.broadband.gov/plan/11-education.
Fibre to the Home Council (2012), 'The FTTH Councils' latest global ranking', available at http://www.ftthcouncil.eu/documents/PressReleases/2012/PR2012_Global_Ranking_Munich_2012_FINAL.pdf (accessed 27 February 2015).
Innovation and Business Skills Australia (2013), 'Post-NBN impact on enterprises and eskills', available at http://www.ibsa.org.au/Portals/ibsa.org.au/docs/Research per cent20& per cent20Discussion per cent20Papers/Post-NBN per cent20impact per cent20on per cent20enterprises per cent20and per cent20e-skills per cent20Exec per cent20Summary.pdf.

Irvine, J. (2013), 'Smart retailers will thrive', [News Limited Network], available at http://www.theaustralian.com.au/news/smart-retailers-will-thrive/story-e6frg6n6-1226604157336 (accessed 9 June 2013).
McKinsey Global Institute (2011), 'Internet matters: the net's sweeping impact on growth, jobs and prosperity', available at http://www.google.com.au/url?sa=t&rct=j&q=&esrc=s&source=web&cd=1&ved=0CCwQFjAA&url=http per cent3A per cent2F per cent2Fwww.mckinsey.com per cent2Ffeatures per cent2F~ per cent2Fmedia per cent2F6BDDD3D756C449C1A9C41633EE5B0732.ashx &ei=mWzKUfDaIcWlkQXhkYHgAQ&usg=AFQjCNHCCSS9R1fSVRBJgf Lmo4gMEA8Ilw&sig2=EhD7Suj-k8QvKP1G7ukM0w&bvm=bv.48340889,d.dGI (accessed 11 March 2013).
Mettler, A. (2012), 'Unleash Europe's digital revolution to kick-start growth', World Economic Forum, available at http://forumblog.org/2012/06/unleash-europes-digital-revolution-to-kick-start-growth/ (accessed 17 January 2013).
Mitchell, S. and L. Pezzullo (2010, July), 'Financial and externality impacts of high-speed broadband for telehealth', available at http://www.archive.dbcde.gov.au/__data/assets/pdf_file/0019/130159/Financialandexternalityimpactsof high-speedbroadbandfortelehealth-311.pdf (accessed 18 January 2013).
National Australia Bank (2013), 'NAB online retail sales index monthly update', available at http://business.nab.com.au/wp-content/uploads/2013/05/nab-online-retail-sales-index-march-2013.pdf (accessed 27 February 2015).
NBN Co (2012), 'Australians demonstrate growing appetite for superfast broadband', available at http://www.nbnco.com.au/about-us/media/news/australians-demonstrate-growing-appetite-for-superfast-broadband.html (accessed 15 April 2013).
Nielson Online Ratings (2012, July), 'Australian online landscape review', available at http://www.nielsen.com/content/dam/corporate/au/en/reports/2012/ Nielsen per cent20Online per cent20Landscape per cent20Review per cent20July per cent202012 per cent20Media per cent20Pack.pdf (accessed 27 February 2015).
PayPal Australia (2012), 'Secure insight/consumer discovery', available at https://www.paypal-media.com/assets/pdf/fact_sheet/Secure_Insight_Consumer_Dis covery_Report.pdf (accessed 27 February 2015).
Ruthven, P. (2012), 'A snapshot of Australia's digital future to 2050', IBIS World, available at http://www-07.ibm.com/au/pdf/1206_AustDigitalFuture_ A4_FINALonline.pdf (accessed 27 February 2015).
Sensis (2012), 'Sensis E-Business Reports: The online experience of SMEs', available at http://about.sensis.com.au/IgnitionSuite/uploads/docs/2012SensisE_ Business_Report.pdf.
Thang Ngo (2013), 'Australian online and social media statistics', available at http://www.thangngo.com/2013/02/online-and-social-media-statistics-in-australia/ (accessed 17 May 2013).
Wolski, G. (2012), 'Australia second most prepared nation to adopt cloud technology', available at http://www.dynamicbusiness.com.au/technology/ australia-second-most-prepared-nation-to-adopt-cloud-technology-27022012.html (accessed 7 April 2013).

15. NGA networks as basis for a sustainable digital society: European country studies (Germany, the UK, the Netherlands and Switzerland)

Ernst-Olav Ruhle, Matthias Ehrler, Natascha Freund and Martin Lundborg

1. INTRODUCTION

The development towards NGA networks in Europe is not only a matter of developments on the national level but also on the European level. The EU and its institutions have targeted the implementation of broadband networks. In a community of currently 28 countries, ICT policies and regulatory approaches are compared to each other thereby trying to avoid trade barriers by harmonising the national legislative approaches in many economic areas. The information and communication technology (ICT) sector is one of those areas where a harmonised approach is key to the overall development. The European Union tries to overcome economic barriers between the countries as it is presumed that, especially in telecommunications markets, economies of scale can be realised. Telecommunications is a business requiring significant investments, and therefore scale economies by conducting investment in several markets in parallel may be very beneficial in the long run.

The European Union has concluded that fostering investments in the ICT sector is important in order to increase broadband coverage and bandwidths all over Europe, which again will contribute to an improvement of Europe's competitive position in the world. Telecommunications networks and services are a facilitator for other industries and thus a valuable enabler for growth.

The developments in the different EU countries may be quite different as regards the establishment and deployment of next generation access networks, via fibre (FTTX), cable or the attempt to make more efficient use of the already currently existing copper infrastructure.

The European Union itself has supported this development by several different policy measures, such as:

- the Digital Agenda, which sets goals for bandwidth and penetration over the next decade in the EU;
- the NGA recommendation, which tries to formulate incentives for investment into fibre-based access networks;
- 'state aid' rules for broadband networks, an aspect that refers to the environment in which it is difficult to realise further benefit from rollout of NGA infrastructure, especially so-called white areas where there is little private economic incentive for investment. State aid in the form of financial support to roll out such networks may be appropriate under certain circumstances. The EU Commission has defined such circumstances in order to also enable users in these areas to benefit from higher bandwidths; and
- the recommendation on cost orientation and non-discrimination, which prescribes costing methodologies for regulators when calculating costs for wholesale access products and introduces the principle of Equivalence of Input (EoI) when imposing remedies on telecommunication operators.[1]

Despite these measures – and others that were taken in late 2013 and early 2014, and that are not discussed here (e.g. the Digital Single Market (DSM) approach) – which have been introduced over a number of years, the EU, in comparison to other regions of the world, has not achieved significant developments towards next generation access networks. In some countries investments have been undertaken, cable networks may be strong and also mobile networks are offering 3G and 4G services; however, incumbent operator investments in new networks are rather low for the time being. It appears to be the case that there is a major attempt to make best possible use of the copper infrastructure as long as new technological features (VDSL, vectoring, bonding, etc.) allow to increase the capabilities on these infrastructures.

In the next sections, we describe the development in selected European countries (EU members and non-members) and try to find similarities and differences in their approaches to building a digital society through more broadband access networks.

2. GERMANY

Germany, as the largest economy in Europe on the one hand, can be seen as a country with remarkable achievements since the beginning of the liberalisation process in 1998, in particular with respect to the inter- and intra-modal competition between the different business models following the ladder of investment. On the other hand, concerning the development of an NGA environment as basis for sustainable digital society policies, the stakeholders encompassing the regulatory authority, institutions and associations as well as infrastructure enablers, network operators and service providers still need to do their homework, at least in the fixed network sector. However, the recent developments within the mobile sector (3G and 4G) and also in the CATV sector can be considered satisfactory.

In February 2009 the Federal Ministry of Economics and Technology released the German Broadband Strategy (BMWi, 2009). The strategy at that time defined two major targets. First, gaps in broadband penetration should be eliminated and capable broadband, defined as at least 1 Mbps, should be made available nationwide by the end of 2010. Second, a total of 75 per cent of all German households should have Internet access with speeds of at least 50 Mbps by the end of 2014. According to the Federal Ministry, in the short run, the approach of the government focused mainly on financial support for local authorities as well as financial options available to companies. In the long run, it focused on incentives within the overall EU regulatory framework. Further it should provide stimulus to promote synergies from infrastructure projects. Specific measures include optimising shared use of existing infrastructure and facilities and compiling a broadband map and a database of existing infrastructure suitable for co-usage as well as of construction sites and work where parallel deployment could be facilitated. In a nutshell, the National Broadband Strategy is based on four pillars (BMWi, 2009, p. 10):

1. using synergies for building up infrastructure through providing access to and using existing public and private facilities;
2. supporting polities for using frequencies, e.g. through exploiting the digital dividend (790 through 862 MHz) for broadband development in rural areas;
3. growth-supporting regulation, emphasising incentives for investments as well as aspects of planning security and supporting cooperative market approaches (e.g. parallel deployment and co-investment); and
4. measures for financial support to connect municipalities that are not equipped with broadband via the market (white spots).

In April 2012 the German government expanded its broadband strategy and defined a nationwide availability for broadband access of at least 50 Mbps by the end of 2018 as an additional target.

Beyond that, the German government has also taken steps to make more spectrum available for mobile broadband by freeing up the 900 MHz band for data and reallocating spectrum in the 800 MHz band.

Taking into consideration this broadband policy approach the German government relies on three NGA technologies fulfilling the aforementioned objectives: 1) fibre optic-based networks in different stages, which means FTTC, FTTB and FTTH; 2) TV cable networks; and 3) mobile technologies, e.g. LTE.

A major step regarding the usage of mobile technologies to provide (adequate) broadband access was the completion of the LTE auction of the 800 MHz spectrum (Digital Dividend) and the new 2600 MHz spectrum in 2010. The auction ended up with three mobile operators (Telekom Deutschland, Vodafone and Telefónica) receiving a licence for the 800 MHz spectrum paying €3.6 billion of the total auction proceeds of €4.4 billion. Only one operator (e-plus) decided not to acquire any 800 MHz spectrum. The allocation in particular of the 800 MHz spectrum was attended with the obligation to cover remote areas before deploying the cities and sub-rural areas. Each federal state had been categorised into four 'priority stages'. Stage 1, for example, consists of towns and districts with fewer than 5000 inhabitants ('white spots'). Operators had to cover 90 per cent of each stage before they could start to deploy 800 MHz in more attractive rural, sub-urban and urban areas. Hence, the operators were keen to meet these obligations quickly. In the meantime the three obliged operators have fulfilled the obligation and have already started deploying their 800 MHz networks in more densely populated areas and in cities (Little, 2012, pp. 1 2). Figure 15.1 shows the LTE availability in Germany in terms of percentage of households.

The second NGA technology with a success story is coaxial cable. In 2006 cable network operators had a market share of only around 3 per cent of broadband (Internet) subscribers. With above-average growth rates since then they raised their market share to around 15 per cent (4.4 million subscribers) at the end of 2012 (Dialog Consult/VATM, 2012). In the past five to seven years cable operators invested significantly into DOCSIS 3.0 ending up with nearly 25 million homes passed currently, which represent almost two-thirds of German households (ANGA, 2013, p. 6). Based on the DOCSIS 3.0 standard cable operators are capable of providing bandwidths of 100 Mbps and more, which is significantly higher compared to the bandwidths offered by xDSL operators. But, the provision of broadband access via cable only makes sense economically if the density of

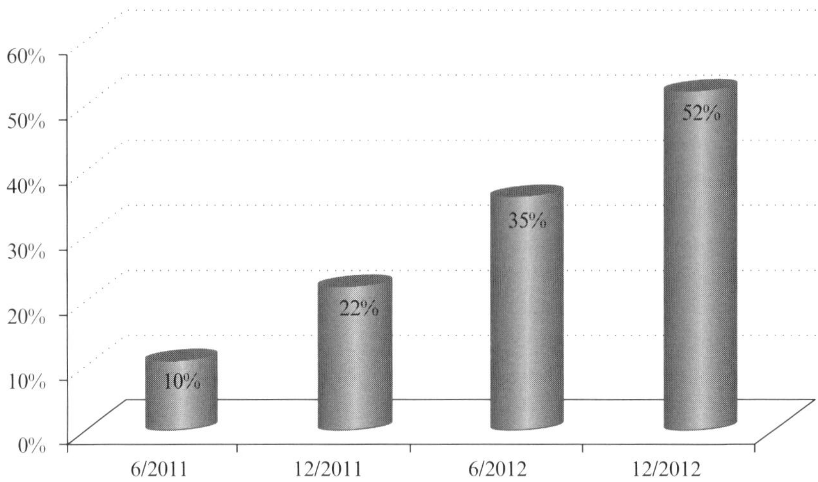

Figure 15.1 LTE availability in Germany (% of households)

potential customers is sufficient to justify the investment for developing a certain region. Thus, cities and urban areas rather tend to be supplied by cable, whereas in remote areas, other (broadband) technologies dominate.[2]

Figures 15.2 and 15.3 show the dynamic development of the German cable market with regard to broadband (Internet) access and telephony.

Compared to the mobile and cable market, the development of the third NGA technology FTTx (FTTC, FTTB and FTTH) is quite different. One major reason is the multi-provider and multi-business model environment that does not exist in a similar way in other European countries. The German FTTx market is highly fragmented. Key players on the market for (active and passive) FTTx access are Telekom Deutschland (TD), alternative established local or regional telco operators (e.g. EWE TEL, inexio, NetCologne, M-net) and public (municipal) utilities or municipalities. Further, private operators are active in remote areas. Other operators like 1&1, QSC, Vodafone and Telefónica are purchasing passive and/or active access products based on other provider infrastructures. All participants are increasingly concerned with the question of whether they should (furthermore) invest in fibre optic (access) infrastructures, and if so, which technology is preferable.

Referring to TD, the incumbent currently pursuing a national FTTC/VDSL strategy and, further, carrying out some FTTH projects in selected cities. Roughly three years ago, TD announced its plan to supply 4 million households with FTTH before the end of 2012. But afterwards TD had to reduce this ambitious target to a few hundred thousand. Most of the

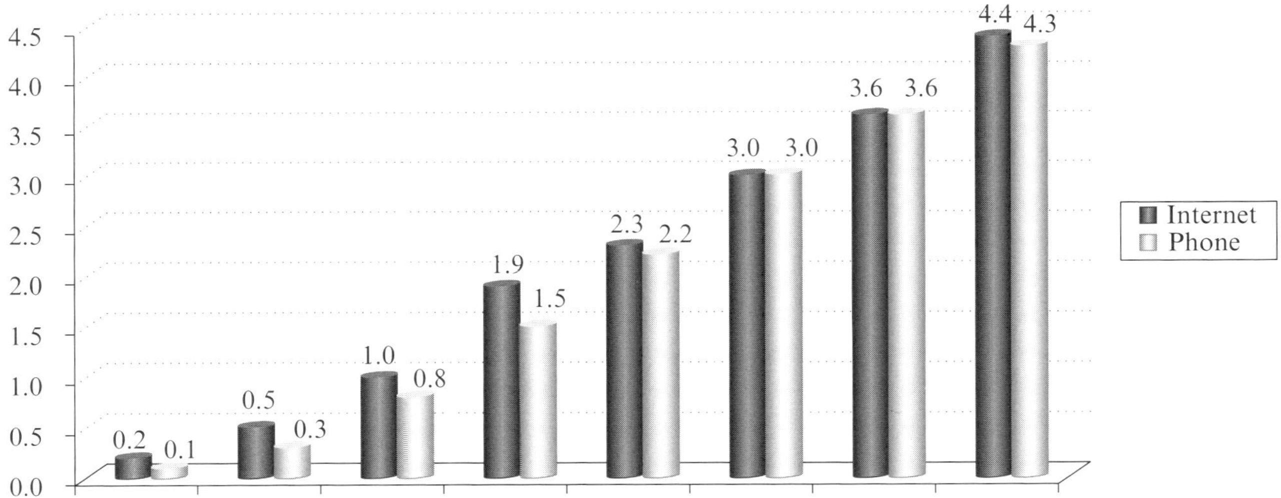

Figure 15.2 *Customer growth of the German cable network operator (millions)*

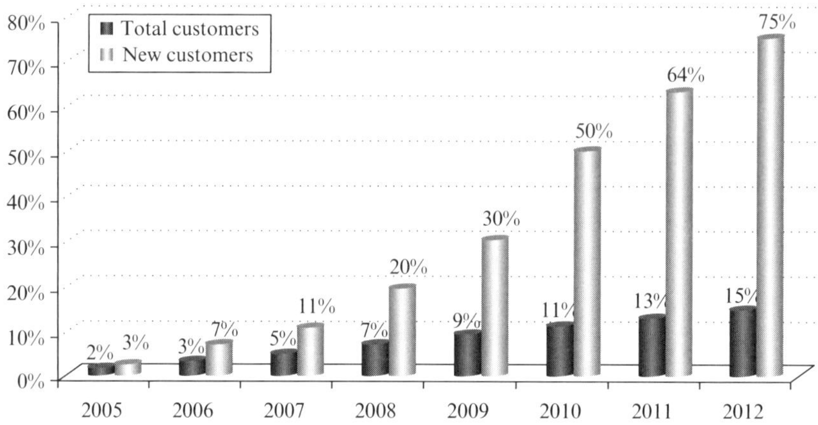

Figure 15.3 German cable network operators' share in broadband customers

alternative operators were deploying FTTB/FTTH networks in cities and upgrading the existing copper networks with FTTC infrastructures in suburban and rural areas. Public utilities and municipalities invest mainly in FTTC.

Another reason for the low penetration rates of FTTB/FTTH is that German consumers are slow in adapting to FTTB/FTTH networks. In cities and regions where FTTH or FTTB is available, only a few households currently subscribe to the new networks. The lack of interest on the part of the consumers is in part owing to existing long-term contracts that bind customers to another operator for up to 24 months and to the fact that many of the new connections are not – or are badly – marketed and communicated to the end users.

At last, connecting all German households with an FTTH infrastructure is very costly. Experts in 2011 came to the conclusion that a nationwide rollout will require investments of about €80 billion (WIK, 2011, p. viii).

Table 15.1 shows the availability of FTTx connections in Germany at the end of 2012.

As a result of the low FTTB/FTTH penetration the discussion regarding the further use of the sub-loop part of the twisted copper pair has intensified in the past 18 months, in particular with respect to the application of the vectoring technology for FTTC infrastructures. But, applying vectoring has implications on the regulation regarding access to the sub-loop because it can only be installed by one operator in the street cabinets – making unbundled access where VDSL technology is being used impossible – and thus, creates a new monopoly. Given that an

Table 15.1 Availability of FTTx technologies in Germany

	Availability (homes passed)
FTTC (VDSL)	approx. 12.5 million
FTTB	approx. 1 million
FTTH	approx. 200,000

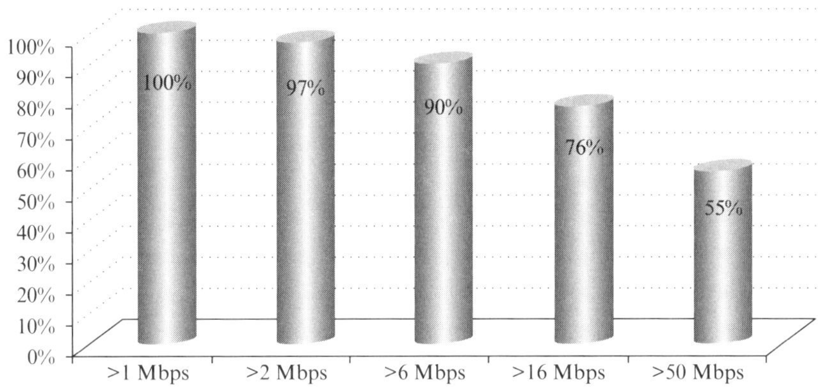

Figure 15.4 Broadband availability in Germany for all technologies (fixed and mobile) (% of households)

adequate wholesale product will be offered by the respective operator, the new technology can be used by all operators.[3]

However, TD has announced its target to cover 24 million households over the next four years with high speed connections. This shall be achieved with a combination of FTTC/VDSL and vectoring as well as LTE connections.

Overall, the German broadband access market is still dominated by xDSL technologies, in particular by ADSL with bandwidths up to 16 Mbps. But the availability of bandwidths up to and over 50 Mbps is constantly growing. Figures 15.4 and 15.5 show the situation on the German market.

As can be seen, the availability of broadband access via cable dominates the access via DSL/FTTx above 16 Mbps connections.

Finally, in Table 15.2 a comparison of the regions with different population densities shows the general availability of broadband access for different speeds.

Regarding the German case the development towards next generation access networks has been driven by: 1) investments of alternative network

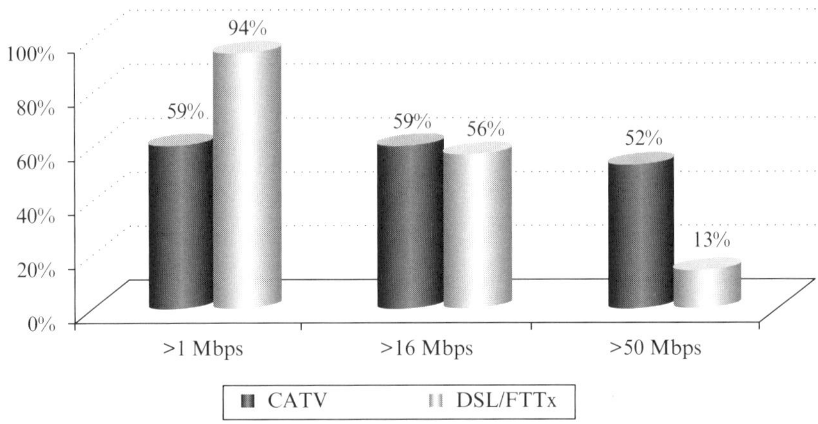

Figure 15.5 *Comparison of CATV and DSL/FTTx availability in Germany (% of households)*

Table 15.2 *Broadband availability in Germany per geographical area (% of households) – all technologies*

	> 1 Mbps	> 2 Mbps	> 6 Mbps	> 16 Mbps	> 50 Mbps
Urban	100.0	99.4	96.8	90.0	77.3
Sub-urban	99.2	96.1	84.5	63.0	32.1
Rural	95.8	90.2	73.8	42.3	10.2

operators, utilities and municipalities (regional authorities) as well as cable network operators; 2) coordination and support through a national broadband strategy by the government and additionally by initiatives of several federal states, an approach that is currently under discussion, the government having initiated a 'network alliance' in late 2013 to accelerate NGA deployment; and 3) the incumbent TD with its more or less consistent FTTC strategy as well as mobile operators, in particular T-Mobile and Vodafone with their fast LTE rollout in the past 18 months.

3. UNITED KINGDOM

The UK is traditionally known as the country that initiates economic and regulatory discussions (e.g. functional separation and equivalence of inputs). Considerable time and effort has been spent in the UK to define the setting of a move towards next generation access networks but in spite

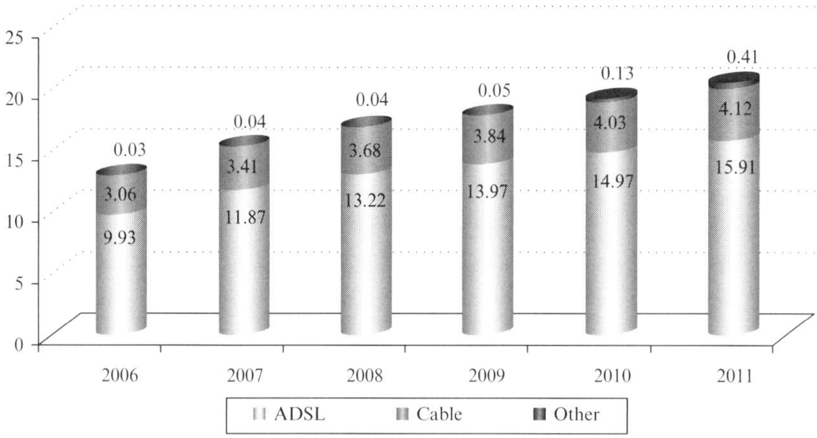

Source: Ofcom (2012), Communications market report.

Figure 15.6 Fixed broadband connections in the UK (millions)

of this, the development towards NGA has been limited so far. The UK hoped to see substantial effects by introducing structural separation and thus to separate the BT retail business from the wholesale business, which is organised under the name OpenReach. Such a network company that provides wholesale products and services to downstream service providers was seen as a key factor for competition on the service side.

However, investment in fibre networks by the incumbent as well as competitors is not impressive. BT has invested £2.5 billion to upgrade the local access network, bringing fibre broadband within reach of two-thirds of UK homes and businesses by the end of spring 2014.[4] Except for BT, most investments are undertaken by Virgin Media.

The fixed broadband penetration in the UK stands at 32.5 per cent of the population and 72 per cent of households, which are predominantly served by ADSL and cable networks and only a minor portion by other broadband such as fibre (Ofcom, 2012a). Figure 15.6 shows the fixed broadband connections in the UK.

As Figure 15.7 shows, household availability of superfast broadband is at 60 per cent (superfast broadband is defined as 30 Mbit/s or higher). In total numbers, there were 1.4 million such connections in March 2012, predominantly from BT and Virgin Media.

Mobile broadband can be categorised in three different ways according to Figure 15.8 and also shows substantial growth. The UK is aware of the importance of a highly capable telecommunication access network. A

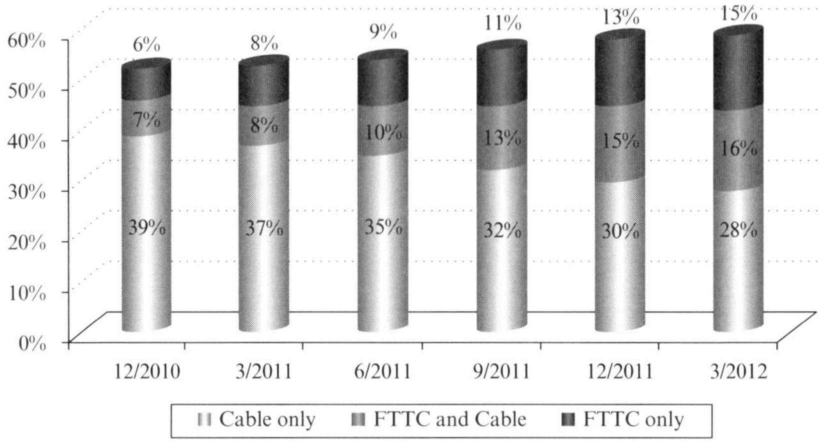

Source: Ofcom (2012), Communications market report, p. 284.

Figure 15.7 Superfast broadband in the UK (% of homes)

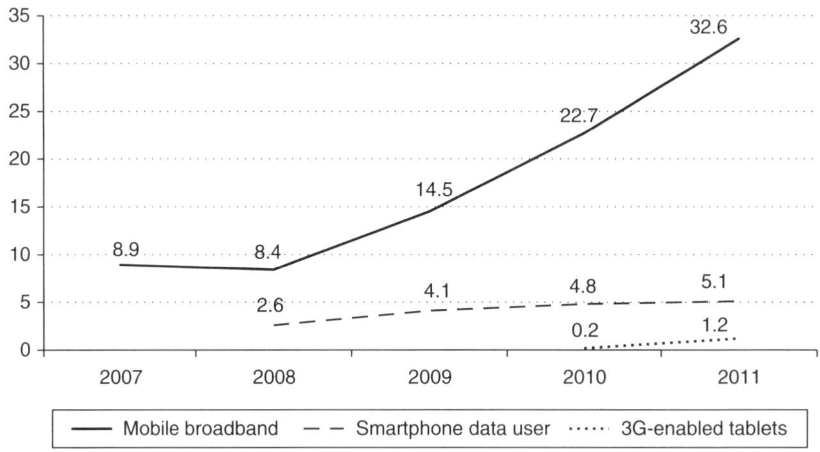

Source: Ofcom (2012), Communications market report, p. 288.

Figure 15.8 UK market for mobile broadband (millions)

national plan has been developed – Broadband Delivery UK – which also contains information about public support for the rollout of next generation access networks.[5]

Municipalities and utilities are also involved, though to a lesser degree than in comparable EU countries. In 2011, the government set aside £100 million for an Urban Broadband Fund (UBF) that will create up to ten 'super-connected' cities across the UK. This was followed in 2012 by a further fund of £50 million for a 'second wave' of cities to benefit from this programme.[6] Energy utilities are rolling out smart meters, too, although this is not dependent on NGA and communication is done via wireless devices.

Access regulation is a key component of the strategy in the UK. Besides structural separation there are also access obligations such as service level agreements and infrastructure access (divided into passive/physical and active access). Fibre access has been included in the market definition for WLA (wholesale local access) but BT is not under any obligation so far to provide wholesale access to unbundled fibre loops. The obligations levied on BT for access products are:

1. Local loop unbundling (LLU), which allows OCPs to physically take over (or share) BT's existing copper lines between the local telephone exchanges and the customer premises.
2. Virtual unbundled local access (VULA), which will have to be provided by BT wherever it deploys its NGA network. VULA will provide a virtual connection that gives OCPs a dedicated link to their customers and substantial control over the services provided.
3. Physical infrastructure access (PIA), which allows OCPs to deploy fibre in the access network using BT's ducts and poles – either to support deployment of fibre-to-the-premises (FTTP) technology, or to support deployment of fibre-to-the-cabinet (FTTC) technology (by enabling a 'backhaul' connection between street cabinets and the OCP's network).[7]

The UK is still heavily using the copper network for NGA purposes and as in many other countries it is being considered whether it is possible to extend the usage of the copper network as this may be cheaper than to deploy new fibre networks. BT has announced the introduction of vectoring technology on its copper network for the end of 2013. The regulatory consequences are being addressed in the wholesale local access review.

In its consultation Ofcom (Ofcom, 2012b, pp. 23–25) has come forward with a number of interesting questions, such as:

> 4.6 Does our general pricing approach to the pricing regulation of VULA remain appropriate, assuming that such a remedy continues to be required? If not, why? Please provide reasons to support your views.

4.8 Have the existing ex ante safeguards against margin squeeze in relation to VULA been effective? If not, what would be an alternative approach? Please provide reasons to support your views.

4.9 What should be the purpose of any ex ante margin squeeze safeguards in relation to VULA (for example, actively promoting expansion by non-BT retailers or simply protecting reasonably efficient retailers) where such safeguards are required? Please provide reasons to support your views.

4.12 Have there been any significant changes since the last market review that mean we should alter our approach to regulating SLU, for example, in response to technological change, assuming that such a remedy continues to be required? If so, please provide reasons supporting your views.[8]

The UK is still a country where discussions about NGA and fibre rollout are intense but in practice the implementation is rather slow. A strong focus can still be found in the regulatory debate with respect to wholesale products to be provided on the basis of the copper network (VULA, PIA).

Regarding the UK case the development towards next generation access networks has been driven by: 1) a duopoly of networks (BT, cable); 2) regulatory wholesale obligations levied by Ofcom; and 3) overarching strategic and policy considerations and plans.

It appears that, in the UK, market activity with respect to NGA rollout is still somewhat behind. However, the discussion is intense. A strong focus on the regulatory measures can be found. Likewise, activities of alternative operators, municipalities, utilities, etc. are rather limited. Politically, NGA is seen as important for Britain's future, yet financing still has to come mostly from private sources.

4. THE NETHERLANDS

The telecommunications markets in the Netherlands have undergone significant change during the last decades. Owing to the introduction of unbundled access to the local loops of the incumbent KPN in 1998 and the introduction of broadband over CATV networks, the market shares in the fixed markets have changed considerably. In 2012, the market share of KPN in the retail markets for broadband decreased to below 45 per cent. The CATV operators UPC and Ziggo have captured about the same fraction of the market, while the operators basing their products on unbundled local loops have captured about 10 per cent of the market.

As can be seen in Figure 15.9, the market is divided between five main operators and an additional number of smaller operators that mainly consist of municipalities, public–private partnerships and others. Of the

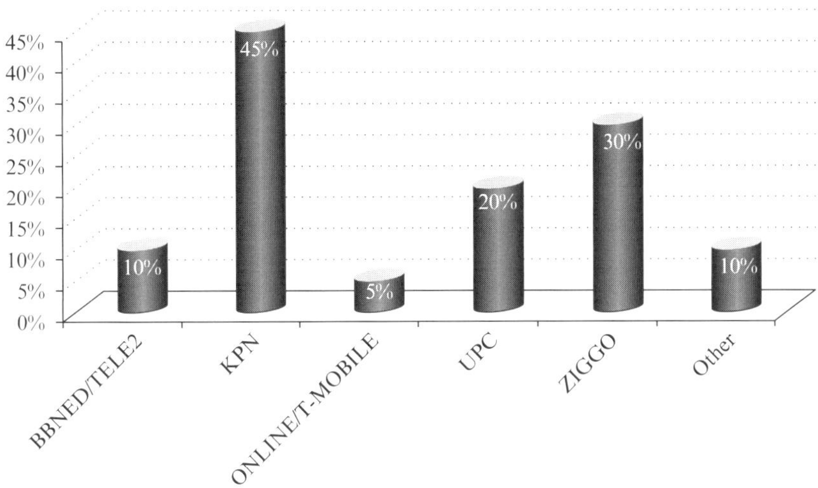

Source: OPTA, Public Report Broadband Q3/2012.

Figure 15.9 Market shares in the Dutch retail broadband markets in Q3/2012

five significant operators, Ziggo and UPC are CATV operators, KPN is the incumbent operator with a legacy copper access network and the other two are operators using the access network of KPN.

When looking at the technologies used, as can be seen in Figure 15.10, the market is almost divided in half by CATV and DSL, while fibre access is limited to only 5 per cent of the market.

The situation, however, is still undergoing continuous change; although the number of FTTH fibre connections is still limited, Figure 15.11 demonstrates that the 305,000 homes connected (not active customers) at the end of 2008 have increased to 1.1 million homes connected in Q2/2012.

Hence, although the investments in FTTH have come rather late, the development is positive, and if it continues, the household coverage of fibre is likely to be higher than 50 per cent in 2020.

Investments in fibre have been driven by the market participants rather than by regulation or government aid as the initial investments in FTTH started before the regulation was adopted to the new technology and before the broadband policy – Digitale Agenda.nl – was published in 2011. It is more likely that it was the threat from the cable operators and the introduction of DOCSIS 2.0 and later DOCSIS 3.0 that has driven the investments in FTTH (De Bijl, 2011, p. 22).

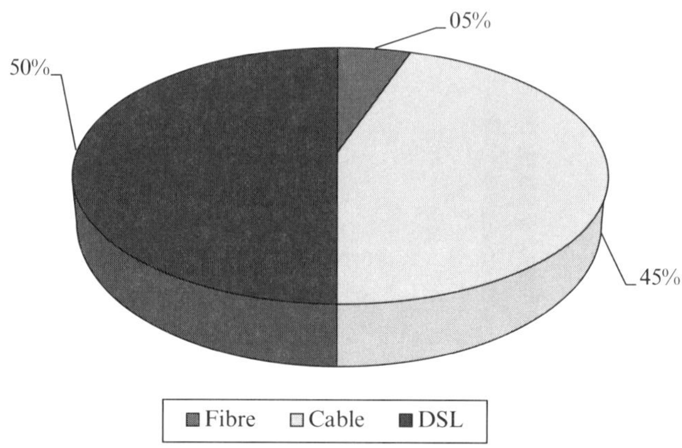

Source: OPTA, Public Report Broadband Q3/2012.

Figure 15.10 Technologies used to provide broadband services in Q3/2012

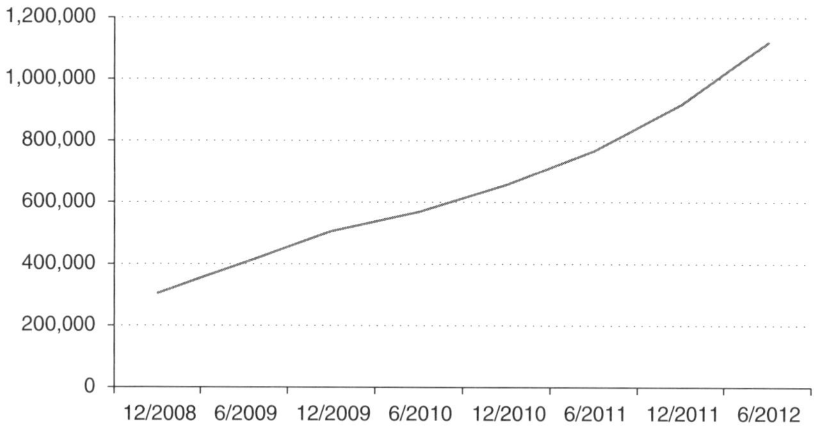

Source: OPTA, Public Report Broadband Q3/2012.

Figure 15.11 FTTH homes connected in the Netherlands in Q3/2012

The regulation was introduced at the same time in the Netherlands as in the rest of Europe. For the broadband markets, especially, the unbundling of the local loop has had a significant impact. At the beginning, the uptake of unbundled lines was limited, but with the introduction and growth of DSL this changed. During 2003 and 2004 the number of lines increased from nearly zero to more than 400,000 lines (De Bijl, 2011, p. 22). The success of local loop unbundling was further perpetuated by a reduction in the regulated prices from €13 in 2002 to only €6 in 2010 (Stratix Consulting, 2010). This was against the initial intention of the regulator that wanted to determine increasing prices in order to create incentives for the alternative operators to invest in their own local loops (De Bijl, 2011, p. 12f).

The outcome in the Dutch market owing to the combined increase in unbundled local loops, the introduction of broadband over CATV networks and DSL in the KPN network has been a market with high Internet/broadband penetration. For example, De Bijl (2011, p. 14) writes: 'At the start of 2009 [. . .] the average broadband penetration rate in the EU was 24%. The Netherlands (together with Denmark) were out in front with a broadband penetration rate of almost 37%.'

The investments in FTTH have been driven on one hand by the incumbent KPN together with Reggeborgh and on the other hand by local initiatives – mostly originating from the municipality in the form of public–private partnerships including e.g. Amsterdam and Rotterdam (see e.g. Sadowski and Nucciarelli, 2008).

KPN started in 2005 to invest in FTTC and FTTH, while Reggeborgh founded Reggefiber to invest in FTTH in e.g. the cities of Deventer, Nuenen and Hillegom. In 2008, KPN and Reggeborgh agreed to consolidate their investments in a joint venture under the name of Reggefiber.[9] At the end of 2012, Reggefiber reported 1.3 million homes passed (up from 0.9 million at the end of 2011) and 419,000 active customers (up from 254,000).[10]

An example of a municipal driven project is Amsterdam Citynet or 'Glasvezelnet Amsterdam' (GNA), which is a consortium of private and public investors. The decision was taken by the city council in 2004 and the joint venture encompassing the city, the housing corporations and private investors was incorporated in 2006. In February 2009, 40,000 homes were connected (FTTH Council Europe, 2013). From 2009 onwards, Reggefiber has increased its ownership in GNA and is now also operating the network.

With regard to the investments in fibre in the Netherlands, the regulation has been more of a hold-up than a promoter. With regard to GNA, the EU Commission started investigations in 2006 and needed a year to take a decision on whether state aid from the municipality of Amsterdam was acceptable or not, delaying the complete project.[11] With regard to the establishment of a joint venture of KPN and Reggeborgh, the national competition authority started investigations and hesitated to give its approval. As it finally did so in 2009, it imposed remedies including functional and structural separation of Reggefiber from KPN and the obligation to provide open access to all access seekers on a non-discriminatory basis.[12]

Two years later, the Dutch Ministry of Economic Affairs, Agriculture and Innovation published the Digitale Agenda.nl. The aim of the agenda is aligned with the aims of the EU Commission in that everyone should have access to 30 Mbit/s in 2020 and 50 per cent should have access to more than 100 Mbit/s. The national broadband initiative has a policy focus on facilitating competition, ensuring open networks, technology neutrality and consumer issues. With regard to fibre networks, the initiative addresses issues regarding reduction of bureaucracy and investment uncertainties, provision of spectrum for mobile services, the introduction of information to stakeholders ('broadband monitor'), regulated access to the networks (open access and non-discrimination) and network neutrality. Further, the initiative includes the establishment of the Taskforce NGN, which has the assignment to help municipals and other stakeholders with the planning, coordinating and facilitating of the network rollout. It also has the assignment to implement the 'broadband monitor'.[13]

The regulatory authority (OPTA) has imposed regulation on to the Digitale Agenda.nl, including regulated access to fibre loops (i.e. unbundling of FTTH networks). Reggefiber is required to provide open access on a non-discriminatory basis and the tariffs should be cost oriented. As FTTH networks are new and investment is still coming through, the regulatory authority decided to assess the costs based on the discounted cash flow (DCF) method. In order to create investments and/or to consider the risks in rolling out a new network, the cost of capital ('discount factor') was set higher than for other regulated products. Further, the costs are indexed to take inflation into account (OPTA, 2013).

As of today we arrive at the conclusion that the markets are making significant progress towards high speed next generation access networks, but that the rollout has already been going on for several years and it will still be a considerable amount of time before FTTH networks become the major technology in the Netherlands.

Regarding the Dutch case the development towards next generation access networks has been driven by: 1) the cable operators investing in broadband and eventually DOCSIS 3.0; 2) the local municipalities investing in fibre in the early years; and 3) Reggefiber and KPN rolling out a national FTTH network as a reaction to the activities of the cable operators and local municipalities.

With regard to the national government, competition authority and the sector-specific regulatory agency (OPTA), the contribution has so far mainly been limited to the regulation of open access, i.e. facilitating the service-based competition. In a couple of cases, these authorities have even rather delayed the progress. This might change, though, as the Digitale Agenda.nl is taking effect in the markets.

5. SWITZERLAND

Switzerland has some unique features as regards the development towards NGA and the digital economy. These features are partly rooted in the fact that not being a member of the European Union has for a long time enabled Switzerland to follow its own path of competition policy in the telecommunications sector. One example of many is that physical unbundling was not introduced any earlier than 2007. Furthermore, there has been an intensive debate and significant efforts to deploy fibre networks by the incumbent Swisscom as well as by alternative network operators, specifically backed by the local and regional utilities. However, so far, broadband connections are still predominantly based on cable networks and DSL.

Broadband Internet access is still on the rise in Switzerland. Despite massive efforts to invest in fibre access networks, e.g. FTTB and FTTH, the two dominant technologies are still DSL and cable, and they are still growing. Meanwhile, as can be seen in Figure 15.12, there are more than 3 million high speed broadband connections in Switzerland.

According to national statistics, the number of fibre access lines is still small.

Switzerland is a leader in broadband connections overall with approximately 40 per cent of households connected. However, with respect to fibre access shares, as Figure 15.13 shows, Switzerland is far behind Japan and Sweden as well as the OECD average. Only Germany, Australia and the United Kingdom show lower shares than Switzerland.

Mobile broadband is also relevant and the usage in Switzerland compares to the international average. Based on the coverage presented in Figure 15.14, the penetration of mobile broadband has already surpassed 50 per cent.

Early on in international comparison, the discussion about fibre access started and this discussion in Switzerland is focused on FTTH. In 2008, local utilities established a joint association under the name 'openaxs' to provide fibre access networks. They felt it to be important that a national fibre network would be introduced that could enable fibre-based access in the whole country. The efforts today encompass local energy utilities, municipalities, service providers and property owners and developers.[14] Today, the activities are also known as SwissFibreNet.

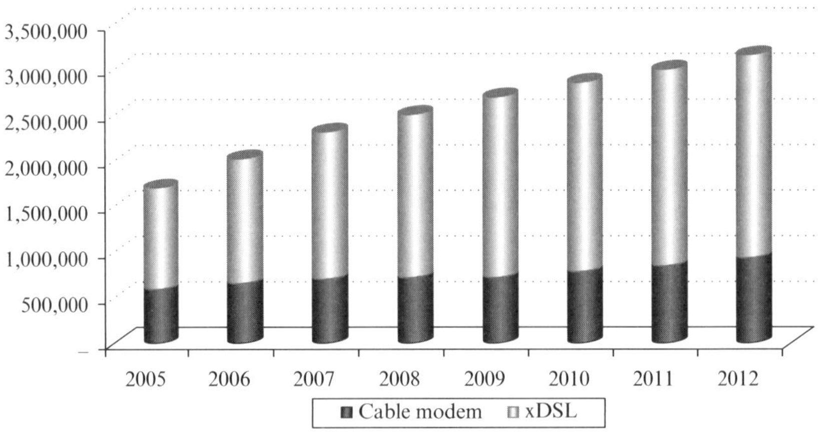

Sources: Statistik Schweiz, the regulatory authority Bakom (www.bakom.admin.ch) and the Eidgendgenoessische Kommmunikationskommission, see http://www.comcom.admin.ch/dokumentation/00439/00565/index.html?lang=de (March 2013).

Figure 15.12 Broadband Internet access in Switzerland

Interestingly, the incumbent operator pursued its own strategy and after the introduction of openaxs launched an FTTH campaign deploying four fibre access lines to households that would allow itself and other providers of communications services to serve the customers. Thus, in principle, Swisscom also followed an open access approach and tried to establish cooperative models.[15] The total amount of investment is unknown, although it will amount to several billion euros in the end.[16]

On the level of policy making and regulation, the trend towards fibre has always been welcomed but it has not been mandated or regulated to any detailed extent. There is a legislature plan for 2011–15 and a strategy for an information society, in which the Federal Council has set the goal to continue to extend a powerful and open telecommunications network. Following this, Bakom has set up the High-Speed Broadband/Next Generation Access Working Group, which aims to map the provision of high-speed broadband services geographically in Switzerland and to make them accessible. This work is also intended to support public authorities.

Starting in 2009, the regulatory authority Bakom[17] has played a crucial role in coordinating the activities in order to find a common approach for the rollout. Switzerland, interestingly, is a country showing great willingness to invest in fibre projects, and a highly capable communications infrastructure is seen as an important factor for both business and society.

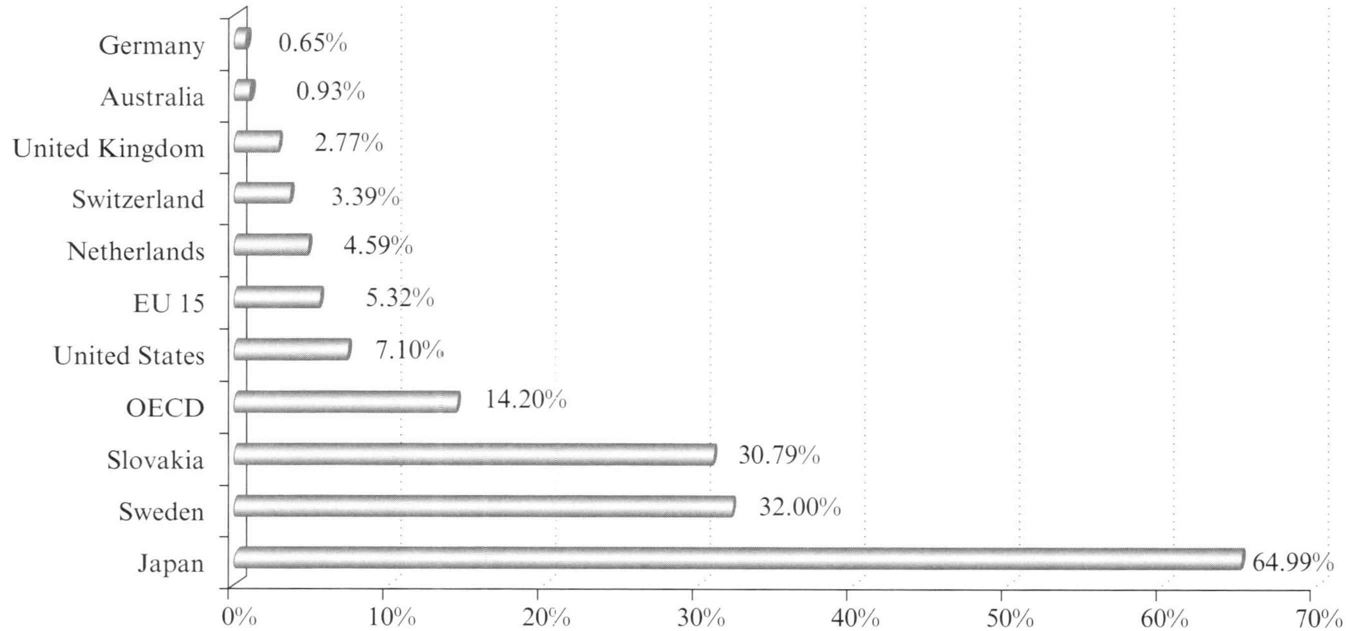

Source: http://www.comcom.admin.ch/dokumentation/00439/00565/index.html?lang=de (March 2013).

Figure 15.13 Fibre access shares

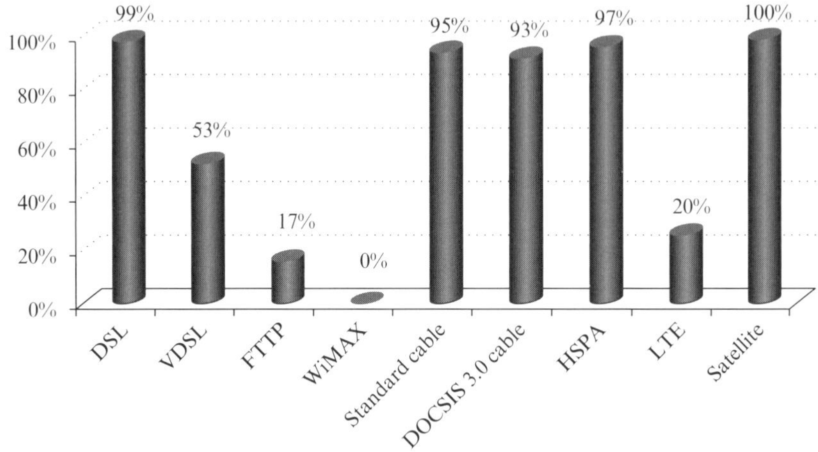

Source: Point Topic (2013), Broadband Coverage in Europe 2012: Coverage for Switzerland, p. 11.

Figure 15.14 Coverage by technology

Under Bakom's leadership, nine round tables have been established to coordinate the work towards a ubiquitous fibre network.

Regarding the Swiss case, the development towards next generation access networks has been driven by: 1) investment of alternative network operators (utilities), the incumbent and cable operators in next generation access; 2) coordination work to enable a nationwide fibre access network; and 3) a high level of willingness to invest in infrastructure combined with a regulatory hands-off approach.

Switzerland is one of the countries that have fostered the open access idea very strongly. It remains to be seen whether the take-up rate for FTTH will move the market forward.

6. COMPARISON

The selected European countries covered in this article display quite different situations and features. With regard to the development towards NGN based on penetration rates and homes passed, the countries covered can be divided into the following groups: 1) countries with limited demand but with a significant development towards NGA, i.e. the Netherlands and Switzerland; and 2) countries lagging behind, i.e. Germany and the UK.

The situation regarding investments is different in the four countries. In Germany the investments are distributed among a number of players in the market while no operator has taken the lead so far. That said, the incumbent Deutsche Telekom has recently announced a nationwide FTTC programme. A similar situation can be found in Switzerland, although here the process started earlier than in Germany. In the Netherlands, the situation is different with the joint venture between the incumbent KPN and Reggefiber taking the lead for the development. In the UK, the progress has been delayed but the incumbent is rolling out infrastructure only in selected areas. In all the countries covered, the CATV operators have so far played a significant role as drivers for more investment into fibre networks, either as investors or as a competitive force threatening the incumbent telecommunication operators:

1. In Germany one can find a multilayer investment situation that is mostly driven by the market rather than induced by the regulatory environment. Several parties are committed to deploying NGA infrastructures. To date significant amounts have already been invested. According to the associations VATM and BREKO, competitors of TD are planning further investments of €3 billion (VATM) and €7.5 billion, thereof €4.2 billion for FTTB/H (BREKO) until 2015.[18] TD has announced it plans to invest up to €6 billion comprising FTTC/VDSL including vectoring and LTE. Cable network operators will complete their networks and, thus, have planned to invest a further €5 billion until 2015.[19] Finally, projects initiated by utilities and municipalities have to be taken into consideration too, as well as the capital expenditure of the mobile operators for LTE.
2. Investment in NGA infrastructure in the UK in most cases is driven by the incumbent and cable network operators. Some investment has also been undertaken by competitors. Hence, BT has invested £2.5 billion to upgrade its local access network, bringing fibre broadband within reach of two-thirds of UK homes and businesses by the end of spring 2014. Except for BT, most investment is undertaken by Virgin Media.
3. The investments in fibre in the Netherlands has been driven by the market participants rather than by regulation or government aid as the initial investments in FTTH started before the regulation was applied to the new technology and before the broadband policy named Digitale Agenda.nl was published in 2011. The investments in FTTH have been driven on one hand by the incumbent KPN together with Reggeborgh, being under competitive pressure from the cable

operators, and on the other hand by local initiatives (municipalities) in the form of public–private partnerships. With regard to the investments in fibre in the Netherlands, the regulation has been more of a hold-up than a promoter.

4. Concerning the deployment of NGA infrastructures the investment discussion in Switzerland is focused on FTTH and cable networks. Investments in fibre access networks (FTTB/FTTH) have already been made in particular by local utilities and municipalities as well as the incumbent Swisscom. Further, service providers and property owners and developers are involved.

In all of the countries covered in this chapter, a national broadband plan has been developed. Research shows that those countries lagging behind have introduced a broadband plan. Whether this is a significant correlation is difficult to say. One could argue that because a country lags behind, it develops a national broadband plan.

1. Germany: In February 2009 the Federal Ministry of Economics and Technology released the German Broadband Strategy based on four pillars (see above). In April 2012 the German government expanded its broadband strategy and defined as an additional target a nationwide availability for broadband access of at least 50 Mbps by the end of 2018.
2. In the UK a national plan has been developed – Broadband Delivery UK – which also contains information about public support for the rollout of next generation access networks.
3. In 2011 the Dutch Ministry of Economic Affairs, Agriculture and Innovation published the Digitale Agenda.nl. The aim of the agenda is aligned with the aims of the EU Commission regarding that everyone should have access to 30 Mbit/s in 2020 and 50 per cent should have access to more than 100 Mbit/s.
4. Switzerland: There is no specific national broadband plan introduced. On the level of policy making and regulation a legislature plan for 2011–15 and a strategy for an information society have been issued, in which the Federal Council has set the goal to continue to extend a powerful and open telecommunications network.

When it comes to regulation, a correlation cannot be observed between the intensity of regulation and the development towards NGA. It is to be seen, though, that the UK with a strict regulation in the form of structural separation and Germany with a successful local loop unbundling regulation (for current generation access networks based on copper, though) have lower penetration rates for fibre offers than the Netherlands and Switzerland.

The Netherlands is an interesting example from a European perspective when it comes to regulation and investments into fibre networks. The regulation was introduced at the same time in the Netherlands as in the rest of Europe and comprises in particular the unbundling of the local loop. With regard to the investments in fibre in the Netherlands, though, the regulation has been more of a hold-up than a promoter.

In Germany, unlike in the other three countries, the local utilities and municipalities have played a major role. Although also relevant in the Netherlands (especially in Amsterdam), in Switzerland (through openaxs) and in the UK, it is only in Germany where local utilities and municipalities are acting as a driving market force. Here, utilities (locally or regional based) have become more and more an integral part concerning NGA investments in the past two to three years, in particular in sub-urban and remote areas of the country. This is because of the highly fragmented market. Further, a comparable role is taken by municipalities and regional authorities respectively. Activities of the latter are partly supported by subsidy programmes and initiatives of the federal states.

The latest development towards NGA networks in Europe is the use of vectoring in order to increase the lifetime and the capabilities of legacy networks. This strategy is now a major cornerstone of the investment programme of Deutsche Telekom in Germany. In the Netherlands and Switzerland, though, which are further advanced, it seems that vectoring is not the most attractive opportunity for the telecommunication providers.

7. CONCLUSIONS

With the country studies presented in this chapter and the different carrier and policy approaches in mind, it is hard to see how Europe will meet the European Commission's broadband targets as set out in the Digital Agenda. The Commission has set targets with at least half of all households in Europe taking a broadband subscription at speeds of 100 Mbps or above. Even with new applications like vectoring, VDSL technologies will struggle to deliver that speed to the majority of consumers.

What can be said from the results from the countries presented is rather what kind of policy/regulatory measurements do not help and not which measurements ensure 'success'. Without the incumbents' willingness to invest, the development towards NGA will remain slow. On the other hand, the UK experience gives room for doubt that relying mainly on the incumbent operator is not the most successful way. Also in Germany, where Deutsche Telekom has been slow in rolling out new infrastructure, the investments by alternative operators have not been sufficient to create a flourishing market for NGA. This leads us to conclude that a competitive environment on the infrastructure level (e.g. intermodal competition) has a positive impact on the willingness to invest in new fibre access networks.

The impact to be derived from a national broadband plan is difficult to assess, especially as those countries with slow uptake were the first ones to implement a broadband plan. Until now, the outcomes of these plans in the countries covered here are limited.

When it comes to sector specific regulation, the results on the development have been limited. It seems as if the structural separation implemented in the UK has not fostered investments into fibre access networks. The German and Dutch regulations have not implemented structural separation but have had a focus on the unbundling of local loops. It is difficult to say, however, whether these countries have benefitted or had any negative impacts of regulation on their investments.

The overall conclusion is that the measurements implemented in the countries covered here have not been sufficient. An improved development towards NGA has generally not been fostered by policies or regulation. Instead it is likely that the countries covered here need to search for measurements other than sector-specific regulation, national broadband plans, structural separation and the like in order to promote investments into NGA networks.

NOTES

1. Most recently, in 2013, the EU Commission started the 'Single Market Initiative' – see http://ec.europa.eu/dgs/connect/en/content/digital-single-market–ttp://ec.europais not the focus of this chapter.
2. However at last with respect to the regulatory environment in Germany it needs to be stated that cable network operators are not burdened with, for instance, (broadband) access obligations.
3. At the end of 2012, TD submitted a request to the federal network agency (BNetzA) to restrict competitors' access options to the sub-loop at the street cabinet, to allow it to implement vectoring in its network. On 10 April 2013 BNetzA published its draft decision on the introduction of vectoring technology in TD's network. The draft states that TD must continue to allow its competitors access to the street cabinet and the sub-loop respectively, from previously unused street cabinets. This means that in future, every company – be it TD or its competitors – can open up the street cabinets for VDSL and consequently for vectoring. However, TD is now allowed to prohibit access to the sub-loop under certain conditions, in order to implement vectoring there itself or to allow another company to do so.
4. http://www.superfast-openreach.co.uk/the-big-build/the-numbers-game.aspx.

5. https://www.gov.uk/broadband-delivery-uk.
6. Ibid.
7. http://stakeholders.ofcom.org.uk/consultations/wla/statement – see section 1.5.
8. http://stakeholders.ofcom.org.uk/consultations/fixed-access-markets.
9. KPN, Press Release, 22 May 2008.
10. http://www.eindelijkglasvezel.nl.
11. http://www.citynet.nl.
12. Nederlandse Mededingingsautoriteit, Besluit nr. 6397/274, '6397/KPN-Reggefiber', 28 July 2009.
13. Ministerie van Ekonomische Zaken, Landbouw en Innovatie (2011), Digitale Agenda.nl.
14. See www.openaxs.ch.
15. See www.openaxs.ch, Broadband Coverortraet/das-netz-von-swisscom/rund-ums-festnetz/glasfaserversorgung.html.
16. See http://www.charder.ch/libraries.files/FTTH_in_Switzerland_v1.pdf for strategic and financial information.
17. See http://www.bakom.admin.ch/themen/technologie/01397/03044/index.html?lang=de.
18. Breitbandbüro des Bundes, Tim Brauckmüller, Zentrale Aspekte des Breitbandausbaus, Presentation on 23 May 2013, p. 7.
19. Breitbandbüro des Bundes, Tim Brauckmüller, Zentrale Aspekte des Breitbandausbaus, Presentation on 23 May 2013, p. 7.

REFERENCES

ANGA (2013), Das deutsche Breitbandkabel, Infrastruktur der Zukunft.

BMWi (2009), Breitbandstrategie der Bundesregierung. Federal Ministry of Economics and Technology: The Federal Government's Broadband Strategy, available at http://www.bmwi.de/Dateien/BBA/PDF/breitbandstrategie-der-bundesregierung,property=pdf,bereich=bmwi,sprache=de,rwb=true.pdf (accessed 22 June 2013).

De Bijl, P.W.J. (2011), 'Broadband policy in the light of the Dutch experience with telecommunications liberalization', CPB Discussion Paper, No. 169.

Dialog Consult/VATM (2012), 12. gemeinsame Marktanalyse 2012.

FTTH Council Europe (2013), 'Case studies collection, January 2013'.

Little, A.D. (2012), 'LTE spectrum and network strategies, strategic options for mobile operators in dynamic 4G mobile markets', *Telecom and Media Viewpoint*, available at http://www.adlittle.com/downloads/tx_adlreports/ADL_LTE_Spectrum_Network_Strategies.pdf (accessed 26 June 2013).

Ministerie van Ekonomische Zaken, Landbouw en Innovatie (2011), Digitale Agenda.nl.

Nederlandse Mededingingsautoriteit, Besluit nr. 6397/274, '6397/KPN-Reggefiber', 28 July 2009.

Ofcom (2012a), 'Communications market report', available at http://stakeholders.ofcom.org.uk/market-data-research/market-data/communications-market-reports/cmr12/uk/ (accessed 26 June 2013).

Ofcom (2012b), 'Fixed access market reviews: wholesale local access, wholesale fixed analogue exchange lines, ISDN2 and ISDN30', available at http://stakeholders.ofcom.org.uk/consultations/fixed-access-market-reviews/ (accessed 26 June 2013).

OPTA (2013), Tariefbesluit ontbundelde glastoegang (FTTH) 2012. OPTA/AM/2013/200686, Den Haag, 22 March 2013.

Sadowski, B. and A. Nucciarelli (2008), *New challenges in municipal broadband network management: from vertical integration to wholesale-retail model*, ITS.
Stratix Consulting (2010), 'FTTH in the Netherlands: 2010 Q1', the Netherlands: Hilversum.
WIK (2011), 'Implikationen eines flächendeckenden: Glasfaserausbaus und sein Subventionsbedarf', No. 359.

16. Understanding market definition, relevant market and significant market power (SMP) frameworks in the EU

Chatchai Kongaut and Erik Bohlin

1. INTRODUCTION

Theoretically, resources are used efficiently in order to produce services or goods in a perfectly competitive market. Therefore the competitive market benefits social welfare overall. Realistically, various markets cannot be perfectly competitive owing to their unique characteristics, such as high investment cost and bottleneck characteristics.[1] The telecommunications sector, for example, is among those markets that cannot be perfectly competitive even though the competition in this sector has been greatly increased since liberalisation. Therefore there is still a need for sector regulation and competition law to encourage a competitive environment and to soften monopoly power. Sector regulation is seen as an ex ante approach more specifically in each sector to facilitate the efficiency of competition when competition law is inadequate. Examples of the ex ante approach are interconnection, unbundling and price regulation. Generally, the ex ante approach is considered as a stronger market intervention compared with the ex post approach, which is applied by competition law. Examples of ex post regulation are antitrust or merger regulation. In addition, regulations generally aim to cope with market structure failure, while competition law mainly focuses on the abusive behaviours of dominant operators. Also, the burden of proof for regulation to intervene is less than in competition law (De Streel, 2008b). Nevertheless, market intervention should be conducted only when there is some failure in the market.

In the telecommunications sector, the common types of market failure are market power and externalities. Market power, which makes imperfect competition, is found when, for example, incumbent operators are the only ones who have infrastructure to provide the whole service while other operators need to buy one or more parts of the services from the incumbents. Without any regulation, the natural market power of incumbents consequently leads to failure of the market, such as high access price or refusing to allow other operators to buy particular services, which can be considered as a market barrier. Examples of this issue are local loop unbundling (LLU) or mobile virtual network operators (MVNO). Another example of market failure in telecommunication services is the externality problem. This problem occurs when two or more networks provide the whole services to their customers but operators still need to buy the bottleneck services from each other. Without intervention, the operator with higher market power can raise its prices for bottleneck services to increase the cost for its rivals. An example of this problem is mobile termination rate. The market power and externalities issues can be seen as a one-way access problem and a two-way access problem, respectively.[2] Therefore before intervention, the regulator needs to specify that there is a problem in the market where there is a need for regulation.

In the European telecommunications sector, market intervention can be simplified as three steps to regulate any inefficient market. The first step is to identify the relevant market. This step aims to assess the competition in the market and relevant services in a particular market. The second step is to test whether there is any significant market power (SMP) in the market. The purpose of this step is to find if there are any market players who are controlling the market or possibly will do so in the future in a particular market. The third step is to apply the appropriate remedy to solve the market failure. This step is where the ex ante regulations are imposed on the market to improve market competition. While several studies mostly focus on the last step, the impact of various regulatory tools – for example, local loop unbundling, termination rate regulation, price regulation or accounting separation – there are still only limited studies regarding market definition and relevant market analysis, which is an important foundation before applying any regulatory tool. While ex ante regulation is mainly analysed and implemented based on economic frameworks, in practice ex ante regulation still relates to legal principles. For instance, relevant market analysis is currently identified based on the principle of competition law. It is not uncommon for economics and law aspects sometimes to go in different directions. This leaves scope for the regulatory body to often make a decision to trade off one option against another. With the telecommunications sector becoming more dynamic and complex, the relationship between economic concepts and competition law principles is becoming more complicated as well. Therefore the purpose of this chapter is to fill the gap in the study of regulation by trying to understand the economic problems in the market definition analysis, relevant market and SMP frameworks in the EU. In addition, this chapter aims to evaluate whether the framework imposed by the European Commission (EC) has so far been successful or has failed as well. After this introductory section, Section 2 provides overviews of Recommendations issued by the European Commission and the SMP framework. In Section 3 economic issues in market definition and the SMP framework in the EU are pointed out, followed by a case study of the broadband market in Section 4 in which more economic issues are explained. Section 5 summarises the public consultation results on the revision of the Recommendations on relevant markets to support the need to adapt the current Recommendations and SMP framework. Lastly, conclusions regarding whether the EU Recommendations have failed or been successful and the direction of future EU Recommendations are discussed in Section 6.

2. EC RECOMMENDATIONS AND THE SMP FRAMEWORK

2.1 Evolution of EC Recommendations and the SMP Framework

After the liberalisation of the telecommunications sector in the EU, the EU regulatory framework was introduced in 1998, for example, the guideline for interconnection and accounting separation, to encourage a more competitive environment in the market. At that time, the national regulatory authorities (NRAs) could define the market players who had significant power, having a market share of more than 25 per cent, as SMP, and consequently impose any ex ante regulation on SMP. The market definition and SMP analysis in the 1998 regulatory framework did not greatly rely on competition law principles. Until 2002, the EC issued guidelines on market analysis and the assessment of significant market power under the Community regulatory framework for electronic communications networks and services. This 2002 regulatory framework is where the principle of competition law started to play an important role in market definition and SMP analysis. The EC consequently issued a Recommendation on relevant product and service markets within the electronic communications sector susceptible to ex ante regulation in 2003 to support the NRAs' decisions on market definition and SMP analysis. With greater competition in the market, the EC revised and reissued the Recommendation on relevant product and service markets susceptible to ex ante regulation in 2007. Recently, with the changes in technology and the development of fibre technology, the EC has issued the Recommendation on regulated access to next generation access networks (NGAs) in 2010, but there has been

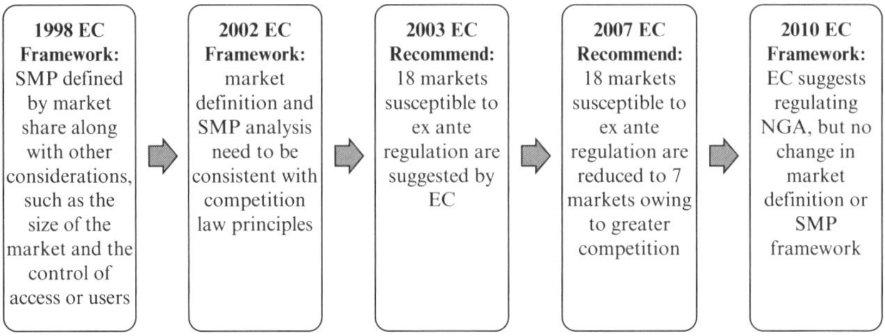

Figure 16.1 The evolution of EC Recommendations on market definition and SMP framework

no change in market definition and SMP analysis since 2007. Figure 16.1 presents the evolution of EC Recommendations on market definition and the SMP framework.

Nevertheless, the EC recently undertook a public consultation in early 2013 on the revision of the Recommendation on relevant markets in which BEREC,[3] the NRAs and several telecommunications operators, such as Telefonica, Telenor and Deutsche Telekom, commented on the questionnaire. The main issues for consultation are relevant trends in the sector that affect the relevant market, the efficiency of the three cumulative criteria test, the relevant markets listed in the 2007 Recommendations, and the markets to be added or removed from the 2007 Recommendation (EC, 2012). A summary of this public consultation is later explained in Section 5.

2.2 The Role of the European Commission (EC) and the National Regulatory Authorities (NRAs) in the Market Definition Procedure

According to Article 7 of Directive 2002/21/EC, NRAs are a main constituent in implementing the telecommunications regulation in each member state. At the same time, the EC acts as a reviewer as well as a decision maker (whether the proposal should be adopted, revised or withdrawn) after NRAs propose their regulatory measure. Importantly, NRAs must be independent of any organisation related to the telecommunications sector, and must use their powers fairly and transparently.

In general, NRAs have several responsibilities such as universal service regulation and radio frequency management. One of the important responsibilities is that NRAs need to regulate any proposition with significant market power. Therefore it is important for NRAs to define

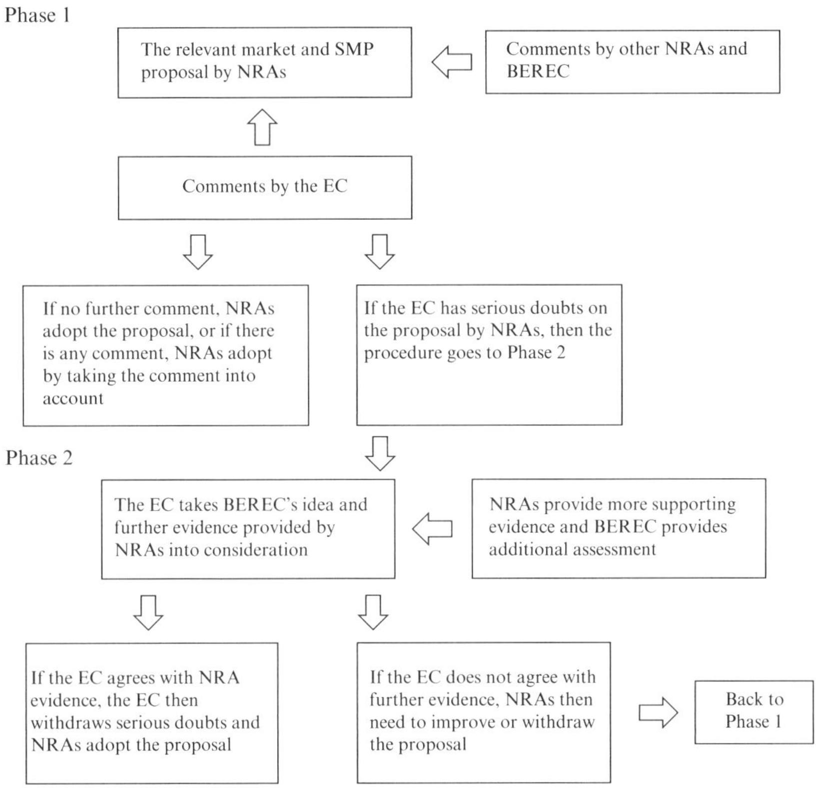

Source: EC (2010b).

Figure 16.2 The procedure for the relevant market and SMP proposal by the EU and NRAs from May 2011

the relevant market and analyse any dominant market player with an SMP framework before imposing any regulation. Generally similar in procedure but differing in purpose from competition law, which aims to prevent anti-competitive behaviour or abuse from dominance, the purpose of the SMP framework instead is to encourage greater competition in the telecommunications market as well as future economic development. The market definition and SMP analysis procedure have been developed over the past decade. Recently, in accordance with Article 7 of the Framework Directive 2009, the procedure of defining the relevant market and SMP in the EU from May 2011 onwards has been as presented in Figure 16.2.

2.3 EC Recommendations

In the EU telecommunications sector, the first step for the EC is to find whether there is an SMP in the market. If so, the chosen remedies are applied to (a) market player(s). Aiming to support the member states, the EC has issued Recommendations to help identify the markets where competition law may be inadequate. In the past decade, the European Commission on Electronic Communications has issued two recommendations, in 2003 and 2007, to identify market definition and SMP. According to the European Commission (2003, 2007), the Annex to the Recommendation on relevant product and service markets within the electronic communications sector in 2003 and 2007 can be summarised as in Table 16.1.

These Recommendations were launched as a guideline for each national regulator of member states to identify the market definition, relevant market and SMP. NRAs can have the same numbers of markets susceptible to ex ante regulation, more or even less than the Recommendations, depending on the market in each member state. According to Table 16.1, from the 2003 Recommendation, there are 18 markets susceptible to ex ante regulation. As the competition in some markets becomes greater over time, the number of markets susceptible to ex ante regulation is reduced to only seven markets in the 2007 Recommendation. It is interesting that in the 2007 Recommendation, while many retail markets have been significantly removed from the 2003 Recommendation, the number of wholesale markets has not greatly changed. Consequently, even though there are some remedies applied at retail level, such as regulating the international roaming price, most of the remedies applied to an SMP in a particular market mainly aim at the wholesale level.

2.4 SMP Framework

This SMP framework is consistent with the definition of relevant market by competition law (see Commission Notice on the definition of relevant market for the purposes of Community competition law, EC 1997). According to BEREC (2012), the SMP framework in the EU can be applied by following three main steps including defining the relevant market, SMP assessment and imposition of ex ante regulation. The three steps of the SMP framework can be simplified as in Figure 16.3.

2.4.1 Defining relevant markets

This first step can be divided into two phases. First, the regulator in each member country assesses the degree of competition in the market. The

Table 16.1 The market susceptible to ex ante regulation according to two Recommendations from 2003 and 2007

Recommendation 2003	Recommendation 2007
Retail level	Retail level
1. Access to the public telephone network at a fixed location for residential customers 2. Access to the public telephone network at a fixed location for non-residential customers	1. Access to the public telephone network at a fixed location for residential customers and non-residential customers (combines 1 and 2 from Recommendation 2003)
3. Publicly available local and/or national telephone services provided at a fixed location for residential customers 4. Publicly available international telephone services provided at a fixed location for residential customers 5. Publicly available local and/or national telephone services provided at a fixed location for non-residential customers 6. Publicly available international telephone services provided at a fixed location for non-residential customers 7. The minimum set of leased lines, comprising the specified types of leased line up to and including 2 M/bs	– (Markets 3–7 from Recommendation 2003 are removed)
Wholesale level	Wholesale level
8. Call origination on the public telephone network provided at a fixed location	2. Call origination on the public telephone network provided at a fixed location
9. Call termination on individual public telephone networks provided at a fixed location	3. Call termination on individual public telephone networks provided at a fixed location
10. Transit services in the fixed public telephone network	– (Market 10 from Recommendation 2003 is removed)
11. Wholesale unbundled access to metallic loops and sub-loops for the purpose of providing broadband and voice services	4. Wholesale (physical) network infrastructure access (including shared or fully unbundled access) at a fixed location (similar to Market 11 in Recommendation 2003)
12. Wholesale broadband access, including bit-stream access	5. Wholesale broadband access (non-physical or virtual network access), including bit-stream access

Table 16.1 (continued)

Recommendation 2003	Recommendation 2007
Wholesale level	Wholesale level
13. Wholesale terminating segments of leased lines	6. Wholesale terminating segments of leased lines, irrespective of the technology used to provide leased or dedicated capacity
14. Wholesale trunk segments of leased lines	– (Markets 14–15 from Recommendation 2003 are removed)
15. Access and call origination on public mobile telephone networks	
16. Voice call termination on individual mobile networks	7. Voice call termination on individual mobile networks
17. The wholesale national market for international roaming on public mobile networks	– (Markets 17–18 from Recommendation 2003 are removed)
18. Broadcasting transmission services to deliver broadcast content to end users	

Source: European Commission (2003, 2007).

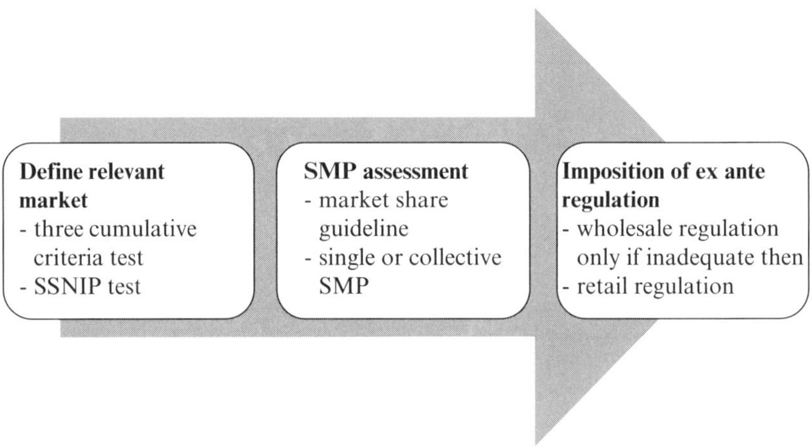

Figure 16.3 SMP framework

degree of competition can be assessed in two aspects: relevant product market and relevant geographic market. In this phase, both demand-side substitutability and supply-side substitutability need to be considered to define relevant market both in the product market and in the geographic market. The analysis that is normally applied at this stage is called the 'small but significant and non-transitory increase in price' (SSNIP) test or hypothetical monopolist test. The SSNIP test is a concept for measuring the relevant market by assuming a hypothetical monopolist and investigating whether a monopolist gains profits when increasing the price of the product by 5 per cent to 10 per cent, while others are constant. If a hypothetical monopolist does not gain any profit by increasing the price, it implies that there could be more substitute services to be included in the relevant market. On the other hand, if a hypothetical monopolist gains some profit, this group of services constitutes the relevant market. The SSNIP test can be calculated by estimating elasticity of demand or, alternatively, critical loss analysis.

The second phase, EC Recommendations 2007, provided that a three cumulative criteria test is applied in order to investigate the market to determine whether there is a need for regulation. In addition, ERG (2008) also provided some guidelines on the three cumulative criteria test as follows:

1. There is any existence of a non-transitory or high barrier to entry. There are several indicators to analyse whether there is any barrier to entering the market, such as structural barriers and legal barriers. Structural barriers are, for example, the existence of sunk costs, vertical integration services diversification, control of infrastructure, etc. Legal barriers include a licence for operating in the market and condition of spectrum use.
2. Market structure has ineffective competition in the relevant period of time. This criterion evaluates the competition in the market in the relevant timeframe. By applying a forward-looking assumption, the indicators for this criterion are, for example, market share, price trend, service diversification, barrier to expansion and potential competition.
3. The ex post treatment by competition law is not enough to solve market failure. This criterion aims to assess the efficiency of competition law in particular services. The indicators of this criterion include degree of generalisation of non-competitive behaviour, degree of difficulty to address non-competitive behaviour, non-competitive behaviour bringing about irreparable damage in related or connected markets, and need for regulatory intervention to ensure the development of effective competition in the long run.

De Streel (2008a), however, claimed that most member states usually only define relevant product market and relevant geographic market with the

SSNIP test while the three cumulative criteria test is ignored since they assume that the EC has already applied and chosen the market according to EC Recommendations. By contrast, Hou (2008) suggested that the first and second criteria are already covered in the SMP assessment procedure and the third criterion can be applied during the imposition of ex ante regulation procedure.

2.4.2 SMP assessment

The SMP position is assessed in this step after the relevant market has been defined. The SMP assessment is generally consistent with the guideline of *dominance* position in competition law. Nevertheless, dominance positions by competition law are not identical to SMP positions by NRAs because the decisions under competition law are not subject to the decisions by NRAs. The market share of market players, either single (individual) dominance or joint (collective) dominance, is normally used as a guideline for single (individual) SMP and joint (collective) SMP, respectively; for example, more than 40 per cent of market share implies a strong indication of dominance while less than 25 per cent market share is usually not considered as a dominant position. Both volume sales and value sales can be used as a measure, depending on the characteristics of the relevant markets. For instance, in call termination in a mobile market, the revenue from call termination is more appropriate to use as a measure than the number of terminating calls, because the calls can have different values, on-net, off-net, or international calls. Call termination revenue is therefore more suitable in this case. Not only is market share used as a measurement to assess SMP, but there are also several criteria to be considered, similar to the three cumulative criteria test, such as control of infrastructure, technological advantage, vertical integration, barrier to new entrances, absent or very low buying power, etc. In addition, since innovation from the telecommunications industry is highly developed, SMP assessment should be based on a forward-looking approach.

Owing to some similarities between the three cumulative criteria test and SMP assessment, there is sometimes no need to perform both tests in specific situations after assessing the degree of competition in the market. ERG (2008) provides the guideline with six scenarios for applying the three cumulative criteria test and SMP assessment relating to 2003 and 2007 Recommendations; however this chapter concludes only four scenarios relating to 2007 Recommendations as in Table 16.2.

2.4.3 Imposition of ex ante regulation

In this step, NRAs decide to apply one or more remedies to the market players who are assessed as SMP to solve market inefficiency as well as to

Table 16.2 Relationship between the three cumulative criteria test and SMP assessment

Scenarios	Three cumulative criteria test	SMP assessment
1. NRAs decide to impose ex ante regulation and the market is in the 2007 Recommendations	Optional (depending on each member state)	Yes
2. NRAs decide to impose ex ante regulation but the market is not in the 2007 Recommendations	Yes	Yes
3. NRAs decide not to impose any ex ante regulation (regardless of whether the market is included in the 2007 Recommendations or not)	Either the three cumulative criteria test or the SMP assessment is sufficient (depending on each member state)	
4. No competition problem or any market failure after assessing the degree of competition in the market	No obligation to review the market	

Source: Adapted from ERG (2008).

encourage more competition in the market. For instance, the regulatory tools to use as remedies can be transparency, non-discrimination, accounting separation, mandatory access, price control or cost accounting. These regulations can be combined or imposed separately, depending on the characteristics of market failure. At the first step, the regulations should first aim to solve market failure at the wholesale level. Only if the remedy at wholesale level is insufficient can NRAs then further regulate and apply remedies in the retail market. Figure 16.4 simplifies the criteria to move from wholesale to retail regulation.

With the SMP framework, the EC has already implemented several remedies at both wholesale level and retail level. Wholesale regulation, for example local loop and bitstream unbundling, has been regulated since 2004, while mobile termination rates have been reduced in several member states since 2008. When wholesale regulation is insufficient, as in the case of international roaming prices, the EU has already regulated the international roaming price for both voice and SMS services. Currently, the focus in the EU is on whether to regulate the next generation network. Understanding the market definition and SMP framework on this issue

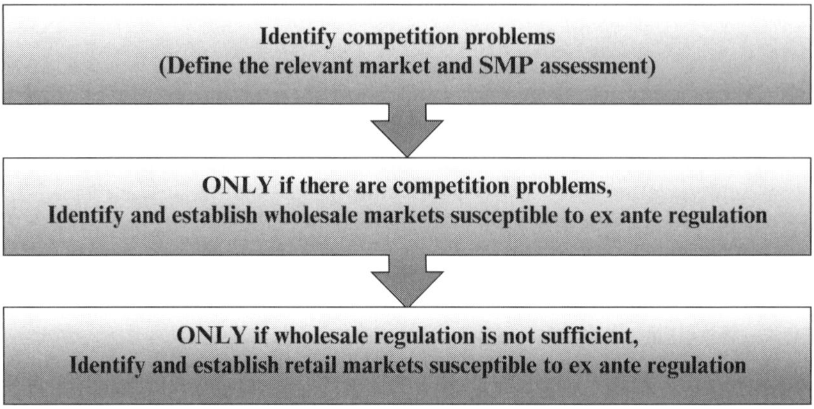

Figure 16.4 The move from wholesale to retail regulation

will benefit future regulation of the next generation network. Despite EC intentions to solve several market failures with regulatory tools, there are still some economic issues that need to be addressed, such as a trade-off between dynamic and static regulation, a two-sided market assumption, platform competition, and substitution between fixed and mobile services. Discarding these economic issues, market intervention can possibly reduce the incentive to invest and even distort the market. In the next section, the economic issues in the SMP framework by the EC are discussed by analysing previous studies.

3. ECONOMIC ISSUES IN THE SMP FRAMEWORK IN THE EU

Based on Article 15(1) of Directive 2002/21/EC, the EC and the NRAs are required to define the relevant market for electronic communications networks and services according to the principles of competition law. Some scholars have criticised that this dependency on competition law causes both the EC and the NRAs to have less flexibility when applying an economic framework to market definition analysis and the SMP framework. For example, De Streel (2008a) mentions four main reasons why the 2003 regulatory framework failed. These reasons include absence of clear objectives, inappropriate coordination with competition law, limited regulatory tools and, lastly, institutional failure. Nevertheless, the author concludes that the reformed 2007 EC proposals are an improvement and a step in the right direction but are still too moderate and need further improvement.

Table 16.3 Previous studies on the problem of SMP frameworks in the EU

Author(s)	Main finding(s)
Larouche and De Visser (2006)	The electronic communication sector has some uncertainty. Policy usually involves a trade-off decision. Therefore, the traditional legal concept cannot be applied without some adjustment.
Schwarz (2007)	The wholesale broadband market can be determined by substitution at the retail level, which consequently leads to a broader relevant market. The EC and NRAs therefore need to review their concept of the SMP framework and market definition analysis by taking this substitution into account.
De Streel (2008a)	Aligning sector-specific regulation such as market definition analysis and the SMP framework with competition law principles (antitrust methodologies) is inefficient owing to the dynamic market of the telecommunications sector.
De Streel (2008b)	Even though the regulation and competition law are converging, they remain significantly different, and it is suggested that sector regulation should have flexibility, objectivity, transparency, harmonisation, proportionality and legal certainty. Also, the sector regulation should be based on an economic concept.
Hou (2008)	In the three cumulative criteria test, the first and second criteria (existence of a non-transitory or high barrier, and market structure has ineffective competition) are already covered by SMP assessment and the third criterion (ex post treatment by competition law is not enough) is better performed in imposing the remedies.
Never and Preissl (2008)	Current telecommunications regulations in the EU have regulated heavily while the three-criteria test can be useful if properly applied.

Selected previous studies relating to the problem in the SMP framework in the EU are summarised in Table 16.3.

On top of these studies, BEREC has also recently provided reports relating to market definition for more understanding on several economic issues; for example, a report on the impact of bundled offers in retail and wholesale market definition (BEREC, 2010), a report on the impact of fixed-mobile substitution in market definition (BEREC, 2011a), a report on relevant market definition for business services (BEREC, 2011b), and a report on co-investment and SMP in NGA networks (BEREC, 2012).

Moreover, Cave et al. (2006) pointed out five important issues related to economic problems when deciding relevant market and market definition in the EU: the hypothetical monopolist test, self-supply, two-sided markets, geographical market definition and next generation networks. These concepts are important to understand when it comes to setting market definition and SMP framework in each member state. First, the hypothetical monopolist test or SNNIP test: this concept is consistent with the degree of competition evaluation in both demand-side substitutability and supply-side substitutability in the defining relevant market process in an SMP framework. Second, the self-supply issue refers to the relevance of self-supply for market definition in wholesale markets. This self-supply problem occurs either when the incumbent is the only operator providing a particular wholesale service or when the incumbent does not provide a particular wholesale service as the service is not subject to ex ante regulation. An example of a self-supply issue is access to local loop. Third, a two-sided market is an economic concept that provides more understanding on the interaction between two or more services. In addition, Evans and Schmalensee (2008) suggest that the market definition analysis needs to be improved when taking account of a two-sided market. Fourth, geographical market definition is also important since one country mostly has several types of market owing to geographical differences. Nevertheless, the regulators in practice determined one service in a whole country as one particular market. Fifth, the next generation networks situation is developing as well as the convergence of technology. With one technology able to substitute for another technology, market definition is more complicated. In the conclusions of Cave et al. (2006), the authors also suggested the market susceptible to ex ante regulation at the end of their reports[4] as a guideline for the EC before they issued the Recommendation in 2007. Many interesting economic concepts from the past decade can be discussed relating to market definition and SMP framework. To further elaborate on economic problems, this section discusses the third issue from Cave et al. (2006), a two-sided market, as well as discussing a trade-off between dynamic and static regulation that is also partly related to the self-supply issue. Moreover, in Section 4, a case study on the broadband market also presents other economic issues, related to the broadband market and next generation networks.

3.1 Is a Two-sided Market Argument Relevant in the Analysis of Market Definition and SMP Framework?

The purpose of market definition is to provide more understanding on the characteristics of telecommunications services before imposing any remedy in the case of market failure. If services are constituted in a

Table 16.4 Examples of a two-sided market in the telecommunications sector

Author(s)	Services in a two-sided market	Supporting comment(s)
Hausman and Wright (2006)	Fixed line callers and mobile subscribers	If the fixed to mobile price is too high compared to the mobile to mobile price, consumers will change to have mobile subscribers instead of calling from fixed telephones.
Valletti (2006)	Mobile telephone callers and mobile telephone receivers	Operators' networks can be seen as a platform for customers to complete their calls and both can gain positive externality the more people they can contact (network effect).
Evans (2007)	Fixed line callers and mobile subscribers	Comparing a telephone network to a shopping mall platform, mobile subscribers are similar to shopping mall customers while fixed line callers represent the retailers.

two-sided market, the imposition of any ex ante regulation needs to be carefully implemented since regulation in one service can affect other services.

Rochet and Tirole (2004) defined two-sided markets as 'markets in which one or several platforms enable interactions between end-users, and try to get the two (or multiple) sides "on board" by appropriately charging each side'. The authors also proposed that markets that can be considered as two-sided have the characteristics of network externalities between users and price discrimination between both sides of the market. According to the definition given by Rochet and Tirole (2004), many scholars have defined some telecommunications services as two-sided markets, such as Hausman and Wright (2006), Valletti (2006) and Evans (2007). Table 16.4 provides examples of a two-sided market in the telecommunications sector from previous studies.

Cave et al. (2006) have also pointed out an example of a dilemma between the SMP framework and a two-sided market. When defining the relevant market, practically and legally, market definition is mostly set from one type of service. The SMP can be found in one particular service, but might not be found as a whole group of services. The remedy is therefore introduced to one particular service to resolve its failure. With a two-sided market, however, the remedy from any particular service can affect another type of service. For example, the proposed remedy in mobile termination (wholesale level) can affect the price in the mobile market (retail level). Eventually, the counter-effects possibly reduce overall welfare. One example of these effects is known as the 'waterbed effect'.

The waterbed effect has been greatly debated with regard to the mobile termination issue. While the operators pointed out that decrease in mobile termination can lead to a rise in mobile retail price or mobile handset subsidy, which ultimately harms consumers, the regulator side found that the waterbed effect is insignificant. In addition, reducing the mobile termination rate would benefit consumers more and partly solve the bottleneck characteristic problem of the mobile termination service. Therefore the EC has recommended NRAs in member states to reduce the mobile termination rate since 2008.[5] Moreover, if SMP is determined by a whole group of services that link together, this might not be practical, especially from the legal point of view, since one service is likely to partly relate to several services. The two-sided market assumption therefore should be considered and carefully taken into account service by service. While it is not uncommon for economics and law sometimes to go in different directions, a two-sided market framework can at least provide more understanding in market analysis, which can be useful, for example, when providing an appropriate remedy to an SMP in any particular market.

3.2 Trade-off between the Dynamic and Static Regulation

Dynamic versus static regulation has been debated in recent years. In the early stage of telecommunications regulation, the infrastructure was already built with the telephone line and the regulation aimed to transform the monopoly incumbent to a competitive market. Conversely, the next generation network is currently only at the stage of building infrastructure, and this requires a large amount of investment. The transformation of the regulatory approach from what is referred to as static to dynamic regulation is needed in order to facilitate the next generation network and benefit overall welfare in the long run (Bauer and Bohlin, 2008). With market definition being the starting point from which to define any particular market for ex ante regulation, it is therefore very important to take the trade-off between the dynamic and static regulations into consideration. According to Grajek et al. (2008), the trade-off between dynamic and static regulations is that more competition in the short term that brings low price and high usage can possibly restrict future crucial investment in the long term. The authors also mention that an understanding of the trade-off between dynamic and static regulation is crucial for a regulatory framework.

An obvious example of the regulatory approach in the EU that aims to bridge the trade-off gap between the static and the dynamic environments is the 'ladder of investment' approach to local loop unbundling (LLU) initiated by Cave and Vogelsang (2003). The idea of a ladder of investment is that it can accommodate both short-term competition and investment in the long term. In the short term, it can stimulate a competitive market at the retail level that supports more services, lower price and better quality for consumers because the ladder of investment approach allows the new entrants to access the incumbent's network at an appropriate price. In the long term, after the new entrants are sufficiently known to consumers, they can gradually invest more parts of networks step by step (similar to a ladder), which encourages investment in the long run. Nevertheless, this approach has been heavily criticised: for example, Bourreau, Dogan and Manant (2010) claimed that the implementation is problematic, especially the appropriate time of decision and the credibility of regulators. Furthermore, later on even Cave (2010) himself suggested that it is not economically feasible for new entrants to completely have their own network at the final step.

The lesson from the ladder of investment is that sometimes the dilemma between dynamic and static cannot be completely solved. Therefore the regulators need to decide whether to take the static side, for example having a lower price with the risk that the future investment might be reduced, or take the dynamic side, which could experience insufficient competition at the current stage. Taking a trade-off between static and dynamic into consideration, the regulator in each country can therefore prioritise the situation case by case when it comes to defining the relevant market and imposing ex ante regulation. NRAs need to decide whether it is more crucial to protect consumers or whether it is more important to invest in infrastructure. Nevertheless, until now, NRAs have had only limited options owing to the strict principles of competition law, as seen in market definition analysis. In the next section, the discussions narrow down to the market definition analysis and SMP framework and its economic problem in the broadband market.

4. A CASE STUDY OF MARKET DEFINITION IN THE BROADBAND MARKET

Many scholars, such as Koutroumpis (2009) and Czernich et al. (2011), have found that broadband is important for economic growth as well as other benefits in several sectors of medicine, education, government, society or even entertainment. Therefore the broadband market has been recently a focus of the European Commission. For instance, the EU Digital Agenda aimed to have broadband coverage for all EU citizens in 2013 and more than 50 per cent of EU citizens should have transmission speeds of more than 30 Mbit/s in 2020. To achieve these ambitious goals, any remedy should be carefully implemented. According to the Recommendation (version 2007) by the EC (refer back to Table 16.1), the broadband market is related to Market 4 (Wholesale physical network infrastructure access) and Market 5 (Wholesale broadband access, non-physical or virtual network access). Currently, there are two major concerns when analysing market definition in the broadband market. First, there is greater competition between different technologies in the fixed broadband market as the development of cable technology and the implementation of fibre technology are becoming more relevant. Second, substitution between fixed and mobile broadband is greatly affecting the broadband market owing to the explosive growth of smartphone usage.

4.1 The Impacts of Competition in Different Platforms in the Fixed Broadband Market on Market Definition

Earlier, when DSL dominated in most member states, the main issue of market definition and the SMP framework in the broadband market focused on wholesale broadband access related to whether NRAs should regulate other infrastructure in the same way as they regulate DSL technology. For example, many NRAs impose unbundled local loop as a remedy in DSL infrastructure but not in cable infrastructure, owing to the difference in characteristics and technology at the wholesale level. In contrast, Schwarz (2007) argued that not only demand-side substitutability and supply-side substitutability at the wholesale level but also retail competitive constraints should be considered during the regulatory process.

Recently, the rolling-out of fibre technology, although still at an early stage, has been further expanded in many member states and is crucial to the future development of ultra-high speed broadband. The issue of market definition and SMP framework in the broadband market has now shifted to whether NRAs should impose the remedy on the fibre market as in the DSL market. The development of fibre technology and the availability of cable technology differ from country to country; so, before intervening in the market, the relevant market of different broadband services therefore needs to be appropriately identified and properly assessed. In Sweden, for instance, the national regulator, PTS, applied the SSNIP test to the wholesale broadband market and decided that DSL, cable and

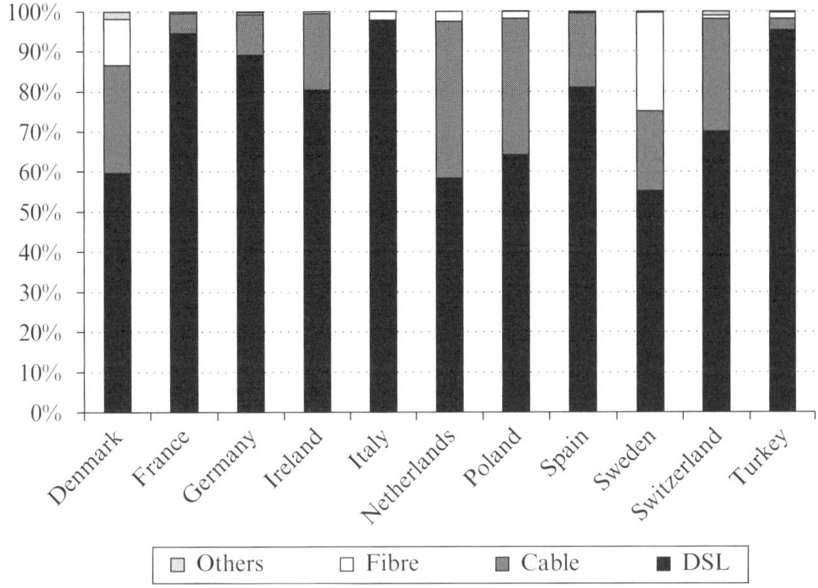

Source: OECD (2011b).

Figure 16.5 Percentage of fixed broadband subscriptions by technology in June 2010

fibre technologies are in the same wholesale market (Srinuan et al., 2012). Figure 16.5 presents the percentage of fixed broadband subscriptions by technology in selected European countries in June 2010.

Even though some countries such as Sweden and Denmark have a high ratio of fibre technology compared with other European countries, unlike copper line local loop (DSL), fibre technology is still at the early stage in other member states; there are therefore some EU member states that are deciding to exclude fibre technology from Markets 4 and 5. BEREC (2012) reported that there are currently 19 NRAs that include fibre networks in the relevant market in Market 4 and 17 NRAs in Market 5. Conversely, there are nine NRAs that exclude fibre networks from Market 4 and 11 NRAs that exclude fibre networks from Market 5. Even though several member states have decided to include fibre technology in Markets 4 and 5, the remedies to the wholesale broadband market in each country differ depending on politics, economics, geography and infrastructure in each member state.

So far, OECD (2011a) has mentioned several remedies that are applied in different member states, such as inter-platform competition, functional separation, an open access requirement, and public–private partnership. However, in many cases NRAs decide not to impose any remedy on fibre technology despite including fibre networks in Market 4 and Market 5 (EC, 2010a).

4.2 The Impacts of the Substitution between Fixed and Mobile Broadband on Market Definition

Impacts on market definition come not only from the competition between different technologies in fixed broadband, but also from the substitution between fixed and mobile broadband. In a similar way to substitution between fixed and mobile in voice services, the significant development of mobile technology is likely to merge the use of fixed and mobile broadband even though the fixed and mobile technologies differ. With currently significant growth in smartphone use, according to OECD data on fixed and mobile broadband subscriptions (June 2012), the number of mobile broadband subscriptions has already surpassed the number of fixed broadband subscriptions in many European countries. Figure 16.6 presents the number of fixed and mobile broadband subscriptions in selected European countries in June 2012.

According to BEREC (2011a), although several NRAs are concerned about the integration between fixed and mobile services, only the Austrian regulator concludes that fixed and mobile services are in the same retail market. The report from BEREC (2011a) also mentions that many NRAs decide not to integrate fixed and mobile services in the same market for the following reasons. In the retail market, NRAs provide several supporting arguments such as: 1) characteristic differences; 2) price differences; 3) immobility of fixed broadband; 4) differences in usage patterns; and 5) preferences of consumers. For the wholesale market, the main reasons are: 1) characteristic differences in the retail market; 2) inadequate indirect competitive constraints from retail substitution; and 3) no direct substitution at the wholesale level.

In addition, in recent years, there have been academic studies analysing the substitution between fixed and mobile broadband; for example, Cardona et al. (2009), Lee et al. (2011), Srinuan et al. (2012) and Wulf et al. (2013). Cardona et al. (2009) is among the early studies to investigate substitution between fixed and mobile broadband by using the Austrian survey data from 2006. The authors found that broadband services (DSL, cable and mobile broadband) are substitutes for and constrain each other in areas in which these services are available. Nevertheless, the Austrian mobile broadband data in 2006 was still quite new. Similarly, Srinuan

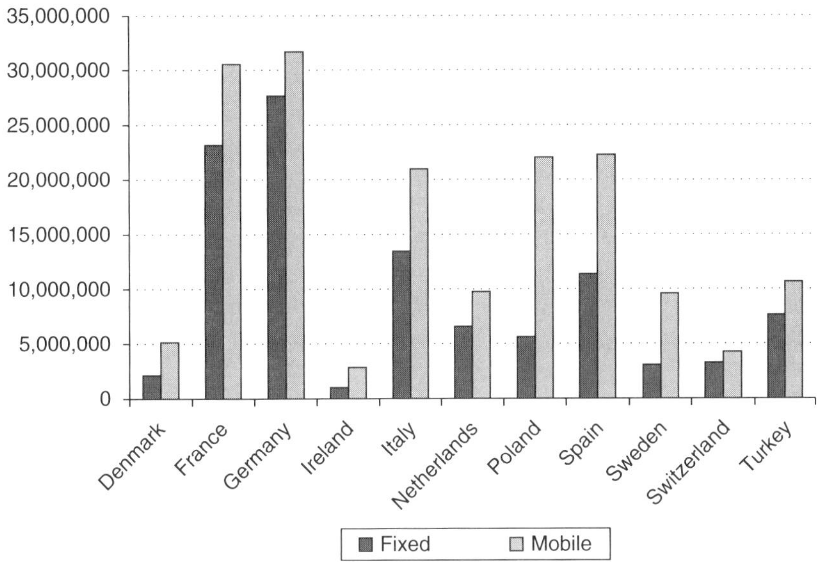

Source: OECD (2012).

Figure 16.6 Fixed and mobile broadband subscriptions in selected European countries in June 2012

et al. (2012) found substitution between broadband services (DSL, LAN/fibre, cable and mobile broadband) in areas where there are more than two broadband services. Conversely, mobile broadband cannot be in the same market as fixed broadband in areas where only DSL and mobile broadband are available. According to Cardona et al. (2009) and Srinuan et al. (2012), the geographical area where broadband services are available is an important factor in defining the relevant market. Conversely, Lee et al. (2011) applied cross-country analysis to an OECD data set from 2003 to 2008 and suggested that mobile broadband is a complement to fixed broadband. Wulf et al. (2013) used logistic regression and the Lotka Volterra Competition (LVC) model on another OECD data set from 2001 to 2009. The authors concluded that the demand for mobile broadband is driven by the high demand for fixed broadband. Consumers therefore usually tend to consider mobile broadband to be a complementary service to fixed broadband. According to these previous studies, it is interesting that when analysing demand elasticity in one specific country (Austria and Sweden as examples), the results tend to show substitution between fixed and mobile broadband services to some extent. On the other hand, analysis across countries provides opposite results, i.e. that mobile broadband is complementary to fixed broadband services.

Nevertheless, BEREC (2011a) concluded that fixed and mobile substitution is highly heterogeneous across European countries. Therefore it is important to consider case by case since one factor might be crucial in one country but not significant in others. In addition, BEREC (2011a) suggested that although mobile and fixed broadband are still not considered to be in the same market in most countries, fixed and mobile substitution analysis can provide insight into a degree of price constraint in several stages of the SMP framework, such as the three cumulative criteria test, SMP assessment and imposition of ex ante regulation.

5. SUMMARY OF THE PUBLIC CONSULTATION RESULTS ON THE REVISION OF THE RECOMMENDATION ON RELEVANT MARKETS

This Public Consultation on the Revision of the Recommendation on Relevant Markets was opened from 16 October 2012 until 8 January 2013. More than 60 contributions from various stakeholders, such as NRAs, telecommunication operators, telecommunication operators' associations and consumer organisations, commented on the Revision of the Recommendation on Relevant Markets. Despite few contributions commenting that the three cumulative criteria test is not the most efficient tool, most contributions believed that the three cumulative criteria test is still an appropriate tool for defining relevant markets. Also, for the stability and certainty of the market, the three cumulative criteria test should be the main regulatory tool for market definition. According to most contributions, they agreed that the market had been developing dynamically, with convergence and bundle services. Changes in the retail situation and the fixed and mobile substitution can affect the market definition; nevertheless, the fixed and mobile substitution is still different between EU member states. While use of the three cumulative criteria test and market trends are going in the same direction according to most stakeholders, the contributions on adding and removing markets from the Recommendation vary. Table 16.5 summarises the supporting and opposing views from stakeholders for adding/keeping/removing in each market that is susceptible to ex ante regulation.

This summary shows that there are still diverse opinions from various stakeholders on economic issues and relevant market. In addition, as stated by BT in the public consultation, market definition should not become an end in itself; instead, it should support the assessment of market

Table 16.5 Summary of supporting and opposing views from stakeholders, according to the Public Consultation Results on the Revision of the Recommendation on Relevant Markets in 2013

Markets susceptible to ex ante regulation	Comments
Current markets susceptible to ex ante regulation according to the Recommendations 2007	
1 (Access to the public telephone network at a fixed location; retail)	While there are some contributions that believe that removing any market might create uncertainty for particular stakeholders, most contributions believe that Market 1 is the most likely candidate to be removed. Greater competition from other services and the decrease in the incumbent's market share imply that there are likely to be insignificant impacts if Market 1 is removed from the list. However, BEREC as well as Telenor, WIND and Vodafone suggested that if Market 1 is removed, the wholesale market regulation should be explicitly ensured.
2 (Call origination on the public telephone network provided at a fixed location)	This is a similar situation to Market 1 in that removing any market might create uncertainty for particular stakeholders. Most contributions, e.g. Deutsche Telekom, FICORA, TeliaSonera and TelekomAustria, suggested that Market 2 should also be removed. The reason for removing it is the changing of the retail markets such as bundled services, OTT service and fixed-mobile substitution; therefore, the fixed telephony regulation should be removed. Nevertheless, as mentioned in Market 1, some contributors, such as BEREC, Telenor, WIND and Vodafone, commented that the wholesale market regulation needs to be ensured if Market 1 is removed. Therefore, Market 2 can possibly be revised to fit the wholesale line rental market instead.
3, 7 (Call termination)	**In favour of keeping call termination (e.g. BEREC)** 1. The termination market is different from the access market where each operator is a de facto monopolist. **In favour of removing call termination (e.g. Deutsche Telekom, GSMA, WIND)** 1. Growing VOIP services are increasing the market competition. 2. With significant termination rate reduction in recent years, competition law and symmetric regulation are sufficient to solve the market failure; therefore, these markets now do not meet the third criterion in the three cumulative criteria test. **Other views** 1. Some operators such as Telenor and TeliaSonera believe that Markets 3 and 7 no longer fulfil the three cumulative criteria test, but if there is a need to be regulated, they should be regulated with reasonable rates.

Table 16.5 (continued)

Markets susceptible to ex ante regulation	Comments
Current markets susceptible to ex ante regulation according to the Recommendations 2007	
	2. Orange France commented that when both fixed and mobile termination rates reach the pure LRIC level, Markets 3 and 7 should be removed from the Recommendation. 3. BEREC and TDC commented that Markets 3 and 7 could be merged into one termination market. Furthermore, TDC believes that SMS can be included with this merged market as well.
4, 5 and 6 (Wholesale broadband related)	The views of Markets 4, 5 and 6 generally focus on the revision of their definition. The main discussion is about technological neutrality and possible consolidation between two or three of them; for example, 1. BEREC suggested that both Markets 4 and 5 should be technologically neutral; e.g. inclusion of cable technology in relevant market. 2. TDC and KPN commented that technology development in the broadband market could lead to a blurring of the wholesale market. Markets 4, 5 and 6 should be consolidated because of supply and demand substitution and they should be based on technological neutrality as well.
Additional markets susceptible to ex ante regulation for the next revised version	
SMS termination	**In favour of adding SMS termination (e.g. DBA, UKE, 3UK)** 1. Similar to voice termination, each mobile operator has 100% of the market shares of the SMS termination wholesale market on its own network. 2. The SMS termination wholesale market met the three cumulative criteria when applying the test. 3. Price discrimination; for example, unlimited on-net SMS. **Against adding SMS termination (e.g. GSMA, Telefonica, EE)** 1. Great increase in the retail level of alternative ways of exchanging messages through mobile phone. 2. Falling price in the retail level. 3. The condition to meet the three cumulative criteria test is not significant enough. **Other views** 1. BEREC suggested that there should not be any additional market, including SMS, but that an explanatory memorandum should be used instead to facilitate the legal aspect. 2. VAT commented that SMS termination might need intervention in some countries, but there is no market failure in most countries in the EU.

Table 16.5 (continued)

Markets susceptible to ex ante regulation	Comments
Additional markets susceptible to ex ante regulation for the next revised version	
Wholesale mobile access and origination (former Market 15 in the 2003 Recommendation)	**In favour of adding wholesale mobile access and origination (e.g. BT, Coop Italia, Noverca)** 1. Market concentration is increasing owing to the recent merger between MNOs and the market shares of large MNOs in the retail market are still high. 2. The decline of MVNOs, especially after Market 15 was deleted in 2007, in Europe has led to less competition in the mobile market. 3. Agreement between MNOs and MVNOs is not based on regulation. Therefore, the hosting operators are in the monopolistic position and MVNOs do not have practical switching capability. **Against adding wholesale mobile access (e.g. GSMA, Telefonica)** 1. The competitive dynamic in the retail mobile market is increasing. 2. The market does not meet the three cumulative criteria test owing to the potential of spectrum trading and allocation. 3. Adding back Market 15 might lead to uncertainty for MVNOs. **Other views** 1. TDC suggested that even though most member states have a competitive wholesale market for mobile access and origination, this market still needs to be carefully monitored.
Others	BEREC suggested that even though any market should be added to the revision of the Recommendation on relevant markets, there are some markets, such as Internet connectivity, backhaul services to reach mobile base stations, and special rate services, that possibly meet the three cumulative criteria test. Nevertheless, these services could be considered case by case by the NRA in each member state or further developed in a revised explanatory memorandum.

Notes: WIND (Wind Telecomunicazioni SpA) is an Italian telecommunications operator; FICORA (Finnish Communications Regulatory Authority); GSMA (Global System Mobile Association) is an association of mobile operators; TDC (TeleDenmark Corporation) is a Danish telecommunications operator; KPN (formerly, Koninklijke PTT Nederland) is a Dutch telecommunications operator; DBA (Danish Business Authority); UKE (Urząd Komunikacji Elektronicznej) is a Polish electronic communications authority; 3UK (Hutchison 3G UK) is a mobile network operator in the UK; EE (formerly, Everything Everywhere) is a mobile network operator in the UK; and VAT (Verband Alternativer Telekom-Netzbetreiber) is an Austrian association of alternative telecommunications operators.

power and the identification of potential market failure. Nevertheless, this public consultation implies that the EC is now concerning itself with the economic issues and dynamic change of the market in order to improve the market definition and SMP framework according to its dynamic and convergent nature. In the upcoming years, while Markets 3 and 7 are still regulated as the present Recommendation, it is likely that we will see the removal of Market 1 and the adjustment of Markets 2, 4, 5 and 6 to better suit the fast development of the broadband market.

6. CONCLUSIONS AND RECOMMENDATIONS

Sections 3 and 4 show that all economic issues need to be addressed when regulators decide to intervene in the market throughout the process, from defining the relevant market to imposing ex ante regulation. To be more precise, platform competition and the substitution between fixed and mobile services need to be more strongly emphasised at the market definition and SMP assessment stages while the two-sided market can be more focused on imposing an ex ante regulation procedure. Importantly, the static and dynamic trade-off should be considered throughout the procedure. In addition, Section 5 provides empirical results of different concerns relating to economic issues from different stakeholders, suggesting that there is a need to revise the Recommendations with these economic issues and the changes in the telecommunications market in mind.

Despite these economic issues, have these Recommendations by the EC so far succeeded or failed? The EC (2010b) reported that although the member states sometimes applied different remedies to solve market failure, the member states overall have better regulations that ultimately benefit consumers. For instance, EU consumer prices for mobile phones for both voice and SMS services currently average 34.5 per cent less than in 2004 while high-speed broadband services are becoming more affordable across member states. With lower prices and greater adoption in the market, these results imply that regulatory tools, such as mobile termination rate reduction and local loop unbundling, benefit consumers by encouraging more competition, more quality and greater variety of services. However, similarly to the trade-off between static and dynamic regulation, studies have suggested that EC Recommendations bring some drawbacks to the industry, especially a decrease on future investment, which is crucial for the next generation network. Several studies show that price regulation and access regulation can reduce operators' incentive to invest. For instance, Weisman (2005) and Roques and Savva (2009) claim that the level price cap regulation can affect incentives to invest for operators; a high price cap can encourage greater investment while too low a price cap is likely to hamper investment incentives. In access regulation, for example, Bourreau and Dogan (2005) and Grajek et al. (2008) mention that access regulation, local loop unbundling, can significantly reduce infrastructure investment incentives that are crucial for future next generation network investment.[6]

Can we conclude that the frameworks established so far by the EC have succeeded or failed? No conclusion can be drawn either way, but the implication from these figures is that it depends on the priorities of each country to choose their own suitable framework and there is also a need to develop a new framework to bridge the gap between static and dynamic trade-off. It can be seen from the revision of Recommendations that the EC has attempted to encourage efficient competition and future investment at the same time. The change in Recommendations 2007 removes several markets susceptible to ex ante regulation, from 18 markets in 2003 to seven markets in 2007. This implies that the EC tends not to intervene in the market if there is enough competition in it. However, there is still some criticism from scholars such as De Streel (2008a) who points out that while the SMP framework in the retail market has decreased, the wholesale level is still strictly regulated. The author recommends that the EC should develop the Recommendation further after the 2007 version to allow more flexibility and future investment. The author further suggests improving the market definition analysis in the Recommendations by instead analysing it from the retail level to assess whether there is any market failure. If there is no SMP, the market analysis should not be continued. On the other hand, if there is a market failure in the retail market, NRAs can then move forward and investigate the wholesale market. Also, Schwarz (2007), who applied the theory of demand elasticity in his study, comments on an issue in the wholesale broadband market, that it depends on the demand and supply-side substitution. The retail competitive constraint therefore should be considered at some stage when analysing market definition and SMP in the broadband wholesale market.

In short, there is still uncertainty in the present-day European telecommunications sector. Not only is there no best solution to resolve the market failure, but regulators also often need to make the decision on the trade-off dilemma (Larouche and De Visser, 2006). Several cases in this chapter, such as two-sided markets, platform competition, substitution between fixed and mobile services and especially the trade-off between static and dynamic regulation, show that it is important for the market definition and SMP analysis to also be concerned with important economic issues. In addition, it is better to balance policy between economics and law, static and dynamic, and consider the issue case by case and country by country depending on the priority and situation in each member state. Importantly, to be able to apply efficient market intervention, the change of EU policy should start at the first stage, which is market definition. Furthermore, the EC released the results of the public consultation on revision of the Recommendation on relevant markets in early 2013. It will be interesting to see what the new revision of the Recommendation on relevant markets will be like in the future. Currently, it can be concluded that regardless of whether the 2003 and 2007 Recommendations by the EC have so far succeeded or failed, the market definition and SMP framework need to be adapted to be more flexible and suitable for the EC framework and the new dynamics of the market.

NOTES

1. A bottleneck in the telecommunications sector can be described as a phenomenon where networks take each other on in a competitive retail market for consumers but still have monopoly power for providing access to subscribers (Armstrong, 2001).
2. See OECD (2004) for a more comprehensive explanation of one-way access and two-way access problems.
3. BEREC is the Body of European Regulators for Electronic Communications, formerly ERG, which assists the EC and the NRAs with the regulatory framework in the electronic communications sector.
4. The recommended market according to Cave et al.'s (2006) report is: 1) Retail fixed access (low-capacity access); 2) Retail fixed access (high-capacity access); 3) Wholesale fixed call origination; 4) Wholesale call termination on individual fixed networks; 5) Wholesale local-tandem transit; 6) Unbundled local loops; 7) Wholesale local-tandem broadband access; 8) Termination of segments of leased lines excluding those of high capacity; 9) Trunk segments of leased lines excluding those of high capacity; and 10) Termination of incoming calls on individual mobile networks.
5. For more waterbed effect studies, for example, see Veronese and Pesendorfer (2009), Growitsch et al. (2010) and Genakos and Valletti (2011).
6. See the more comprehensive reviews of broadband investment and regulation in Cambini and Jiang, 2009.

REFERENCES

Armstrong, M. (2001), 'The theory of access pricing and interconnection'. MPRA Paper 15608, University Library of Munich, Munich.

Bauer, J.M. and E. Bohlin (2008), 'From static to dynamic regulation: recent developments in US telecommunications policy', *Intereconomics*, **43** (1), 38–50.

BEREC (2010), Report on the impact of bundled offers in retail and wholesale market definition, BEREC, BoR(10)64.

BEREC (2011a), Report on the impact of fixed-mobile substitution in market definition, BEREC, BoR(11)54.

BEREC (2011b), Report on relevant market definition for business services, BEREC, BoR(10)46 Rev1b.

BEREC (2012), Draft report on co-investment and SMP in NGA networks, BEREC, BoR(11)69.

Bourreau, M. and P. Dogan (2005), 'Unbundling the local loop', *European Economic Review*, **49** (1), 173–99.

Bourreau, M., P. Dogan and M. Manant (2010), 'A critical review of the "ladder of investment" approach', *Telecommunications Policy*, **34** (11), 683–96.

Cambini, C. and Y. Jiang (2009), 'Broadband investment and regulation: a literature review', *Telecommunications Policy*, **33** (10–11), 559–74.

Cardona, M., A. Schwarz, B.B. Yurtoglu and C. Zulehner (2009), 'Demand estimation and market definition for broadband Internet services', *Journal of Regulatory Economics*, **35** (1), 70–95.

Cave, M. (2010), 'Snakes and ladders: unbundling in a next generation world', *Telecommunications Policy*, **34** (1–2), 80–85.

Cave, M. and I. Vogelsang (2003), 'How access pricing and entry interact', *Telecommunications Policy*, **27** (10–11), 717–27.

Cave, M., U. Stumpf and T. Valletti (2006), 'A review of certain markets included in the Commission's Recommendation on Relevant Markets subject to ex ante Regulation: an independent report for the European Commission', available at http://ec.europa.eu/ information_society/policy/ecomm/doc/library/ext_studies/review_experts/review_regulation.pdf (accessed 15 April 2013).

Czernich, N., O. Falck, T. Kretschmer and L. Woessmann (2011), 'Broadband infrastructure and economic growth', *The Economic Journal*, **121** (552), 505–32.

De Streel, A. (2008a), 'Current and future European regulation of electronic communications: a critical assessment', *Telecommunications Policy*, **32** (11), 722–34.

De Streel, A. (2008b), 'The relationship between competition law and sector specific regulation: the case of electronic communications', *Reflets et Perspectives de la Vie Économique*, **47** (1), 55–72.

ERG (2008), Report on guidance on the application of the three criteria test, ERG(08)21.

European Commission (1997), Commission Notice on the definition of relevant market for the purposes of Community competition law, C372/5.

European Commission (2003), Commission Recommendation of 11 February 2003 on relevant product and service markets within the electronic communications sector susceptible to ex ante regulation in accordance with Directive 2002/21/EC of the European Parliament and of the Council on a common regulatory framework for electronic communication networks and services, 2003/311/EC.

European Commission (2007), Commission Recommendation on relevant product and service markets within the electronic communications sector susceptible to ex ante regulation in accordance with Directive 2002/21/EC of the European Parliament and of the Council on a common regulatory framework for electronic communications networks and services, C(2007) 5406 rev 1.

European Commission (2010a), Telecoms: Commission report on national telecoms regulation shows more competition but no single market, IP/10/644.

European Commission (2010b), Telecoms: how the Article 7 consultation and notification mechanism works: frequently asked questions, MEMO/10/226.

European Commission (2012), 'Questionnaire for the public consultation on the revision of the Recommendation on relevant markets', available at https://ec.europa.eu/digital-agenda/en/news/public-consultation-revision-recommendation-relevant-markets (accessed 13 May 2013).

Evans, D.S. (2007), 'Do mobile operators have a dominant position in a market for the wholesale termination of calls from fixed to mobile?', *The Economics of Mobile Prices*, **7**, Vodafone Policy Paper Series, Vodafone Group Plc.

Evans, D.S. and R. Schmalensee (2008), 'Markets with two-sided platforms', in *Competition Law and Policy (ABA Section of Antitrust Law) 1*, Chapter 28, pp. 667–93.

Genakos, C. and T. Valletti (2011), 'Testing the "waterbed" effect in mobile telecommunications', *Journal of the European Economic Association*, **9**, 1114–42.

Grajek, M., H. Friederiszick and L.H. Roller (2008), 'Analyzing the relationship between regulation and investment in the telecom sector'. ESMT White Paper No. WP-108-01, Berlin.

Growitsch, C., J.S. Marcus and C. Wernick (2010), 'The effects of lower mobile termination rates (MTRs) on retail price and demand', *Communications and Strategies*, **80**, 119–40.

Hausman, J. and J. Wright (2006), 'Two-sided markets with substitution: mobile termination revisited', available at http://econ-www.mit.edu/files/1038 (accessed 2 May 2013).

Hou, L. (2008), 'The assessment of collective SMP: lessons learned from the first round of market review', *Proceeding of the First Annual Conference on Competition and Regulation in Network Industries (CRNI)*, 28 November, Brussels.

Koutroumpis, P. (2009), 'The economic impact of broadband on growth: a simultaneous approach', *Telecommunications Policy*, **33** (9), 471–85.

Larouche, P. and M. De Visser (2006), 'The triangular relationship between the Commission, NRAs and national courts revisited', *Communications and Strategies*, **64**, 125–45.

Lee, S., M. Marcu, S. Lee et al. (2011), 'An empirical analysis of fixed and mobile broadband diffusion', *Information Economics and Policy*, **23** (3–4), 227–33.

Never, H. and B. Preissl (2008), 'The three-criteria test and SMP: how to get it right', *International Journal of Management and Network Economics*, **1** (1), 100–127.

OECD (2004), Access pricing in telecommunications, Paris.

OECD (2011a), Next generation access networks and market structure, Paris.

OECD (2011b), OECD communications outlook 2011, Paris.

OECD (2012), 'Fixed and mobile broadband subscription by country (June 2012)', available at http://www.oecd.org/internet/broadband/oecdbroadbandportal.htm (accessed 20 April 2013).

Rochet, J. and J. Tirole (2004), 'Two-sided markets: an overview', mimeo, IDEI University of Toulouse.

Roques, F.A. and N. Savva (2009), 'Investment under uncertainty with price ceilings in oligopolies', *Journal of Economic Dynamics and Control*, **33** (2), 507–24.

Schwarz, A. (2007), 'Wholesale market definition in telecommunications: the issue of wholesale broadband access', *Telecommunications Policy*, **31** (5), 251–64.

Srinuan, P., C. Srinuan and E. Bohlin (2012), 'Fixed and mobile broadband substitution in Sweden', *Telecommunications Policy*, **36** (3), 237–51.

Valletti, T. (2006), 'Mobile call termination: a tale of two-sided markets', *Communications and Strategies*, **61**, 61–77.

Veronese, B. and M. Pesendorfer (2009), *Wholesale termination regime, termination charge levels and mobile industry performance*, London: Ofcom.

Weisman, D. (2005), 'Price regulation and quality', *Information Economics and Policy*, **17** (2), 165–74.

Wulf, J., S. Zelt and W. Brenner (2013), 'Fixed and mobile broadband substitution in the OECD countries: a quantitative analysis of competitive effects'. *Proceeding of the System Sciences (HICSS), 2013 46th Hawaii International*, 7–10 January, Hawaii.

Index